TECHNOLOGICAL INTERVENTIONS IN MANAGEMENT OF IRRIGATED AGRICULTURE

Innovations in Agricultural and Biological Engineering

TECHNOLOGICAL INTERVENTIONS IN MANAGEMENT OF IRRIGATED AGRICULTURE

Edited by

Megh R. Goyal, PhD, PE

Susmitha S. Nambuthiri, PhD

Richard Koech, PhD

Apple Academic Press Inc.
3333 Mistwell Crescent
Oakville, ON L6L 0A2 Canada

Apple Academic Press Inc.
9 Spinnaker Way
Waretown, NJ 08758 USA

First issued in paperback 2021

ISBN-13: 978-1-77463-643-5 (pbk)
ISBN-13: 978-1-77188-592-8 (hbk)

Library and Archives Canada Cataloguing in Publication

Technological interventions in management of irrigated agriculture / edited by Megh R. Goyal, PhD, PE, Susmitha S. Nambuthiri, PhD, Richard Koech, PhD.

(Innovations in agricultural and biological engineering)

Includes bibliographical references and index.

Issued in print and electronic formats.

ISBN 978-1-77188-592-8 (hardcover).--ISBN 978-1-315-20430-7 (PDF)

1. Irrigation--Technological innovations. 2. Irrigation--Management. 3. Water in agriculture--Technological innovations. 4. Agricultural innovations. I. Goyal, Megh Raj, editor II. Series: Innovations in agricultural and biological engineering

S613.T43 2017 631.5'87 C2017-906509-2 C2017-906510-6

CIP data on file with US Library of Congress

Apple Academic Press also publishes its books in a variety of electronic formats. Some content that appears in print may not be available in electronic format. For information about Apple Academic Press products, visit our website at **www.appleacademicpress.com** and the CRC Press website at **www.crcpress.com**

CONTENTS

LIST OF CONTRIBUTORS

K. R. Anil, MTech
Ph.D. Research Scholar, Faculty of Civil Engineering, Cochin University of Science and Technology, Cochin-22 and Director, National Coir Research Management Institute, Trivandrum, Kerala, India, E-mail: kranil2003@yahoo.com

Bavita Asthir, PhD
Senior Biochemist-cum-Head, Department of Biochemistry, Collage of Basic Sciences and Humanities, Punjab Agriculture University, Ludhiana, Punjab 141004, India. E-mail: b.asthir@rediffmail.com; basthir@pau.edu

Michael Eshetu Bisa, MSc
Lecturer, Department of Water Resource and Irrigation Engineering, Arba Minch University, POB 21, Arba Minch, Ethiopia. E-mail: michaeshe@gmail.com

Elena Bresci, PhD
Professor, Department of Agricultural, Food and Forestry Systems, University of Firenze, Via San Bonaventura 13, Firenze 50145, Italy. E-mail: elena.bresci@unifi.it

Jingzheng Chen, MSc
Ph.D. Candidate and Research Scientist, Hunan Academy of Forestry, 658 South Shaoshan Rd., Changsha, Hunan 410004, China. E-mail: chenjingzhen621@sina.com

Gaurav S. Dave, PhD
Assistant Professor, Department of Biochemistry, Saurashtra University, Rajkot 360005, Gujarat, India. E-mail: gsdspu@gmail.com

Pieter Van der Zaag, PhD
Professor and Head, IWSG Department, Water Resources Section, UNESCO-IHE, Delft University of Technology, The Netherlands. E-mail: p.vanderzaag@unesco-ihe.org

Fitsume Yemenu Desta, MSc
Ph.D. Student, School of Environmental and Rural Sciences, University of New England, POB 2350, 137 Mann St. Armidale, Australia. E-mail: fitsum_yimenu@yahoo.com; fdesta@myune.edu.au

Savan D. Fasara, MSc
Research Scholar, Department of Biochemistry, Saurashtra University, Rajkot, Gujarat, India. E-mail: sdfasara@gmail.com

Ceyhun Göl, PhD
Associate Professor, Department of Forest Engineering, Çankırı Karatekin University, Çankırı, 18200, Turkey.

Megh R. Goyal, PhD, PE
Retired Professor in Agricultural and Biomedical Engineering, University of Puerto Rico–Mayaguez Campus; and Senior Technical Editor-in-Chief in Agriculture Sciences and Biomedical Engineering, Apple Academic Press Inc., PO Box 86, Rincon, PR 00677, USA. Email: goyalmegh@gmail.com

Asheber Haile, MSc
Associate Researcher, Ethiopian Institute of Agricultural research, Debre Zeit Agricultural Research Center, POB 32, Debre Zeit, Ethiopia. E-mail: ashu_haile@yahoo.com

Eric W. Harmsen, PhD
Professor, Department of Agricultural and Biosystems Engineering University of Puerto Rico, Mayaguez, Puerto Rico 00681, USA. E-mail: eric.harmsen@upr.edu; harmsen1000@gmail.com; Website: www.pragwater.com

Divya Jain, MSc
Ph.D. Research Scholar, Department of Biochemistry, Collage of Basic Sciences and Humanities, Punjab Agriculture University, Ludhiana, Punjab 141004, India. E-mail: divya.jain6464@gmail.com

Lijuan Jiang, PhD
Professor, Central South University of Forestry and Technology, 498 South Shaoshan Rd., Changsha, Hunan 410004, China. E-mail: znljiang2542@163.com

Girma Kassa, MSc
Associate Researcher, Ethiopian Institute of Agricultural research, Debre Zeit Agricultural Research Center, POB 32, Debre Zeit, Ethiopia. E-mail: girmakaza@gmail.com

Chhaya R. Kasundra, MSc
Student, Department of Biochemistry, Saurashtra University, Rajkot, Gujarat, India. E-mail:chhayachemist@gmail.com

Bhavesh D. Kevadiya, PhD
Postdoctoral Fellow, University of Nebraska Medical Center, Omaha, NE, USA. E-mail: bbhavesh-patel@gmail.com

R. K. Koech, PhD
Program Leader, Agricultural Water Management, School of Environmental and Rural Science, University of New England, Armidale NSW 2351, Australia. E-mail: richardkoech@hotmail.com

Prashant D. Kunjadia, PhD
Assistant Professor, B. N. Patel Institute of Paramedical and Science, Bhalej Road, Anand, Gujarat, India. E-mail: pdkunjadia@yahoo.com

Philip K. Langat, PhD
Chief Manager Technical Services and Operations, Tana and Athi Rivers Authority and part-time Lecturer, Land Resource Management & Agricultural Technology, University of Nairobi, Queensway House 7th Floor, Kaunda Street, P.O. Box 47309-00100 GPO Nairobi, Kenya. E-mail: philkibet@yahoo.com

Changzhu Li, PhD
Professor, Hunan Academy of Forestry, 658 South Shaoshan Rd., Changsha, Hunan 410004, China. E-mail: lichangzhu2013@aliyun.com

Peiwang Li, MS
Associate Professor, Hunan Academy of Forestry, 658 South Shaoshan Rd., Changsha, Hunan 410004, China. E-mail: lindan523@163.com

Qiang Liu, MSc
Ph.D. Candidate, Central South University of Forestry and Technology, 498 South Shaoshan Rd., Changsha, Hunan 410004, China. E-mail: liu.qiangcs@163.com

Joe Masabni, PhD
Assistant Professor, Texas A&M AgriLife Research and Extension at Overton, Texas A&M University, 1710 FM 3053 N, Overton, TX 75684, USA. E-mail: joe.masabni@ag.tamu.edu

Chris J. Matocha, PhD
Associate professor, Department of Plant and Soil Sciences University of Kentucky, Lexington, KY, 40546-0091, USA. E-mail: cjmato2@uky.edu

Blazan Mijatovic, MSc
Research analyst, Department of Plant and Soil Sciences University of Kentucky, Lexington-KY, 40546-0091, USA. E-mail: blazan@uky.edu

Tom G. Mueller, PhD
Research Scientist, Decision Science and Modeling, John Deere & Co. Urbandale, IA, 50263 USA. E-mail: muellerthomasg@johndeere.com

Susmitha S. Nambuthiri, PhD
Researcher, Washington state Department of Health, Washington, USA 98512; 5126 Slate Ct SE, Olympia, WA 98501; E-mail: susmitha.agri@gmail.com

Genhua Niu, PhD
Associate Professor, Texas A&M AgriLife Research Center at El Paso, 1380 A&M Circle, El Paso, TX 79927, USA. E-mail: gniu@ag.tamu.edu

Sonal V. Panara, MSc
Student, Department of Biochemistry, Saurashtra University, Rajkot, Gujarat, India. E-mail: sonalipanara3@gmail.com

Eduardo A. Rienzi, PhD
Research Fellow, Department of Plant and Soil Sciences University of Kentucky, Lexington, KY, 40546-0091, USA. E-mail: Eduardo.rienzi@uky.edu

Gaurav V. Sanghvi, PhD
Assistant Professor, Department of Pharmaceutical Sciences, Saurashtra University, Rajkot 360005, Gujarat. E-mail: sanghavi83@gmail.com

Hubert H. G. Savenije, PhD
Professor and Head, Water Resources Section, UNESCO-IHE, Delft University of Technology, The Netherlands. E-mail: H.H.G.Savenije@tudelft.nl

Frank J. Sikora, PhD
Director, Regulatory Services Soil Testing Laboratory, 103 Bruce Poundstone Regulatory Service, Lexington, KY 40546-0275, USA. E-mail: frank.sikora@uky.edu

Youping Sun, PhD
Assistant Professor, Department of Plants, Soils and Climate Utah State Uni 4820 Old Main Hill Logan UT 84322, USA, E-mail: youping.sun@usu.edu

Gebeyehu Tegenu, BSc
Junior Researcher, Ethiopian Institute of Agricultural research, Debre Zeit Agricultural Research Center, POB 32, Debre Zeit, Ethiopia. E-mail: gebeyehutegenu@gmail.com

Deepak Kumar Verma, MSc
Ph.D. Research Scholar, Agricultural and Food Engineering Department, Indian Institute of Technology, Kharagpur 721302, West Bengal, India. Email: deepak.verma@agfe.iitkgp.ernet.in; rajadkv@rediffmail.com

Subha Vishnudas, PhD
Associate Professor, Faculty of Civil Engineering, Cochin University of Science and Technology, Cochin-22, India. E-mail: v.subha@cusat.ac.in

Filiz Yüksek, PhD
Scientist, Pazar Forest Management Directorate, 53300-Pazar, Rize, Turkey

Turan Yüksek, PhD
Professor & Head, Faculty of Fine Arts, Design and Architecture, Department of Landscape Architecture, Recep Tayyip Erdogan University Rize-53100, Turkey. E-mail: turan53@yahoo.com; turan.yuksek@erdogau.edu.tr

Erdogan Yüksel, PhD
Assistant Professor, Department of Forest Engineering, Artvin Çoruh University, Artvin, 08000, Turkey

LIST OF ABBREVIATIONS

ABS	Australian Bureau of Statistics
AC	alder coppice
AE	application efficiency
AIP	alkaline inorganic pyrophosphatase
AlaAT	alanine amino transferases
ANOVA	analysis of variance
APCC	Asian and Pacific Coconut Community
BIS	Bureau of Indian Standards
CBV	coir bhoovastra
CD	coir dust
CF-IRMS	continuous-flow isotope ratio mass spectrometer
CG	coir geotextiles
CGC	coir geotextile with crop
CP	control plot
CWR	crop water requirement
DBE	debranching enzyme
Dp	degree of polymerization
DU	distribution uniformity
DZARC	Debrezeit Agriculture Research Center
EC	electrical conductivity
ECBs	erosion control blankets
EFB	empty fruit bunch
EIRA	Ethiopian Institute of Agriculture Research
ET	evapotranspiration
FAO	Food and Agriculture Organization
FC	field capacity
GBFMP	German Bundersanstalt fur Material Prufung
GIR	gross irrigation requirements
GLM	general linear model
GMID	Goulburn Murray Irrigation District
GOT	glutamate oxaloacetate transaminase
GPT	glutamate pyruvate transaminase
HMW-GS	high-molecular-weight glutenin subunits
IPP	inorganic pyrophosphate

IPRID	International Programme for Technology Research in Irrigation and Drainage
ISCS	International Soil Classification System
IWMI	International Water Management Institute
KSCRI	Konni Soil Conservation Research Station
LMW-GS	low-molecular-weight glutenin subunits
LS	loamy sand
MAD	management allowed deficit
MIR	MId-infrared
NADH	nicotinamide adenine dinucleotide
NAPT	North American Proficiency Testing
NCEA	National Centre for Engineering in Agriculture
NCRMI	National Coir Research Management Institute
NMI	National Measurement Institute
NUE	nitrogen-use efficiency
NUpE	nitrogen uptake efficiency
NUtE	nitrogen utilization efficiency
OC	organic carbon
OHP	overhead projector
PAW	plant available water
Pi	inorganic phosphate
PLS	partial least-squares
Pn	photosynthesis
PRESS	predicted residual sum of squares
PST	pressure-sensitive transducer
PTB	pipes through the bank
PTP1B	protein tyrosine phosphatase-1B
PWP	permanent wilting point
RCBD	randomized complete block design
RE	requirement efficiency
RLGW	Rannar, Lindgren, Geladi, and Wold
RP-HPLC	reversed-phase high-performance liquid chromatography
SCL	sand-clay loam
SCRS	Soil Conservation Research Station
SD	standard deviation
SIRMOD	surface irrigation simulation, evaluation, and design
SISCO	surface irrigation simulation, calibration, and optimization
SOC	soil organic carbon
SOM	soil organic matter
SPR	soil penetration resistance

SWS	soil water storage
TC	tea cultivation
TGW	1000-grain weight
TN	total nitrogen
UAV	unmanned aerial vehicle
VIP	variables important for prediction
vis-NIR	visible-near-infrared
WAS	weeks after sowing
WC	water content
WMO	World Meteorological Organization
WSA	water-stable aggregates
WUE	water-use efficiency

PREFACE 1 BY MEGH R. GOYAL

Conservation of natural resources (land, water, air, humanity) is a burning issue among world communities. Although we have been able to harness these resources for our benefit, we are also responsible for making our planet a miserable place to live: contamination of land, water, and air; disparity in distribution of shelter, fiber, and food causing death of millions due to hunger and poverty; diminishing human ethical, moral, and spiritual values; wars for no reasons; global warming, among others. Agricultural engineers in cooperation with others can help to alleviate these issues.

My son, Vinay K. Goyal, and his wife, Stacey Carpenter graciously blessed us by inviting my wife and me to join them for an educational tour to California, Nevada, Utah, Colorado, and Arizona to visit the widest and tallest trees, hottest point in Death Valley, a wind-mill energy park, irrigation fields, forest lands under the US National Park Service, Colorado River, Hoover Dam, Lakes, marine life, Grand Canyons, among others. We also visited Mt. Hood in Oregon, glaciers, the Artic Circle, Denali National Park, Mount McKinley, wildlife parks, the University of Alaska in Alaska, and the University of British Columbia in Vancouver, BC.

The General Sherman tree is a giant sequoia (*Sequoiadendron giganteum*) tree located in the Sequoia National Park in California. With a height of 286 ft (87 m) or more, a circumference of 113 ft (34 m) or more, an estimated bole volume of up to 52,500 ft^3 (1487 m^3), and an estimated age of 1800–2700 years, this is among the tallest, widest, and longest lived of all organisms on Earth. Death Valley National Park is the lowest and hottest place located east of the Sierra Nevada. The park protects the northwest corner of the Mojave Desert and contains a diverse desert environment of salt-flats, sand dunes, badlands, valleys, canyons, and mountains. It is the largest national park in the lower 48 states and has been declared an International Biosphere Reserve. In Death Valley, Badwater Basin is noted as the lowest point in North America, with an elevation of 86 m below sea level and the hottest point with a record historical temperature of 134°F. Zion Canyon is located in the Zion National Park near Springdale, Utah, which is home to 289 species of birds, 75 mammals (including 19 species of bats), and 32 kinds of reptiles inhabiting the four life zones: desert, riparian, woodland, and coniferous forest. Zion National Park includes mountains, canyons, buttes, mesas, monoliths, rivers, slot canyons, and natural arches. The Grand Canyon is a steep-sided canyon carved by the Colorado River in the state of Arizona in the United States. It is contained within Grand Canyon National Park and is 277 miles (446 km) long, up to 18 miles (29 km) wide, and attains a depth of over a mile (6093 ft or 1857 m).

The Colorado River with 2330 km drains an expansive, arid watershed that encompasses parts of seven US and two Mexican states. Colorado River water allocations consist of 9.1% for Mexico and 90.9% for the USA, which includes 26.7% for CA, 23.5% for CO, 17.0% for AZ, 14.4% for UT, 6.4% for WY, 5.1% for NM, and 1.8% for NV. The Colorado River system is a vital source of water for agriculture and urban areas in much of the southwestern desert lands of North America. The river and its tributaries are controlled by an extensive system of dams, reservoirs, and aqueducts, which in most years divert its entire flow to furnish irrigation and municipal water supply for almost 40 million people both inside and outside the watershed. The Colorado's large flow and steep gradient are used for generating hydroelectric power and its major dams regulate peaking power demands. According to the US Bureau of Reclamation <usbr.gov>:

> The Colorado River emphasizes its crucial role as the "lifeblood" sustaining millions of Americans across dozens of cities and countless farms in the American West. For the seven states—Arizona, California, Colorado, Nevada, New Mexico, Utah and Wyoming—the Colorado River has stimulated growth and opportunity for generations. Today it is as important as ever for residents of the Colorado River Basin region of the USA. The Colorado River Compact of 1922 established that each basin (upper and lower) is entitled to 7.5 million acre-feet of Colorado River water annually. It also grants priority entitlement to the lower basin. Additionally, a 1944 international treaty guarantees 1.5 million acre-feet of Colorado River water to Mexico each year. Congress authorized several projects to build dependability into the river's resource and reduce the risk from its erratic and destructive flows. By the early 1950s, many federal projects were in place in the lower basin—including the All-American Canal, Laguna Dam, Imperial Dam, Parker Dam, Davis Dam and the iconic Hoover Dam. In 1956, Congress authorized one of the most extensive and complex river resource development projects in the world, the Colorado River Storage Project (CRSP), to allow upper basin states to develop their Colorado River water apportionments. With 26.2 million acre feet of capacity, Lake Powell accounts for more than 86% of the 30.6 million acre feet of total storage capacity across

CRSP's four main units. That storage is key to ensure that the upper basin can meet its annual delivery obligation to the lower basin without creating shortages for upper basin states. Additionally, CRSP facilities and participating projects provide other valuable benefits such as hydroelectric power, flood control, agricultural irrigation and recreation. The US Bureau of Reclamation manages CRSP and other Colorado River projects to develop and protect water and related resources in an environmentally and economically sound manner for the American public. CRSP project facilities can generate enough electricity for nearly 5.8 million customers in the above seven Western states. In fact, each CRSP project is self-sustaining; costs for facilities within each generating unit are paid by that unit, not shared or covered by other units in the CRSP. Power generation revenues also support recovery and environmental programs within the basin, reduce salinity in the river and rehabilitate local irrigation systems. Facilities like the Glen Canyon Dam have been integral to development across the seven Colorado River Basin states and they will continue to play a vital role in the future of the US West. Storage provided by Glen Canyon Dam in particular has enabled the upper basin to prolong drought successfully, while making consistent full water deliveries to the lower basin without creating shortages for upper basin states. As western populations continue to grow, so do the challenges and complexities associated with water management [http://www.usbr.gov/newsroom/newsrelease/detail.cfm?RecordID=55987].

The website <http://www.usbr.gov/projects/Facility.jsp?fac_Name=Hoover+Dam&groupName=General> indicates that

Hoover Dam and Lake Mead, spanning the Arizona–Nevada state line, are located in the Black Canyon of the Colorado River about 35 miles southeast of Las Vegas, Nevada. It is a concrete thick-arch structure, 726.4 feet high and 1,244 feet long. The dam contains 3.25 million cubic yards of concrete; total concrete in the dam and appurtenant works is 4.4 million cubic yards ….

<http://www.usbr.gov/lc/hooverdam/history/storymain.html> writes that Hoover Dam is a testimony to construct monolithic projects in the midst of adverse conditions. Built during the Depression; thousands of men and their families came to Black Canyon to tame the Colorado River. It took less than five years, in a harsh and barren land, to build the largest dam of its time. Now, years later, Hoover Dam still stands as a world-renowned structure. The Dam is a National Historic Landmark and has been rated by the American Society of Civil Engineers as one of America's Seven Modern Civil Engineering Wonders.

According to USBR, the dam was completed on September 30, 1935. Hydrological data of the Hoover Dam is given below.

Total water storage at elevation	28,945,000 acre-ft at 1221.4
Maximum water surface elevation	1232 ft
Spillway type	Concrete-lined, side channel, gate-controlled overflow weir
Spillway capacity at elevation	270,000 cfs at 1232 ft
Outlet works capacity at elevation	52,200 cfs at 1219.6 ft
Drainage area	167,800 sq mi
Probable maximum flood (PMF) report	1980 PMF August general storm with 100-year snowmelt

Thus, my trip to the western United States testifies the importance of conservation of land and water resources.

Colorado River

The Colorado River is the sole source of water for the Imperial Valley in southern California

Lake Mead in 2010, showing the "bathtub ring" [Top]

Colorado River in the Grand Canyon seen from Pima Point [Left]

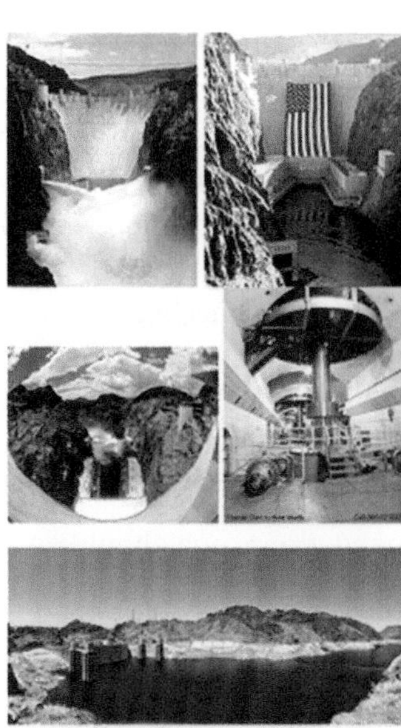

Generators on the Arizona side of Hoover Dam

http://www.usbr.gov/lc/hooverdam/
[Hoover Dam images from USBR]

Hoover Dam panoramic view

Desert

Badwater Basin elevation sign:
Hottest point.

General Sherman is a giant sequoia (Sequoiad endron giganteum) tree
located in the Giant Forest of Sequoia National Parkin Tulare Country,
in the U.S. state of California [Right top]
Source: National Park Service (US Govt.)

Apple Academic Press, Inc. published my first book on *Management of Drip/Trickle or Micro Irrigation*, as well as a 10-volume set under the book series *Research Advances in Sustainable Micro Irrigation*, in addition to other books in the focus areas of agricultural and biological engineering. The mission of this book series is to introduce the profession of agricultural and biological engineering.

At the 49th annual meeting of the Indian Society of Agricultural Engineers at Punjab Agricultural University during February 22–25 of 2015, a group of ABEs convinced me that there is a dire need to publish book volumes on the focus areas of agricultural and biological engineering (ABE). This is how the idea was born on the new book series titled *Innovations in Agricultural and Biological Engineering*.

The contributions by all cooperating authors to this book volume has been most valuable in the compilation. Their names are mentioned in each chapter and also in the list of contributors. This book would not have been written without the valuable cooperation of these investigators, many of whom are renowned scientists who have worked in the field of ABE throughout their professional careers.

Dr. Susmitha S. Nambuthiri and Dr. Richard Koech join me as coeditors of this book volume. Both are frequent contributors to my book series and are staunch supporters of my profession. Their contributions to the contents and quality of this book have been invaluable.

I thank the editorial staff, Sandy Jones Sickels, Vice President, and Ashish Kumar, Publisher and President at Apple Academic Press, Inc., for making every effort to publish the book when the diminishing water resources are a major issue worldwide. Special thanks are due to the AAP Production Staff. I request the reader to offer your constructive suggestions that may help to improve the next edition.

I express my deep admiration to my family for understanding and collaboration during the preparation of this book. I dedicate this book volume to Stacey Carpenter and Vinay K. Goyal (my son), who were our tour guides during our visits to the Hoover Dam, Colorado River, and U.S. National Parks in California, Nevada, Utah, Arizona; and largest irrigation valley in California. Their dedication helped me to appreciate the wonders of nature and human beings in these States. Who will not be astonished and marveled to see the tallest/widest trees and hottest point in the Western USA? After seeing them off in San Jose, CA, we continued our journey to Oregon and Washington state, Vancouver, BC; and Alaska, again for appreciating nature and to observe how the human activities have diminished the forest lands (Lungs of Nature) and Glaciers. At all these places, I could listen to the cries and rivers flowing from the glowing eyes of our Mother Planet, asking for help to conserve our natural resources of land, water, and air. I hear similar concerns from Jim Greenwald and Avani Desai. Jim in his book, *Tears for Mother Planet*, (ISBN-13:978-1607038160 PublishAmerica, November 17, 2008, pages 82), emphasizes his concern, my concern, and your concern: We should all share in keeping this planet, our home, alive for future generations to enjoy. It is about saving our planet, raising awareness, creating a concern within each person to do something rather than sitting back and wishing. Jim says, "I hope that something I wrote spurs you to action, to become involved in the saving of this planet. It would be hard to imagine a person not concerned enough to act to save what they have for their children, who deserve no less. That oft used phrase of "what can I do, I am only one person rings hollow." The poem titled *Make it Green* by Mrs. Avani Desai (June of 2014), <http://www.familyfriendpoems.com/poem/make-it-green>, recites the similar cry.

As an educator, there is a piece of advice to one and all in the world: "Permit that our Almighty God, our Creator, allow us to appreciate and conserve His Gifts without contaminating our planet. Do you want to join me to wipe tears of our Mother Planet? I invite my community in agricultural engineering to contribute book chapters to the book series by getting married to my profession." I am in total love with our profession by length, width, height, and depth. Are you?

— Megh R. Goyal, PhD, PE
Senior Editor-in-Chief

PREFACE 2 BY SUSMITHA S. NAMBUTHIRI

According to the Food and Agricultural Organization of the United Nations, water scarcity is a growing concern across the globe and is projected to become more severe due to increases in population growth, urbanization, and per capita consumption as well as changing water availability due to climate change. Irrigation withdrawals account for over 70% of all freshwater used and produce over 40% of the world's food supply. Irrigation engineering offers solutions to address water management challenges by playing significant roles in areas such as access to providing efficient supply of water for irrigation and by provides protection against soil and water erosion, allowing sustainable crop cultivation.

The book, titled *Technological Interventions in Management of Irrigated Agriculture*, discusses the development of some useful models and their applications in the field of soil and water conservation. The book contains four parts with 13 chapters. Farmer-friendly irrigation scheduling methods, model-based analysis of crop water requirement, ways to optimize surface irrigation systems, hydraulic design, and management of surface water systems are discussed through various book chapters in Part I of the book. Part II highlights ways to improve soil properties by taking into account spatial, temporal, and spectral variability in soil properties.

Part III of the book covers various innovative research studies in the field of soil and water conservation. Studies on soil and water productivity of vegetable cultivation under water-stressed areas, application of coir geotextiles, and the role of biofertilizers in controlling soil degradation and maintaining fertile top soil are discussed in this part. Part IV provides discussion on crop management strategies to enhance efficient use of marginal and saline lands for non-conventional crops.

The book will serve as an invaluable resource for graduate and undergraduate students in the field of agriculture, agricultural, biological, and civil engineering and also other branches of natural resources sciences. The book will be helpful for all academicians, research investigators, field engineers, agronomists, soil scientists, and extension personnel who directly or indirectly deal with soil and water engineering. The contributions by the authors of different chapters of this book are very valuable which are duly acknowledged. The authors are well experts in their fields and have long years of experience in these areas. It is needless to mention that without their support, this book would have not been published successfully. Their names are mentioned in each chapter and also separately in the list of contributors.

I take the opportunity to offer my heartfelt obligations to Prof. Dr. Megh R. Goyal, "*Father of Irrigation Engineering of 20th Century in Puerto Rico*" and editor of this book series, who has benevolently given me an opportunity to serve as the co-editor of this book. Through his arduous task of editing various books in agricultural engineering, Dr. Goyal has benefitted educators, planners, decision makers, and farmers throughout the world. My special thanks to all the editorial staff of Apple Academic Press, Inc., for making every effort to publish this volume.

Readers are encouraged to translate the knowledge and techniques discussed in this book to address the real-world opportunities and challenges. The readers are also requested to offer constructive suggestions that may help to improve the next edition. I express my deep appreciation to my family, friends, and colleagues and contributors for their help and moral support during the preparation of the book.

— **Susmitha S. Nambuthiri, PhD**
Co-editor

PREFACE 3 BY RICHARD KOECH

Agriculture is vital for the provision of food and fiber to humankind. In many parts of the world, especially in developing countries, the majority of people depend on agriculture for their livelihoods. While thousands of years ago agriculture was largely practiced using traditional means characterized by high labor requirement, in the modern times and especially in the developed world, it has been revolutionized through the use of new technologies and practices. In the recent past, however, agriculture has come under increased pressure from among other factors, increasing human population and climate variability.

In particular, competition for the scarce freshwater resources has risen tremendously in the recent decades. Agriculture in many countries is the major consumer of freshwater, most of which is used for the irrigation of crops, fibers, and pastures. The increasing human population and improving lifestyles has necessitated that more water be allocated for domestic consumption. On the other hand, with increased environmental awareness, there has been a move toward ensuring that water resources are used in a sustainable manner, with sufficient quantities set aside for ecological purposes. In some countries, for instance, Australia, deliberate efforts have been taken to reallocate more water for environmental purposes.

Therefore, to increase the production of food, fibers, and pastures through irrigation under the conditions of increased competition for water resources, research and development must be undertaken to uncover effective, innovative, and sustainable solutions. Possible strategies to this problem may be broadly categorized into agronomic and engineering approaches. Agronomic approaches focus on the improvement of the crop, fiber, or pasture, for instance, through better variety selection, plant breeding and nutrition/fertilization, and disease/pest protection. On the other hand, engineering strategies focus on the technologies, mechanisms, water-use efficiencies, and management of irrigation systems.

This book is divided into four parts: (i) engineering interventions in irrigation management, (ii) technological interventions in management of soil properties, (iii) technological inventions for soil and water conservation, and (iv) crop management for non-conventional use. Both engineering and agronomic aspects of agricultural management are discussed using a total of 13 chapters contributed by experienced authors who are recognized as experts in various fields. It is anticipated that the book will be invaluable to academicians, researchers, engineers, agronomists, extension officers, students, and farmers in the broad discipline of agricultural and biological engineering.

I have a background in agricultural engineering with special interests in irrigation engineering, optimization, and management. I am also interested in the dissemination of research findings through writing and publishing of technical materials, including books and articles. In this regard, I wish to wholeheartedly thank Prof. Megh R. Goyal, the Senior Editor-in-Chief of this book and whom I have never met in person, for graciously giving me the opportunity to serve as Co-editor. I also thank Apple Academic Press, Inc., for publishing the book and hence making it available to many readers across the globe. Last, but by no means the least, I thank my family and friends for their moral support during the preparation of this book.

— **Richard Koech, PhD**
Co-editor

WARNING/DISCLAIMER

Read Carefully

The goal of this compendium, *Technological Interventions in Management of Irrigated Agriculture*, is to guide the world engineering community on how to efficiently process agricultural products. The reader must be aware that dedication, commitment, honesty, and sincerity are most important factors, in a dynamic manner, for a complete success. It is not a one-time reading of this compendium.

The editors, the contributing authors, the publisher, and the printer have made every effort to make this book as complete and as accurate as possible. However, there still may be grammatical errors or mistakes in the content or typography. Therefore, the contents in this book should be considered as a general guide and not a complete solution to address any specific situation in irrigation. For example, each case study in irrigated agriculture is unique, and results are valid only for the specific locality.

The editors, the contributing authors, the publisher, and the printer shall have neither liability nor responsibility to any person, any organization, or entity with respect to any loss or damage caused, or alleged to have caused, directly or indirectly, by information or advice contained in this book. Therefore, the purchaser/reader must assume full responsibility for the use of the book or the information therein.

The mention of commercial brands and trade names are only for technical purposes. It does not mean that a particular product is endorsed over another product or equipment not mentioned. The author, cooperating authors, educational institutions, and the publisher Apple Academic Press, Inc., do not have any preference for a particular product.

All web links that are mentioned in this book were active on March 31, 2017. The editors, the contributing authors, the publisher, and the printing company shall have neither liability nor responsibility, if any of the web links is inactive at the time of reading of this book.

ABOUT SENIOR EDITOR-IN-CHIEF

Megh R. Goyal, PhD, PE, is a Retired Professor in Agricultural and Biomedical Engineering from the General Engineering Department in the College of Engineering at University of Puerto Rico–Mayaguez Campus; and Senior Acquisitions Editor and Senior Technical Editor-in-Chief in Agriculture and Biomedical Engineering for Apple Academic Press Inc.

He has worked as a Soil Conservation Inspector and as a Research Assistant at Haryana Agricultural University and Ohio State University. He was first agricultural engineer to receive the professional license in Agricultural Engineering in 1986 from the College of Engineers and Surveyors of Puerto Rico. On September 16, 2005, he was proclaimed as "Father of Irrigation Engineering in Puerto Rico for the twentieth century" by the ASABE, Puerto Rico Section, for his pioneering work on micro irrigation, evapotranspiration, agroclimatology, and soil and water engineering. During his professional career of 45 years, he has received many prestigious awards. A prolific author and editor, he has written more than 200 journal articles and textbooks and has edited over 50 books. He received his BSc degree in engineering from Punjab Agricultural University, Ludhiana, India; his MSc and PhD degrees from Ohio State University, Columbus; and his Master of Divinity degree from Puerto Rico Evangelical Seminary, Hato Rey, Puerto Rico, USA.

ABOUT THE CO-EDITORS

Dr. Susmitha Nambuthiri

Susmitha Nambuthiri, PhD, is a former postdoctoral scientist in the Horticulture Department of Purdue University, West Lafayette, Indiana, USA. She has served as an agronomic resource professional with the state agriculture department of the Government of Kerala in India and has conducted several irrigation management projects across various crop production systems. She published more than 20 peer-reviewed journal articles, four book chapters, and more than 50 conference proceedings and abstracts in the area of optimizing soil and water management resources for sustainable crop production. Dr. Susmitha received her Master's degree in soil science from Tamil Nadu Agricultural University, India, and her Doctorate in soil science from the University of Kentucky, Lexington, Kentucky, USA.

Dr. Richard Koech, PhD

Richard Koech, PhD, is an Agricultural Engineer with diverse research interests that include irrigation engineering, agricultural water management, hydraulics and fluid mechanics, flow measurement and instrumentation, computational fluid dynamics, and, spatial analysis and GIS modeling. He is presently a Program Leader (Agricultural Water Management) at the University of New England, Armidale, Australia. His duties include research, teaching, and supervision of undergraduate and postgraduate research students. Dr. Koech has authored about 20 journal articles and conference papers. He has also written several book chapters and is a reviewer for many journals in his area of expertise. He worked as an Agricultural Engineer in the Ministry of Agriculture in Kenya for nine years and later as a Lecturer at Masinde Muliro University of Science and Technology (MMUST), also in Kenya. Before joining the University of New England, Dr. Koech worked as a Research Engineer at the University of South Australia in Adelaide, Australia. In his career, Dr. Koech has received scholarship awards from the Japan International Cooperation Agency (JICA) to attend a short course at Obihiro University of Agriculture and Veterinary Medicine in Japan; Australia Aid for International Development (AusAID) to pursue an MSc degree; and the University of Southern Queensland (USQ) to undertake PhD research.

Dr. Koech obtained a BSc degree in Agricultural Engineering from Egerton University (Kenya) and his master's and PhD degrees from the University of Southern Queensland, Toowoomba, Australia.

OTHER BOOKS ON AGRICULTURAL & BIOLOGICAL ENGINEERING BY APPLE ACADEMIC PRESS, INC.

Management of Drip/Trickle or Micro Irrigation
Megh R. Goyal, PhD, PE, Senior Editor-in-Chief

Evapotranspiration: Principles and Applications for Water Management
Megh R. Goyal, PhD, PE, and Eric W. Harmsen, Editors

Book Series: Research Advances in Sustainable Micro Irrigation
Senior Editor-in-Chief: Megh R. Goyal, PhD, PE

Volume 1: Sustainable Micro Irrigation: Principles and Practices
Volume 2: Sustainable Practices in Surface and Subsurface Micro Irrigation
Volume 3: Sustainable Micro Irrigation Management for Trees and Vines
Volume 4: Management, Performance, and Applications of Micro Irrigation Systems
Volume 5: Applications of Furrow and Micro Irrigation in Arid and Semi-Arid Regions
Volume 6: Best Management Practices for Drip Irrigated Crops
Volume 7: Closed Circuit Micro Irrigation Design: Theory and Applications
Volume 8: Wastewater Management for Irrigation: Principles and Practices
Volume 9: Water and Fertigation Management in Micro Irrigation
Volume 10: Innovation in Micro Irrigation Technology

Book Series: Innovations and Challenges in Micro Irrigation
Senior Editor-in-Chief: Megh R. Goyal, PhD, PE

Volume 1: Principles and Management of Clogging in Micro Irrigation
Volume 2: Sustainable Micro Irrigation Design Systems for Agricultural Crops: Methods and Practices
Volume 3: Performance Evaluation of Micro Irrigation Management: Principles and Practices
Volume 4: Potential of Solar Energy and Emerging Technologies in Sustainable Micro Irrigation
Volume 5: Micro Irrigation Management Technological Advances and Their Applications

Volume 6: Micro Irrigation Engineering for Horticultural Crops: Policy Options, Scheduling, and Design
Volume 7: Micro Irrigation Scheduling and Practices
Volume 8: Engineering Interventions in Sustainable Trickle Irrigation

Book Series: Innovations in Agricultural and Biological Engineering
Senior Editor-in-Chief: Megh R. Goyal, PhD, PE

- Modeling Methods and Practices in Soil and Water Engineering
- Food Engineering: Modeling, Emerging issues and Applications.
- Emerging Technologies in Agricultural Engineering
- Dairy Engineering: Advanced Technologies and Their Applications
- Food Process Engineering: Emerging Trends in Research and Their Applications
- Soil and Water Engineering: Principles and Applications of Modeling
- Developing Technologies in Food Science: Status, Applications, and Challenges
- Agricultural and Biological Engineering Practices
- Soil Salinity Management in Agriculture: Emerging Technologies and Applications
- Engineering Practices for Agricultural Production and Water Conservation: An Interdisciplinary Approach
- Flood Assessment: Modeling and Parameterization
- Food Technology: Applied Research and Production Techniques
- Processing Technologies for Milk and Milk Products: Methods, Applications, and Energy Usage
- Engineering Interventions in Agricultural Processing
- Technological Interventions in Processing of Fruits and Vegetables
- Technological Interventions in Management of Irrigated Agriculture
- Engineering Interventions in Foods and Plants
- Technological Interventions in Dairy Science: Innovative Approaches in Processing, Preservation, and Analysis of Milk Products
- Novel Dairy Processing Technologies: Techniques, Management, and Energy Conservation
- Sustainable Biological Systems for Agriculture: Emerging Issues in Nanotechnology, Biofertilizers, Wastewater, and Farm Machines
- State-of-the-Art Technologies in Food Science: Human Health, Emerging Issues and Specialty Topics
- Scientific and Technical Terms in Bioengineering and Biological Engineering

EDITORIAL

Apple Academic Press Inc., (AAP) is publishing various book volumes on the focus areas under book series titled *Innovations in Agricultural and Biological Engineering*. Apple Academic Press, Inc., is publishing volumes in the specialty areas defined by *American Society of Agricultural and Biological Engineers* (http://asabe.org).

The mission of this series is to provide knowledge and techniques for agricultural and biological engineers (ABEs). The series aims to offer high-quality reference and academic content in Agricultural and Biological Engineering (ABE) that is accessible to academicians, researchers, scientists, university faculty, and university-level students and professionals around the world. The following material has been edited/modified and reproduced from: "Megh R. Goyal, 2006. Agricultural and biomedical engineering: Scope and opportunities. Paper Edu_47 Presentation at the Fourth LACCEI International Latin American and Caribbean Conference for Engineering and Technology (LACCEI' 2006): Breaking Frontiers and Barriers in Engineering: Education and Research by LACCEI University of Puerto Rico – Mayaguez Campus, Mayaguez, Puerto Rico, June 21–23."

WHAT IS AGRICULTURAL AND BIOLOGICAL ENGINEERING (ABE)?

"*Agricultural Engineering (AE) involves application of engineering to production, processing, preservation and handling of food, fiber, and shelter. It also includes transfer of technology for the development and welfare of rural communities*," according to http://isae.in. "*ABE is the discipline of engineering that applies engineering principles and the fundamental concepts of biology to agricultural and biological systems and tools, for the safe, efficient and environmentally sensitive production, processing, and management of agricultural, biological, food, and natural resources systems*," according to http://asabe.org. "*AE is the branch of engineering involved with the design of farm machinery, with soil management, land development, and mechanization and automation of livestock farming, and with the efficient planting, harvesting, storage, and processing of farm commodities*," definition by: http://dictionary.reference.com/browse/agricultural+engineering.

"*AE incorporates many science disciplines and technology practices to the efficient production and processing of food, feed, fiber and fuels. It involves disciplines like mechanical engineering (agricultural machinery and automated machine systems), soil science (crop nutrient and fertilization, etc.), environmental sciences (drainage and irrigation), plant biology (seeding and plant growth management), animal science (farm animals and housing), etc.,*" (Source: http://www.bae.ncsu.edu/academic/agricultural-engineering.php)

According to https://en.wikipedia.org/wiki/Biological_engineering: "*Biological engineering (BE) is a science-based discipline that applies concepts and methods of biology to solve real-world problems related to the life sciences or the application thereof. In this context, while traditional engineering applies physical and mathematical sciences to analyze, design and manufacture inanimate tools, structures and processes, biological engineering uses biology to study and advance applications of living systems.*"

SPECIALTY AREAS OF ABE

Agricultural and Biological Engineers (ABEs) ensure that the world has the necessities of life including safe and plentiful food, clean air and water, renewable fuel and energy, safe working conditions, and a healthy environment by employing knowledge and expertise of sciences, both pure and applied, and engineering principles. Biological engineering applies engineering practices to problems and opportunities presented by living things and the natural environment in agriculture. BA engineers understand the interrelationships between technology and living systems, have available a wide variety of employment options. The http://asabe.org indicates that "*ABE embraces a variety of following specialty areas.*" As new technology and information emerge, specialty areas are created, and many overlap with one or more other areas.

1. **Aqua Cultural Engineering**: ABEs help design farm systems for raising fish and shellfish, as well as ornamental and bait fish. They specialize in water quality, biotechnology, machinery, natural resources, feeding and ventilation systems, and sanitation. They seek

ways to reduce pollution from aqua cultural discharges, to reduce excess water use, and to improve farm systems. They also work with aquatic animal harvesting, sorting, and processing.

2. **Biological Engineering** applies engineering practices to problems and opportunities presented by living things and the natural environment.

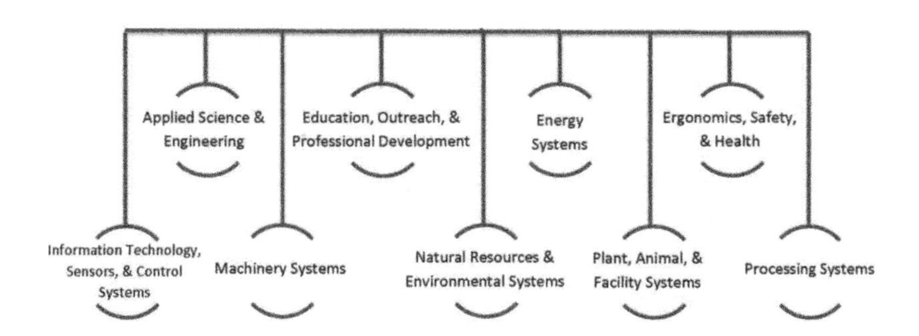

3. **Energy:** ABEs identify and develop viable energy sources – biomass, methane, and vegetable oil, to name a few – and to make these and other systems cleaner and more efficient. These specialists also develop energy conservation strategies to reduce costs and protect the environment, and they design traditional and alternative energy systems to meet the needs of agricultural operations.
4. **Farm Machinery and Power Engineering**: ABEs in this specialty focus on designing advanced equipment, making it more efficient and less demanding of our natural resources. They develop equipment for food processing, highly precise crop spraying, agricultural commodity and waste transport, and turf and landscape maintenance, as well as equipment for such specialized tasks as removing seaweed from beaches. This is in addition to the tractors, tillage equipment, irrigation equipment, and harvest equipment that have done so much to reduce the drudgery of farming.
5. **Food and Process Engineering:** Food and process engineers combine design expertise with manufacturing methods to develop economical and responsible processing solutions for industry. Also food and process engineers look for ways to reduce waste by devising alternatives for treatment, disposal and utilization.
6. **Forest Engineering**: ABEs apply engineering to solve natural resource and environment problems in forest production systems and related manufacturing industries. Engineering skills and expertise are needed to address problems related to equipment design and manufacturing, forest access systems design and construction; machine-soil interaction and erosion control; forest operations analysis and improvement; decision modeling; and wood product design and manufacturing.
7. **Information and Electrical Technologies Engineering** is one of the most versatile areas of the ABE specialty areas, because it is applied to virtually all the others, from machinery design to soil testing to food quality and safety control. Geographic information systems, global positioning systems, machine instrumentation and controls, electromagnetics, bioinformatics, biorobotics, machine vision, sensors, spectroscopy: These are some of the exciting information and electrical technologies being used today and being developed for the future.

8. **Natural Resources:** ABEs with environmental expertise work to better understand the complex mechanics of these resources, so that they can be used efficiently and without degradation. ABEs determine crop water requirements and design irrigation systems. They are experts in agricultural hydrology principles, such as controlling drainage, and they implement ways to control soil erosion and study the environmental effects of sediment on stream quality. Natural resources engineers design, build, operate and maintain water control structures for reservoirs, floodways and channels. They also work on water treatment systems, wetlands protection, and other water issues.
9. **Nursery and Greenhouse Engineering**: In many ways, nursery and greenhouse operations are microcosms of large-scale production agriculture, with many similar needs – irrigation, mechanization, disease and pest control, and nutrient application. However, other engineering needs also present themselves in nursery and greenhouse operations: equipment for transplantation; control systems for temperature, humidity, and ventilation; and plant biology issues, such as hydroponics, tissue culture, and seedling propagation methods. And sometimes the challenges are extraterrestrial: ABEs at NASA are designing greenhouse systems to support a manned expedition to Mars!
10. **Safety and Health:** ABEs analyze health and injury data, the use and possible misuse of machines, and equipment compliance with standards and regulation. They constantly look for ways in which the safety of equipment, materials and agricultural practices can be improved and for ways in which safety and health issues can be communicated to the public.
11. **Structures and Environment:** ABEs with expertise in structures and environment design animal housing, storage structures, and greenhouses, with ventilation systems, temperature and humidity controls, and structural strength appropriate for their climate and purpose. They also devise better practices and systems for storing, recovering, reusing, and transporting waste products.

CAREERS IN AGRICULTURAL AND BIOLOGICAL ENGINEERING

One will find that university ABE programs have many names, such as biological systems engineering, bioresource engineering, environmental engineering, forest engineering, or food and process engineering. Whatever the title, the typical curriculum begins with courses in writing, social sciences, and economics, along with mathematics (calculus and statistics), chemistry, physics, and biology. Student gains a fundamental knowledge of the life sciences and how biological systems interact with their environment. One also takes engineering courses, such as thermodynamics, mechanics, instrumentation and controls, electronics and electrical circuits, and engineering design. Then student adds courses related to particular interests, perhaps including mechanization, soil and water resource management, food and process engineering, industrial microbiology, biological engineering or pest management. As seniors, engineering students' work in a team to design, build, and test new processes or products.

For more information on this series, readers may contact:

Ashish Kumar, Publisher and President
Sandy Sickels, Vice President
Apple Academic Press, Inc.
Fax: 866-222-9549
E-mail: ashish@appleacademicpress.com
http://www.appleacademicpress.com/publishwithus.php

Megh R. Goyal, PhD, PE
Book Series Senior Editor-in-Chief
Innovations in Agricultural and Biological Engineering
E-mail: goyalmegh@gmail.com

PART I

Engineering Interventions in Irrigation Management

CHAPTER 1

SIMPLE SPREADSHEET METHOD FOR SCHEDULING IRRIGATION

ERIC W. HARMSEN*

Department of Agricultural and Biosystems Engineering, University of Puerto Rico, Mayaguez, Puerto Rico 00681, USA

**Corresponding author. *E-mail: eric.harmsen@upr.edu; harmsen1000@gmail.com*

CONTENTS

ABSTRACT

A simple spreadsheet method for scheduling irrigation is presented. The water balance method used is based on the methodology of the United Nations Food and Agriculture Organization. Soil moisture is depleted from the soil profile by evapotranspiration. By maintaining the soil moisture content between the field capacity and the threshold moisture content, water stress can be avoided and 100% of the potential yield can be achieved, ignoring reductions in yield due to other factors such as fertility, disease, and salinity. The method provides an estimate of the crop stress factor, from which the relative seasonal yield can be estimated for many crops. Several graphs are provided in the spreadsheet, which help the user evaluate their real-time water management. In addition to estimates of the relative yield, the spreadsheet also provides an estimate of the lost irrigation (i.e., irrigation lost to surface runoff or deep percolation). Two examples are provided for the bell peppers located on the southern coast of Puerto Rico.

1.1 INTRODUCTION

Over-application of irrigation can result in loss of water, fuel, and chemicals and may contaminate surface or groundwater. Under-application of water can result in a reduction in crop yield.[4] Figure 1.1 shows an example of the relationship between the relative crop yield and the relative seasonal crop water requirement. The curve is based on yield and water-use data for corn, wheat, cotton, barley, pepper, and rice.[7–10] Assuming that the regression equation is applicable to crops grown in the Caribbean region and that the loss in net revenue for a crop is directly proportional to the relative yield reduction, Table 1.1 shows the loss, in dollars per acre, for 17 crops grown in the Caribbean region. The net potential revenue can be obtained when 100% of the crop water requirement is applied. The model budget information for the crops listed in Figure 1.2 was obtained from the *Conjunto Tecnológico* publications of the University of Puerto Rico Agricultural Experiment Station.[12] As an example, suppose that the seasonal water requirement for a 10-acre farm of peppers (pimientos) was 400 mm, but only 320 mm was applied and no additional rainfall was received during the season. The relative amount of water applied then was 320 mm/400 mm = 0.8 or 80%. From Table 1.1, the dollar loss for a pepper crop is \$129/acre × 10 acres = \$1290 for 10 acres (4 ha). The purpose of this exercise is to illustrate the importance of applying the correct amount of water. The values (in dollar) in Table 1.1 are based on model budgets that are several years old. Therefore, the values in the table should be used for illustration purposes only.

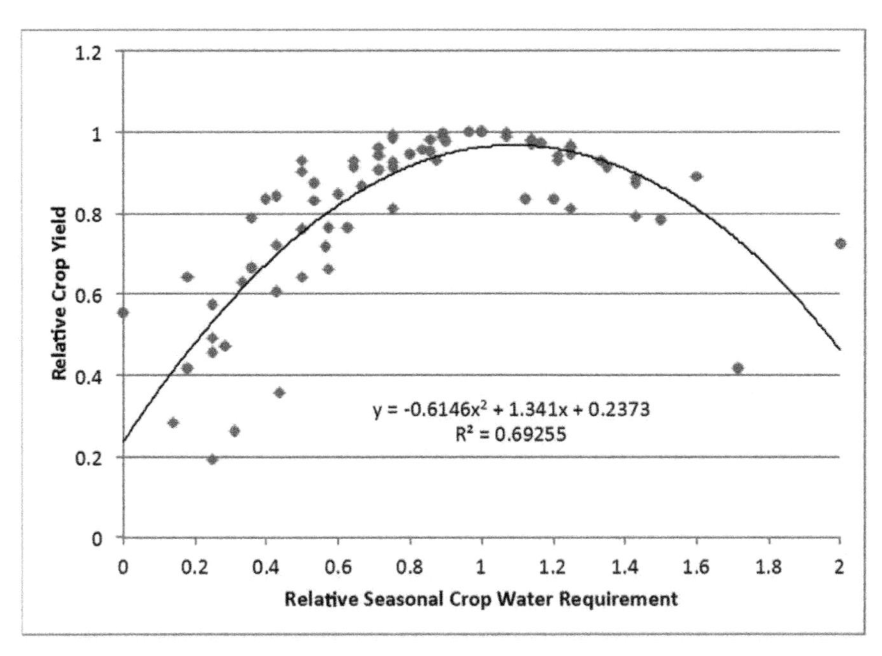

FIGURE 1.1 The relationship between the relative crop yield and the relative seasonal crop water requirement.

	Percentage of Crop Water Requirement Applied						
	40	50	80	100	130	150	180
CROP*	$ Lost / Acre						
Gandules	47	32	10	0	12	35	69
Pepinillo	111	76	25	0	15	56	124
Repollo	256	174	57	0	21	103	247
Sandia	293	199	65	0	23	114	277
Platanos y Guineos, Plantilla	318	216	71	0	24	122	299
Calabaza	390	265	87	0	27	146	359
Cebolla	543	369	121	0	34	195	490
Pimiento	578	393	129	0	36	206	519
Barenjena	757	514	169	0	44	264	670
Platanos y Guineos, Reton~o	1,006	684	225	0	76	388	945
Melon, Cantaloupe y Honeydew	1,027	698	229	0	56	352	899
Raices y Tuberculos	1,041	707	232	0	57	356	911

FIGURE 1.2 Amount, in dollars, lost per acre as a percentage of crop water requirement applied (*based on model budget data from the *Conjunto Tecnológico, UPR Experiment Station*[12]).

Various methods for scheduling irrigation exist, including evapotranspiration, soil moisture, and water balance methods. The method discussed in this chapter is a combination of the evapotranspiration and water balance methods. The advantage of this method is that soil samples do not need to be taken to obtain the soil moisture content. Harmsen[4] presented an evapotranspiration approach based on daily maps of reference evapotranspiration for Puerto Rico. The operational algorithm provides estimates of daily reference evapotranspiration for these three methods.[5,6]

This chapter presents simple spreadsheet method for scheduling irrigation.

1.2 METHODS

In this section, a simple spreadsheet method for scheduling irrigation is described. The methodology used in the spreadsheet is based on the United Nations Food and Agriculture Organization (FAO), Irrigation and Drainage Paper 56, *Crop evapotranspiration—Guidelines for computing crop water requirements*.[1] The spreadsheet can be downloaded from the Internet free of charge from: https://pragwater.com/2011/12/17/a-simple-irrigation-scheduling-spreadsheet-program/. Figures 1.3 and 1.4 show columns A–H

Enter	Enter	Enter		Enter	Enter		
Date	Field Capacity	Wilting Point	Total Available Water	Root Depth	Management Allowed Deficit	Readily Available Moisture Content	Threshold Moisture Content
	FC	WP	TAW	RD	MAD	RAW	θ_t
	%	%	%	m	fraction	%	%
3/14/16	37	20	17	0.25	0.3	5.1	31.9
3/15/16	37	20	17	0.26	0.3	5.1	31.9
3/16/16	37	20	17	0.27	0.3	5.1	31.9
3/17/16	37	20	17	0.28	0.3	5.1	31.9
3/18/16	37	20	17	0.29	0.3	5.1	31.9
3/19/16	37	20	17	0.30	0.3	5.1	31.9
3/20/16	37	20	17	0.31	0.3	5.1	31.9
3/21/16	37	20	17	0.32	0.3	5.1	31.9
3/22/16	37	20	17	0.33	0.3	5.1	31.9
3/23/16	37	20	17	0.34	0.3	5.1	31.9
3/24/16	37	20	17	0.35	0.3	5.1	31.9
3/25/16	37	20	17	0.36	0.3	5.1	31.9
3/26/16	37	20	17	0.37	0.3	5.1	31.9
3/27/16	37	20	17	0.38	0.3	5.1	31.9
3/28/16	37	20	17	0.39	0.3	5.1	31.9
3/29/16	37	20	17	0.40	0.3	5.1	31.9
3/30/16	37	20	17	0.41	0.3	5.1	31.9
3/31/16	37	20	17	0.42	0.3	5.1	31.9
4/1/16	37	20	17	0.43	0.3	5.1	31.9

FIGURE 1.3 Columns A–H of the irrigation scheduling spreadsheet (Soil Moisture Worksheet Tab).

and I–Q of the *Soil Moisture Worksheet*, respectively, and Table 1.1 lists the definition of these columns. Figure 1.5 shows columns A–H for *Application Rate Worksheet*. The user only needs to input data into the yellow highlighted cells. Values in all other cells are calculated automatically. The spreadsheet can be filled-in on a daily basis or could be updated once in a week. It can also be used to evaluate hypothetical scenarios to evaluate the potential

I	J	K	L	M	N	O	P	Q
Enter on First Date		Enter Daily				Enter		
Moisture Content	Crop Stress Factor	Average Crop Evapotranspiration	Average Evapotranspiration Adjusted for Stress	Soil Water Deficit	Irrigation needed	Applied Irrigation or Rainfall	Did Stress Occur?	Lost Irrigation
θ	K_s	ET_c	$ET_{c\,adj}$					
%		mm	mm	%	mm	mm		mm
34.00	1.00	3.80	3.80	3.0	8	0	NO	0
32.54	1.00	3.90	3.90	4.5	12	0	NO	0.00
31.09	0.93	3.80	3.54	5.9	16	0	YES	0.00
36.97	1.00	4.00	4.00	0.0	0	20	NO	0.00
35.59	1.00	4.20	4.20	1.4	4	0	NO	0.00
34.19	1.00	3.90	3.90	2.8	8	0	NO	0.00
32.93	1.00	3.90	3.90	4.1	13	0	NO	0.00
31.72	0.98	4.20	4.13	5.3	17	0	YES	0.00
37.00	1.00	4.20	4.20	0.0	0	22	NO	0.13
35.76	1.00	4.10	4.10	1.2	4	0	NO	0.00
34.59	1.00	4.30	4.30	2.4	8	0	NO	0.00
33.40	1.00	4.20	4.20	3.6	13	0	NO	0.00
32.26	1.00	4.30	4.30	4.7	18	0	NO	0.00
31.13	0.94	4.40	4.12	5.9	22	0	YES	0.00
35.72	1.00	4.50	4.50	1.3	5	22	NO	0.00
34.59	1.00	4.60	4.60	2.4	10	0	NO	0.00
33.47	1.00	4.70	4.70	3.5	14	0	NO	0.00
32.35	1.00	4.80	4.80	4.6	20	0	NO	0.00
31.24	0.94	4.80	4.53	5.8	25	0	YES	0.00

FIGURE 1.4 Columns I–Q of the irrigation scheduling spreadsheet (Soil Moisture Worksheet Tab).

TABLE 1.1 Definitions for the Spreadsheet Columns.

Worksheet	Column	Column label	Definition
Soil Moisture Worksheet	A	Date	Month/day/year
	B	FC	Field capacity (%)
	C	WP	Wilting point (%)
	D	TAW	Total available water (%)
	E	RD	Root depth (m)
	F	MAD	Management allowed deficit
	G	RAW	Readily available water (%)
	H	θ_t	Threshold moisture content (%)
	I	θ	Moisture content (%)
	J	K_s	Crop stress factor
	K	ET_c	Crop evapotranspiration (mm)

TABLE 1.1 *(Continued)*

Worksheet	Column	Column label	Definition
	L	$ET_{c\ adj}$	Adjusted crop evapotranspiration (mm)
	M		Soil water deficit (%)
	N		Irrigation needed (mm)
	O		Applied irrigation or rainfall (mm)
	P		Did stress occur (yes or no)
	Q		Lost irrigation (mm)
Application Rate	A	Date	Month/day/year copied from Soil Moisture Worksheet
	B		Irrigation needed (mm) copied from Soil Moisture Worksheet
	C		Field area (acre)
	D		Percent wetted area (%)
	E		Irrigation efficiency (%)
	F		Volume of water to apply (gallons)
	G		Pump manifold flow rate (gallons per min)
	H		Time to apply irrigation (h)

	A	B	C	D	E	F	G	H
1			Enter				Enter	
2	Date	Applied Irrigation or Rainfall	Field Area	Percent Wetted Area	Irrigation Efficiency	Volume of Water to Apply	Pump Manifold Flow Rate	Time to Apply Irrigation
3								
4		mm	Acres	%	%	gallons	Gallons per Minute	Hours
5	3/14/16	0	5	100	90	0	500	0.0
6	3/15/16	0	5	100	90	0	500	0.0
7	3/16/16	0	5	100	90	0	500	0.0
8	3/17/16	20	5	100	90	118777	500	4.0
9	3/18/16	0	5	100	90	0	500	0.0
10	3/19/16	0	5	100	90	0	500	0.0
11	3/20/16	0	5	100	90	0	500	0.0
12	3/21/16	0	5	100	90	0	500	0.0
13	3/22/16	22	5	100	90	130655	500	4.4
14	3/23/16	0	5	100	90	0	500	0.0
15	3/24/16	0	5	100	90	0	500	0.0
16	3/25/16	0	5	100	90	0	500	0.0
17	3/26/16	0	5	100	90	0	500	0.0
18	3/27/16	0	5	100	90	0	500	0.0
19	3/28/16	22	5	100	90	130655	500	4.4
20	3/29/16	0	5	100	90	0	500	0.0
21	3/30/16	0	5	100	90	0	500	0.0
22	3/31/16	0	5	100	90	0	500	0.0
23	4/1/16	0	5	100	90	0	500	0.0

FIGURE 1.5 Columns A–H of the irrigation scheduling spreadsheet (Soil Moisture Worksheet Tab).

crop stress, estimate relative yield, water use, and so on. Owing to space limitations, Figures 1.3–1.5 show only the first 24 days of the crop season.

1.2.1 SPREADSHEET DATA ENTRY

1. In the Soil Moisture Worksheet, enter the date of the first day of the crop season.
2. Enter soil field capacity and wilting point (in %). If user knows the soil texture, one can obtain an estimate of the field capacity and wilting point online from various sources or from textbooks.[2]
3. The rooting depth needs to be entered for each day of the season. If this information is not available, the user can use the planting depth and the maximum root depth, obtained from the literature, and then creates a linear transition from the planting depth to the maximum depth at the end of the season. The maximum rooting depths for various crops are given in Table 22 of Allen et al.[1]
4. Management allowed deficit (MAD) is given in *FAO 56 Table 22*[1]; however, the FAO refers to it as the depletion fraction (*P*). The *P* and MAD refer to the same parameters.
5. Enter soil volumetric moisture content (in %) on the first day of the season. After the first day, the soil moisture will be estimated by the spreadsheet.
6. Enter the evapotranspiration (ET_c) for each day. Many meteorologic stations will provide estimates of daily reference evapotranspiration (ET_o). User will need to multiply the ET_o by the appropriate crop coefficient (K_c) for the crop under consideration to obtain ET_c. Growth stage lengths and K_c values for number of crops are available in Tables 11 and 12, respectively, in FAO Paper 56.[1]
7. Enter the rainfall or irrigation each day.
8. In the next Worksheet Tab (Application Rate), enter the date of the first day of the crop season.
9. In the next Worksheet Tab (Application Rate), enter the area of the field in acres.
10. Enter the percent wetted area. If drip irrigation is used, only a portion of the field may become wet.
11. Enter the percent efficiency of the irrigation system.
12. Enter the pump flow rate in gallons per minute.

1.2.2 USING THE SPREADSHEET TO SCHEDULE THE IRRIGATION

The objective of the analysis is to minimize crop stress as indicated by the crop stress factor (K_s). If this parameter is <1, then the crop is in a state of water stress. To simplify the analysis, the spreadsheet provides the column "Did Stress Occur?" (Column P, Soil Moisture Worksheet). When stress occurs, as indicated by "Yes," then the user should irrigate the crop. To see how much the user should irrigate, see the column called "Irrigation Needed" (Column N, Soil Moisture Worksheet). Enter the amount the user wants to irrigate in the column called "Applied Irrigation or Rainfall" (Column O, Soil Moisture Worksheet). Finally, go to the Application Rate Worksheet Tab to see how many hours the user should run the irrigation system to put on the desired amount of water.

1.3 GRAPHICAL RESULTS

1.3.1 SOIL MOISTURE GRAPH WORKSHEET TAB

If the soil moisture content falls below the green line (threshold moisture content, θ_t), it indicates that the crop is in stress. Throughout the crop season, the graph should never fall below this line, if possible. It is also very important that the soil moisture does not exceed the field capacity because this water will be wasted, either through deep percolation or surface runoff.

1.3.2 $ET_{c\ adj}$ WORKSHEET TAB

The evapotranspiration is shown in this worksheet. $ET_{c\ adj}$ is the crop ET_c multiplied by the crop stress factor and represents the real ET_c. ET_c is also shown in the graph for comparison.

1.3.3 CROP STRESS FACTOR WORKSHEET TAB

This graph shows the crop stress factor as a function of time for the entire crop season. This line should be as close as possible to 1 throughout the season. Any value lower than 1 indicates that stress occurred. Note that in this worksheet, a value of the relative seasonal crop yield is also provided. The yield response factor (K_y) required to calculate the relative yield can be obtained from Table 24 of Allen et al.[1] for various crops.

1.3.4 CUMULATIVE ET VERSUS IRRIGATION WORKSHEET TAB

The purpose of this graph is to show how close the applied irrigation was to the ET_c. Throughout the season and at the end of the season, the value of the cumulative irrigation should be as close as possible to the cumulative ET_c. (Note in this graph that the ET used is the ET_c and not the $ET_{c\ adj}$).

1.3.5 APPLICATION EXAMPLES

1.3.5.1 EXAMPLE 1

In this example, we will schedule irrigations for a bell peppers. If the crop stress occurred for 1 day, then irrigation was applied the next day at more or less the value of the deficit ("Irrigation needed"). The following information applies to this example:

1. The soil is an silty clay loam having a field capacity and wilting point of 37 and 29%, respectively. The initial soil moisture content is 30%.
2. MAD for peppers is 0.3.
3. Initial soil moisture content is 34% for the irrigated area of 5 acre; the percent wetted area is 100%, irrigation efficiency is 90%; and the irrigation pump is capable of pumping 500 gallons per minute.
4. The location is Puerto Rico's southern coast.
5. The length of the growing season is 85 days (March 14–June 6).
6. No rainfall occurred during the growing season.
7. The change in rooting depth with time was obtained from the field.
8. ET_c was obtained from the equation $\mathrm{ET_c = K_c \times ET_o}$, where, ET_o is the reference evapotranspiration and the K_c values for peppers were obtained from the publication Allen et al.[1] ET_o was obtained from a weather station. Nowadays, weather stations routinely estimate ET_o.

Figure 1.6 shows the soil moisture content during the growth of the bell peppers. Note that irrigation was applied when the soil moisture became close to the threshold moisture content. And the amount of irrigation applied was just enough to increase the soil moisture content to the field capacity. Figure 1.7 shows the $ET_{c\ adj}$ and the ET_c. In this case, they are essentially identical because no stress was allowed to occur (i.e., $K_s = 1$ during the entire season; Fig. 1.8). Figure 1.9 shows the cumulative $\mathrm{ET_c}$ versus irrigation.

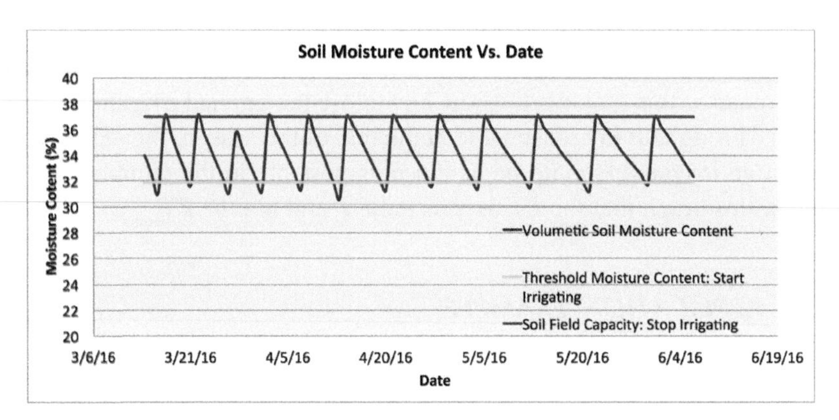

FIGURE 1.6 Soil moisture content during the growing season for Example 1.

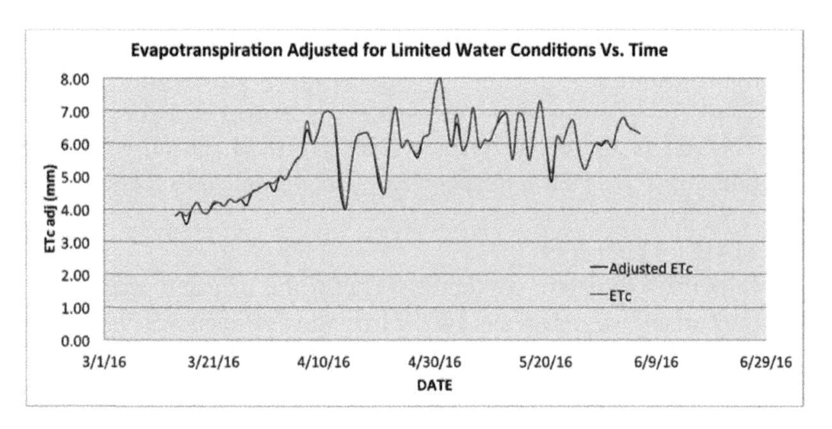

FIGURE 1.7 $ET_{c\,adj}$ and ET_c during the growing season for Example 1.

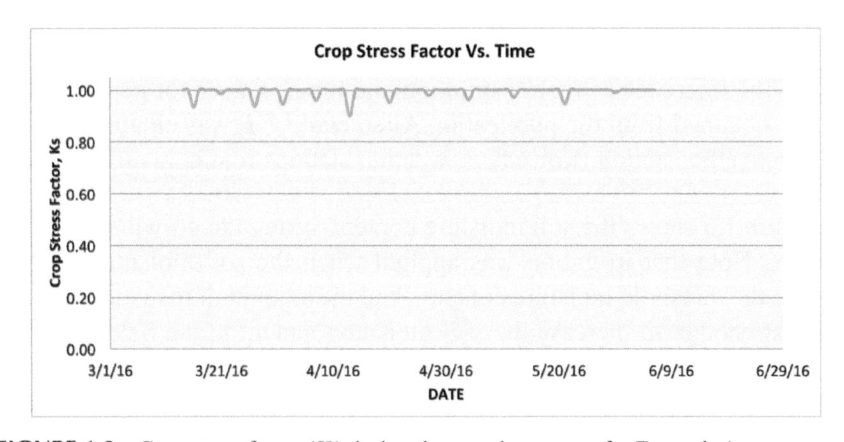

FIGURE 1.8 Crop stress factor (K_s) during the growing season for Example 1.

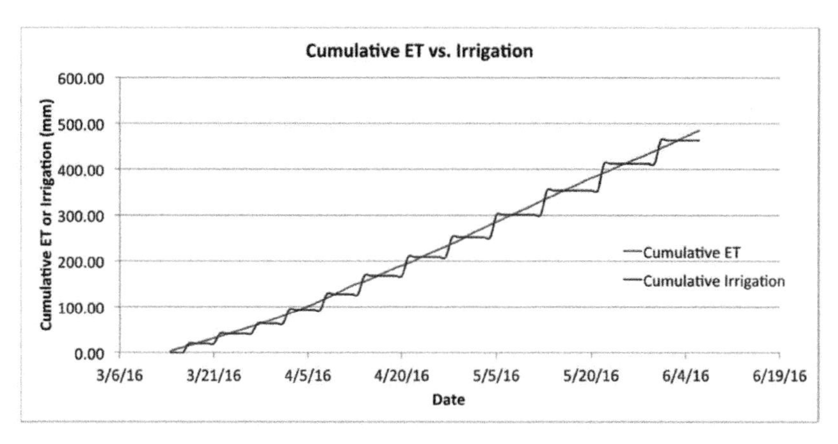

FIGURE 1.9 Cumulative ET_c and irrigation (plus rain) during the growing season for Example 1.

The estimated relative yield was 99.3% based on the value of the seasonal response function (K_y) of 1.1 for pepper.[1] The total water applied was 464 mm with no loss from surface runoff or deep percolation, and the pump was run for a total of 91.9 h during the season.

1.3.5.2 EXAMPLE 2

In this example, the user will apply 25.4 mm (1 in.) per week, which is a common practice. All other data from Example 1 are the same. Figures 1.10–1.13 show the soil moisture content, the $ET_{c\ adj}$ and ET_c, the crop stress factor, and the cumulative irrigation and ET_c during the season.

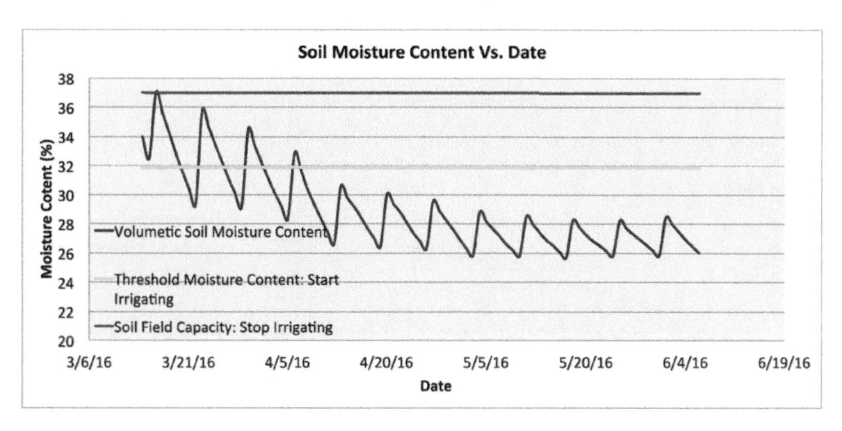

FIGURE 1.10 Soil moisture content during the growing season for Example 2.

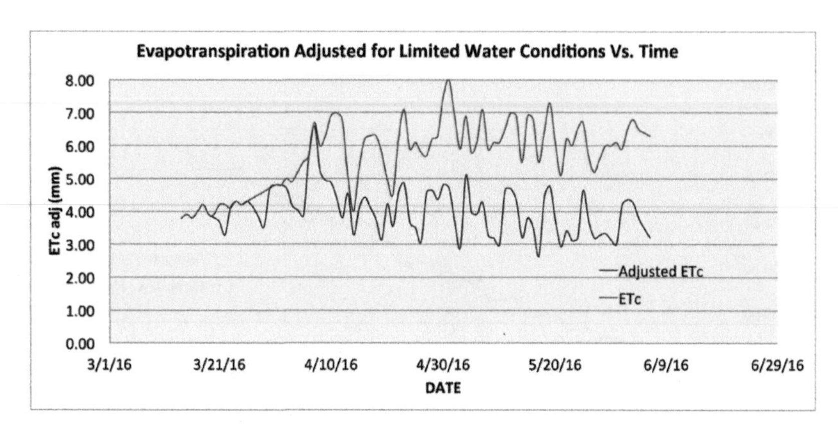

FIGURE 1.11 $ET_{c\ adj}$ and ET_c during the growing season for Example 2.

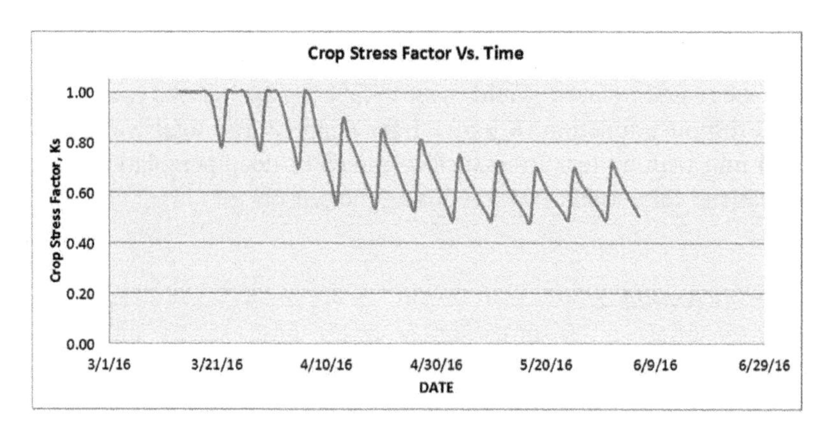

FIGURE 1.12 Crop stress factor (K_s) during the growing season for Example 2.

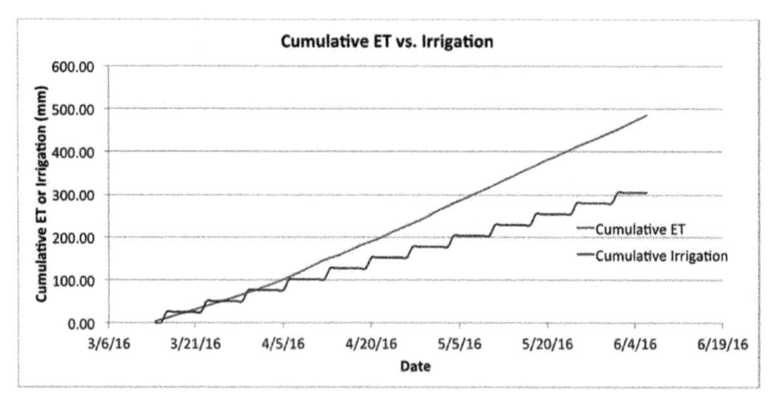

FIGURE 1.13 Cumulative ET_c and irrigation (plus rain) during the growing season for Example 2.

The estimated relative yield is 69.5%. Interpolating the dollar loss per acre from Figure 1.2 gives $300.60 per acre or $1503 for 5 acre. The total water applied was 305 mm, with 3.5 mm irrigation lost to surface runoff and/or deep percolation and the pump was run for a total of 60.3 h during the season. This example clearly shows the potential for yield loss for the commonly used "1-inch per week" method.

1.4 DISCUSSION

The spreadsheet method can provide useful real-time information to an irrigator during the growing season. The method can also be used as an education tool to better understand the implications of different irrigation scheduling schemes. One of the challenges of using the spreadsheet is the requirement to enter the daily crop evapotranspiration. Operational models of ET_o are commonly available on the Internet, for example, the website <https://pragwater.com/daily-reference-evapotranspiration-eto-for-puerto-rico-hispanio-laandjamaica/> provides high-resolution daily ET_o for Puerto Rico, the US Virgin Islands, Dominican Republic, Haiti, Jamaica, and Cuba. The methodologies, which include ET_o estimates using the Hargreaves–Samani,[3] Penman–Monteith,[1] and Priestly–Taylor[11] methods, are described in Harmsen et al.[5]

1.4.1 METHOD LIMITATIONS

Using a spreadsheet method similar to the one described in this chapter is easier than done. If the spreadsheet is not updated frequently, it is quite possible that the crop will begin to experience stress without the grower knowing it. Updating the spreadsheet on a daily basis may not be possible if the source of the ET data is a weather station that needs to be physically downloaded. Weather stations that automatically upload their data to the Internet at least once a day are now commonly available.

Many of the input values used in the spreadsheet may be estimates, for example: field capacity, wilting point, initial soil moisture content (which varies spatially in the field), rooting depth, evapotranspiration, and so on. To the degree that these parameters are in error, then the results from the spreadsheet will also be in error (garbage-in garbage-out).

The spreadsheet graphs may need to be adjusted when the spreadsheet is used with new data. In general, the graphs have been designed with relative horizontal and vertical scales, which change with the data being used.

However, there may be some data that cause the graphical output to appear incorrectly, in which case the user should make the necessary adjustments to the graphs.

1.5 SUMMARY

Water, fuel, chemicals, and money can be lost if an irrigation system is not managed properly. In this chapter, a simple spreadsheet was presented that can help a farmer to schedule irrigation and achieve good water management. The method is based on the United Nations Food and Agriculture Organization Paper No. 56. An example problem was presented in which the soil moisture was maintained between the field capacity and the threshold moisture content, which resulted in no stress and no yield loss. A second example was presented in which 1 in. of water was applied each week to the pepper crop and the resulting yield and dollar losses were 69.5% and $1503, respectively.

ACKNOWLEDGMENT

Financial support for this study was provided by NOAA CREST under NOAA/EPP grant #NA11SEC4810004 and USDA Hatch Project H-402.

KEYWORDS

- **evapotranspiration**
- **irrigation management**
- **Penman–Monteith**
- **soil water deficit**
- **water stress**

REFERENCES

1. Allen, R. G.; Pereira, L. S.; Raes, D.; Smith, M. *Crop Evapotranspiration. Guidelines for Computing Crop Water Requirements*; FAO Irrigation and Drainage Paper No. 56, Rome, Italy, 1998, 300 p.

2. Hargreaves, G. H.; Samani, Z. A. Reference Crop Evapotranspiration from Temperature. *Appl. Eng. Agric. ASAE* **1985,** *1*(2), 96–99.
3. Harmsen, E. W. Technical Note: A Simple Web-based Method for Scheduling Irrigation in Puerto Rico. *J. Agric. Univ. P.R.* **2012,** *96*, 3–4.
4. Harmsen, E. W.; Mecikalski, J.; Mercado, A.; Tosado Cruz, P. In *Estimating Evapotranspiration in the Caribbean region Using Satellite Remote Sensing*, Proceedings of the AWRA Summer Specialty Conference, Tropical Hydrology and Sustainable Water Resources in a Changing Climate, San Juan, Puerto Rico, August 30–September 1, 2010; pp 1–10.
5. Harmsen, E. W.; Mecikalski, J.; Cardona-Soto M. J.; Rojas Gonzalez, A.; Vasquez, R. Estimating Daily Evapotranspiration in Puerto Rico Using Satellite Remote Sensing. *WSEAS Trans. Environ. Dev.* **2009,** *6*(5), 456–465.
6. Irmak, S.; Rathje, W. R. *Plant Growth and Yield as Affected by Wet Soil Conditions Due to Flooding or Over-irrigation*; NebGuide, University of Nebraska–Lincoln Extension, Institute of Agriculture and Natural Resources, 2008, 4 p.
7. Jalota, S. K.; Sood, A. J.; Vitale, D.; Srinivasan, R. Simulated Crop Yields Response to Irrigation Water and Economic Analysis: Increasing Irrigated Water Efficiency in Punjab. *Agron. J.* **2007,** *99*(July–August), 1073–1084.
8. Ottman, M.; Husman, S. H. *Solum Barley*; Publication No. 194030. Cooperative Extension, University of Arizona, College of Agriculture, Tucson, Arizona, 1994, 4 p.
9. Oweis, T.; Hachum, A. Water Harvesting and Supplemental Irrigation for Improved Water Productivity of Dry Farming Systems in West Asia and North Africa. *Agric. Water Manag.* **2006,** *80*, 57–73.
10. Priestly, C. H. B.; Taylor, R. J. On the Assessment of Surface Heat Flux and Evaporation Using Large Scale Parameters. *Mon. Weath. Rev.* **1972,** *100*, 81–92.
11. Fangmeier, D. D.; Elliot, W. J.; Workman, S. R.; Huffman, R. L.; Schwab, G. O. *Soil and Water Conservation Engineering*, 5th Edition; Thomson Delmar Learning: Clifton Park, NY, 2006, 552 p.
12. University of Puerto Rico Agricultural Experiment Station, *Package Practices (Conjunto Tecnológico) Model Budget.* For Pumpkin, Avocado, Cabbage, Coriander, Cucumber melon, Water melon, Onions, Plantain, Banana, Roots and Tubers. http://www.eea.uprm.edu/publicaciones/conjuntos-tecnologicos/ (accessed January 31, 2016).

CHAPTER 2

ESTIMATION OF CROP WATER REQUIREMENT AND SEASONAL IRRIGATION WATER DEMAND FOR *ERAGROSTIS TEF*: MODEL-BASED ANALYSIS

FITSUME YEMENU DESTA[1,*], MICHAEL ESHETU BISA[2], GEBEYEHU TEGENU[3], ASHEBER HAILE[3], and GIRMA KASSA[3]

[1]*School of Environmental and Rural Sciences, University of New England, 137 Mann St. Armidale, Armidale POB 2350, Australia*

[2]*Department of Water Resource and Irrigation Engineering, Arba Minch University, POB 21 Arba Minch, Ethiopia*

[3]*Ethiopian Institute of Agricultural Research, Debre Zeit Agricultural Research Center, POB 32 Debre Zeit, Ethiopia*

[*]*Corresponding author. E-mail: fitsum_yimenu@yahoo.com; fdesta@myune.edu.au*

CONTENTS

ABSTRACT

This chapter has the objective of estimating the water requirements and seasonal irrigation water demand of tef at selected sites in the central parts of Ethiopia, based on CropWat 8.1 MODEL in case of tef (*Eragrostis tef*).

The analysis showed that the seasonal water requirement (mm/season) for tef was 481 for Quncho, 448 for DZ-01-974, and 408 for DZ-01-976. The seasonal irrigation water demand for Quncho variety at Debre Zeit was 681 and 833 mm for heavy and light soils, respectively. At Alem Tena, it was 836 mm.

2.1 INTRODUCTION

Tef (*Eragrostis tef*) is a stable grain crop in Ethiopia and is grown diversely over different agro-ecologies of the country. Gondar, Shewa, and Gojam have been reported to be the major growing areas of the crop. The National Research Council[4] mentioned that tef is a reliable cereal for unreliable climates, especially those with dry seasons of unpredictable occurrence and length. Naturally, the crop withstands low moisture, waterlogging, and anoxic conditions better than maize, wheat, or sorghum. Its production is currently expanding to most of waterlogging-prone vertisol areas replacing wheat and barley[1] and to drought-affected areas[6] replacing maize and sorghum production due to unpredictable and unreliable rainfall.

Although tef has contributed to food security in Ethiopia, its productivity still is below 1 ton/ha and the total production cannot meet the demand and the price has also been increasing by more than fivefold during the last decade.[7] The major reason for low productivity of tef is lodging and the production entirely depends on rain-fed agriculture. The attempt to improve the yield per hectare of the crop is solely due to breeding research. Since research on tef begun in 1950s, more than 30 varieties of tef have been released and some agronomic management options have also been developed.[3] The average grain yield of improved varieties ranges from 1400 to 2200 kg/ha under appropriate cultural practices on the farmers' field.[2,5] This would still make tef less favorite for farmers as the other cereals are yielding more than 3000 kg/ha in rain-fed agriculture. However, from the field experiments conducted at Melkassa, irrigated tef can produce more than threefold of the rain-fed crop.[7] Therefore, irrigated tef production should be a crucial component for improving the productivity and production at national level.

This chapter has been made efforts to determine the tef water requirement and provide preliminary information on irrigation scheduling for tef for selected areas in East Shoa, central parts of Ethiopia.

2.2 METHODOLOGY

This research was conducted in selected districts of East Shoa (Ada, Akaki, and Alem Tena) in Ethiopia (Fig. 2.1). The soil types in the study area were: vertisols for Ada and Akaki, and light soils for Alem Tena. Long-term weather data has been processed to estimate the reference evapotranspiration data. The soil physical properties were adopted from CropWat 8.1 for light, heavy, and medium soils. From the long-term records, the dependable values based on 80% probability of occurrences were selected for further analysis. The crop coefficient of tef was adopted from studies by Yenesew[7] and the length of the growing period for the different varieties of tef has been obtained from the report by Kebebew et al.[3] The length of each growing stage of tef was interpolated from the study by Yenesew,[7] who reported 80 days of growing season for tef. The illustration for crop coefficients versus growth stage can be used from FAO publications, in the case of other crops.

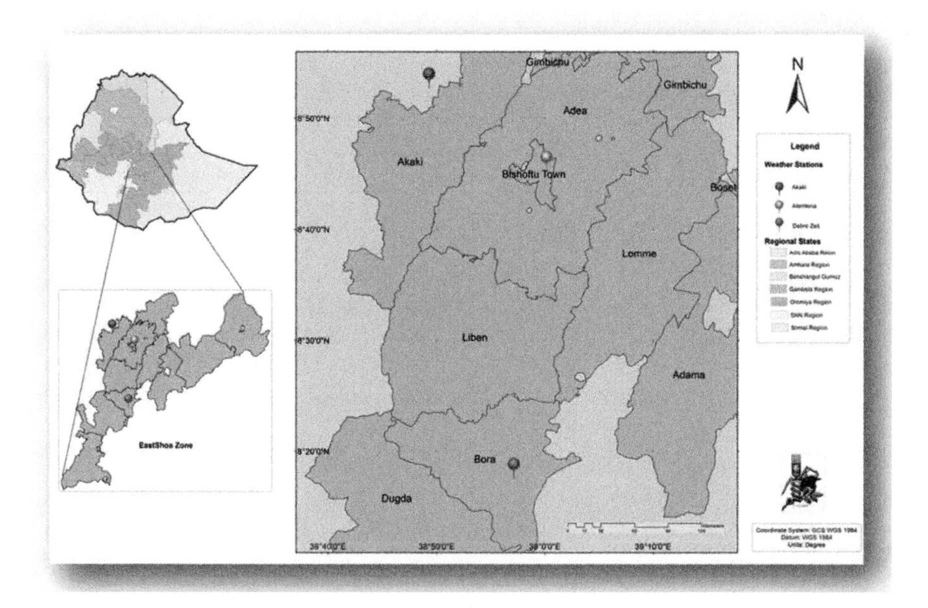

FIGURE 2.1 Location map of the study sites.

The crop water requirement for each of the tef varieties was estimated based on the reference evapotranspiration, length of each of growth stage, and crop coefficient of the stage. The critical depletion level of the crop was estimated for sorghum, as there were no such data for tef until this study was undertaken. The irrigation scheduling was also developed using the CropWat 8.1 software. The ideal planting date was taken as January 1 and the rainfall was purposely set to zero, in order to know the total water demand without any intermittent rainfall.

2.3 RESULTS AND DISCUSSION

The estimated seasonal crop water demand for each variety at Debre Zeit[2,3,6] was 481 mm for Quncho, 448 mm for DZ-01-974, and 401 mm for DZ-01-976 (Table 2.1). The differences in the seasonal water demand are entirely attributed to the variation in the length of the growing season (days to maturity) of each variety. The highest crop water demand of the crop was observed at mid-stage and was 57 (184 mm), 53 (171 mm), and 48 days (152 mm) after planting for Quncho, DZ-01-974, and DZ-01-976, respectively. The high crop water requirement for the initial stage than that for the development stage is difference of 20 days between these two stages; the former has 30 days of stay period.

TABLE 2.1 Crop Water Requirement of Tef for Three Varieties.

Parameter	Qunicho				
	Initial	**Development**	**Mid**	**Late**	**Total seasonal demand**
K_c	0.6	0.8	1.2	0.8	–
ET_o	4.6	4.3	5.3	5.6	–
No. of days	43	14	29	29	–
ET_c	118.68	48.16	184.44	129.92	481.2
DZ-01-974					
K_c	0.6	0.8	1.2	0.8	–
ET_o	4.6	4.3	5.3	5.6	–
No. of days	40	13	27	27	–
ET_c	110.4	44.72	171.72	120.96	447.8
DZ-01-976					
K_c	0.6	0.8	1.2	0.8	–
ET_o	4.6	4.3	5.3	5.6	–
No. of days	36	12	24	24	–
ET_c	99.36	41.28	152.64	107.52	400.8

Table 2.2 shows the lowest values of reference evapotranspiration for Akaki, followed by DebereZeit and Alem Tena. The highest monthly ET_o for Akaki was observed in April (5.24 mm/day), while the lowest was detected in July (4.01 mm/day, which corresponds to the mid of the growing season in the area). Similarly at Alem Tena, the highest was in May (5.6 mm/day) and the lowest value (4.53 mm/day) was for July. The highest value of 5.93 mm/day was observed in March. August had the lowest (3.75 mm/day) reference evapotranspiration value in Debre Zeit but the highest value of 5.76 mm/day was observed both in May and June for this particular site.

TABLE 2.2 Reference Evapotranspiration (mm/day) Values at Different Sites.

Months	Sites		
	Akaki	Alem Tena	Debre Zeit
January	4.87	4.70	4.73
February	4.71	5.14	5.40
March	4.87	5.56	5.37
April	5.24	5.48	5.60
May	4.68	5.60	5.76
June	4.79	5.33	5.76
July	4.01	4.53	3.80
August	4.07	4.59	3.75
September	4.58	4.81	4.13
October	4.84	4.96	5.10
November	4.09	4.82	4.64
December	4.76	4.54	4.49
Average	4.63	5.01	4.88

The gross irrigation water requirement (mm/season) of tef at Debre Zeit (heavy soil) was 571, while that of the light soil was 71. In other words, if one wants to irrigate hectare of land (Quncho variety as a reference), the total seasonal irrigation water need will be 571 and 717 m^3 for heavy and light soils, respectively.

The preliminary irrigation scheduling at Debre Zeit shows that the total number irrigation cycles was 5 for one season for tef on vertisols. Once the soil is set to field capacity before planting (January 1), the next subsequent irrigation application can be February 1, February 27, March 16, and April 6 for heavy soils, based on the results of the model (Fig. 2.2).

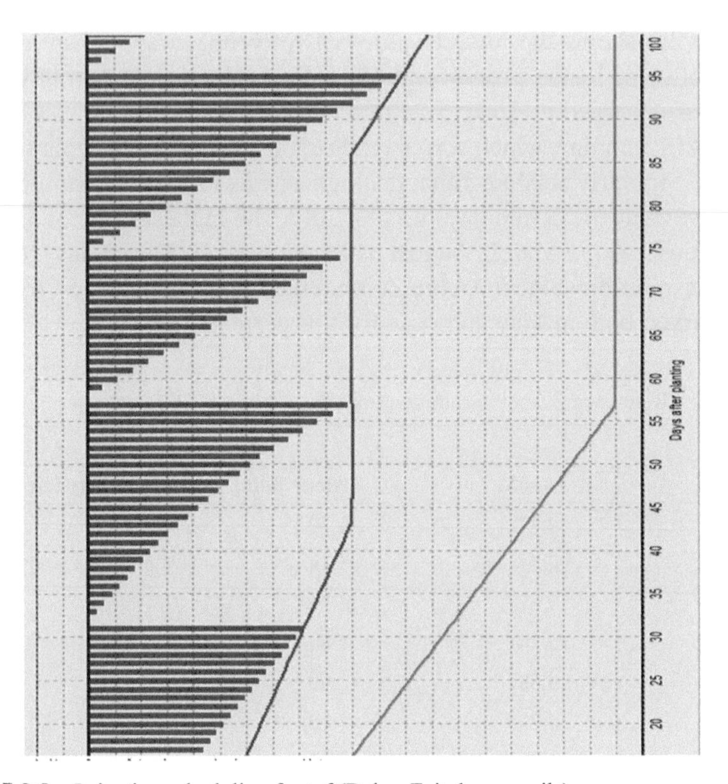

FIGURE 2.2 Irrigation scheduling for tef (Debre Zeit, heavy soils).

For light soils, the number of irrigation cycles was 16 (Fig. 2.3), once set to field capacity at planting (January 1), the next subsequent soil moisture replenishment should commence on January 6, January 12, January 20, January 29, February 8, February 17, February 24, March 2, March 8, March 13, March 18, March 24, March 30, April 6, April 14, and April 24.

The average irrigation interval for light soils is about 7 days. In the first two stages (initial and development), the soils should not be depleted beyond 25 mm of moisture for light soils which have 60 mm available soil water-holding capacity, whereas in heavy soils the depletion could reach to 90 mm before the next irrigation application.

The seasonal gross irrigation water requirement (mm) of tef at Alem Tena was 833 mm (Table 2.3) that consisted of 166.7 for initial, 109.9 for development, 281 for mid, and 275 for late stage, respectively. The total irrigation cycle as estimated by the model was 16, where the frequency of irrigation was about 8 days for initial stage, 5 days for development, 6 for mid, and 10 days for late stage, respectively.

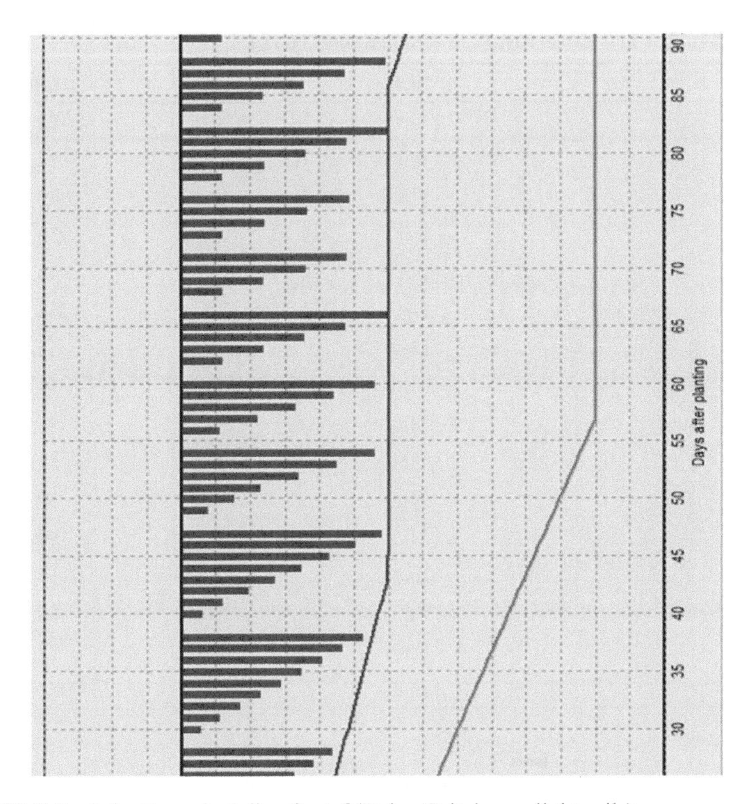

FIGURE 2.3 Irrigation scheduling for tef (Debre Zeit, heavy light soils).

During the first 25 days after planting, the soil moisture depletion should never go beyond 25 mm of the available soil water, because the crop will suffer critical soil moisture stress. Similarly, for the period between 30 and 85 days after planting, the depletion level could be kept until the soil has depleted 30 mm of the available soil moisture that has been replenished to the field capacity. In the late stage, the crop can perform well even when the soil moisture has depleted 35 mm of the available soil moisture.

As Akaki has the same soil type as Debre Zeit, the irrigation scheduling is closely similar in pattern and the seasonal gross irrigation is 590 mm, which is 19 mm higher than that for Debre Zeit. Although annual daily reference evapotranspiration for Debre Zeit is little higher than that for Akaki, yet the months selected for irrigated cropping of tef should have higher values of the reference evapotranspiration and hence the gross irrigation water requirement. Irrigation scheduling program is shown in Figure 2.4 at Alem Tena and Figure 2.5 at Akaki.

TABLE 2.3 Seasonal Gross Irrigation Requirements (GIR).

Day	Stage	GIR
5	Init	23.2
11	Init	27.9
18	Init	32.9
26	Init	38.5
35	Init	44.2
45	Dev	54.8
53	Dev	55.1
59	Mid	53.7
65	Mid	58.6
71	Mid	59.1
76	Mid	50.1
82	Mid	59.8
88	End	59.5
95	End	66.4
103	End	72.6
112	End	77

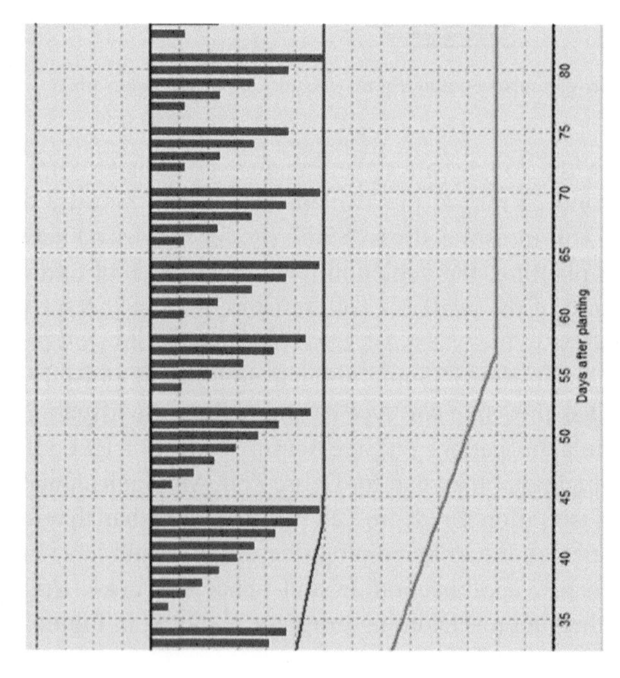

FIGURE 2.4 Irrigation scheduling for tef at Alem Tena (light soils).

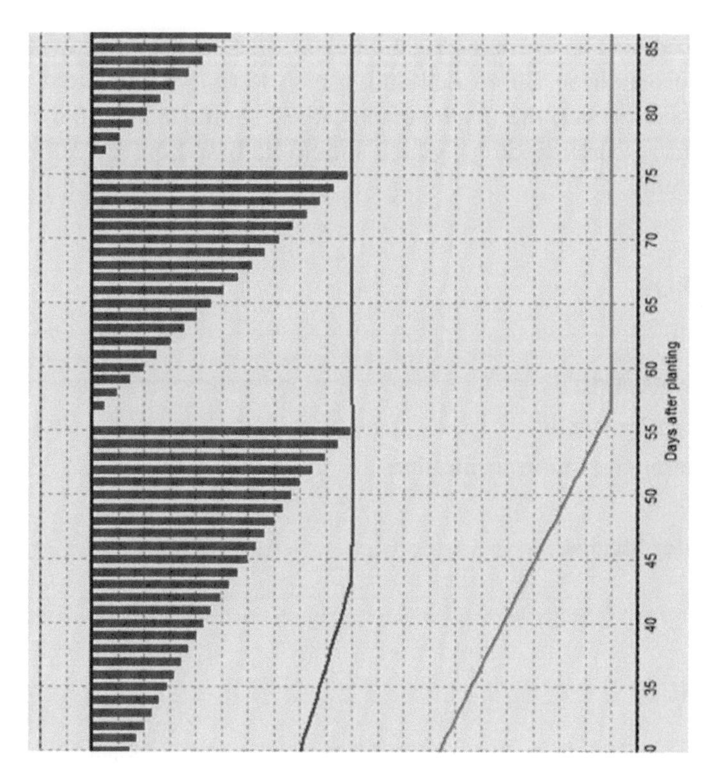

FIGURE 2.5 Irrgation scheduling for tef at Akaki.

2.4 SUMMARY

Tef (*E. tef*) is the stable grain in Ethiopia and is diversely grown over different agro-ecologies of the country. Recently, the increasing demand for the grain has attracted attention for improving the productivity and production at a national level. The irrigated tef is the simplest strategy to increase the production and improve the productivity. For this, the first study is to estimate the crop water requirement of tef and subsequently determine the irrigation scheduling for the areas where the crop could be grown.

The present study had the objective of estimating the water requirements and seasonal irrigation water demand of tef at selected sites in the central parts of Ethiopia, based on CropWat 8.1 MODEL.

The analysis showed that the season seasonal water requirement (mm/season) for tef was 481 for Quncho, 448 for DZ-01-974, and 408 for DZ-01-976. The seasonal irrigation water demand for Quncho variety at Debre Zeit was 681 and 833 mm for heavy and light soils, respectively. At Alem Tena, it was 836 mm. The average irrigation cycles for the whole growing season for vertisols (both at Debre Zeit and Akaki) were five compared to 10 for light soils. However, the results of this study should be substantiated with subsequent research findings, which should be based on real soil moisture measurements.

KEYWORDS

- **CropWat 8.1**
- **irrigation cycle**
- **reference evapotranspiration**
- **tef**
- **water logging**

REFERENCES

1. Abuhay, T.; Teshale, A. Review of Tef Research in the Marginal Rainfall Areas of Ethiopia: Past and Future Prospects. In *Proceedings of the 25th Anniversary of Nazreth Agricultural Research Center: 25 Years of Experience in Lowland Crops Research*, Assefa, H., Ed.; Nazreth Agricultural Research Center: Nazreth, Ethiopia, 1995; pp 57–67.
2. Hailu, T.; Mulu, A.; Kebebew, A. Improved Varieties of Tef (*Eragrostis tef* (Zucc.) Trotter). Releases of 1970–1995, Research Bull. No. 1, Debre Zeit Agricultural Research Center, Alemaya University of Agriculture, Addis Abeba, Ethiopia, 1995, 100 p.
3. Kebebew, A.; Sherif, A.; Getachew, B.; Gizaw, M.; Hailu, T.; Sorrells, M. E. Quncho: The First Popular Tef Variety in Ethiopia. *Int. J. Agric. Sustain.* **2011,** *9*(1), 11–20.
4. National Research Council. *Lost Crops of Africa, Volume I: Grains*; National Academy Press: Washington, DC, USA, 1996; 200 p.
5. Seyfu, K. Cropping Systems, Production Technology, Pests, Diseases, Utilization and Forage Use of Millets with Special Emphasis on Tef in Ethiopia. In *Small Millets in Global Agriculture*; Seetharam, A. K. W., Ed.; Oxford and IBH Publishing Co. Pvt. Ltd.: New Delhi, India, 1989; pp 303–314, Proceedings of the First International Small Millets Workshop, Bangalore, India, October 29–November 2, 1986.
6. Seyfu, K. *Tef (Eragrostis tef) Breeding, Genetic Resources, Agronomy, Utilization and Role in Ethiopian Agriculture*. IAR: Addis Ababa, 1993; 115 p.
7. Yenesew, M. *Agricultural Water Productivity Optimization for Irrigated Tef (Eragrostic tef) in Water Scarce Semi-Arid Region of Ethiopia*. Taylor & Francis Group, 2015; 224 p, ISBN 978-1-138-02766-4.

CHAPTER 3

RECENT TRENDS IN WATER-USE OPTIMIZATION OF SURFACE IRRIGATION SYSTEMS IN AUSTRALIA

R. K. KOECH[1,*] and P. K. LANGAT[2]

[1]*Agricultural Water Management, School of Environmental and Rural Science, University of New England, Armidale NSW 2351, Australia*

[2]*Land Resource Management & Agricultural Technology, University of Nairobi, Queensway House 7th Floor, Kaunda Street, P.O. Box 47309-00100 GPO Nairobi, Kenya.*

[*]*Corresponding author. E-mail: richardkoech@hotmail.com*

CONTENTS

ABSTRACT

The surface system has seen improvements in the recent past ranging from infrastructural upgrades or automation to changes in the management practices and regulatory regimes. Data presented have demonstrated that these improvements have contributed to better water use efficiency of the surface system. Past designs of automatic and real-time control systems have been excessively complex and costly, and hence no significant adoption is evident. A new, simple automatic real-time control system for furrow irrigation is described in this chapter. Field trials have demonstrated potential for water savings, and further development and commercialization of the system is underway.

3.1 INTRODUCTION

Surface irrigation involves conveying water over the field surface by gravity. In surface irrigation systems, the soil surface acts both as a means of conveying water from one end to the other and also the surface through which infiltration occurs. The most common configurations of the surface system are basin, bay, and furrow. Basin as the name suggests is level in all directions and is bounded by earthen embankments with a gate for water inlet. A basin can either be rectangular or square in shape. This system is popular for the irrigation of rice, which is usually grown under ponded conditions. Bay (or border) consists of sloping, rectangular blocks of land with free draining conditions at the lowest end. In furrow irrigation, water is conveyed through small channels with a gentle slope toward the downstream end. The spacing of these channels or furrows generally correspond to the spacing of the crop to be established.

Surface irrigation is the main irrigation system not only in Australia, but also throughout the world. In 2013–2014, the system accounted for 59% of the total irrigated land in Australia, the majority of which is located in the Murray Darling Basin.[1] The system is particularly suited to irrigation of broad-acre crops and those that need to be grown in a pool of water. Pasture for grazing and crops such as cotton and rice are mainly grown using surface systems. Surface systems, in general, are simple and have lower energy and initial capital requirements compared to the conventional pressurized systems.

In the context of surface irrigation, water-use efficiency (WUE) is defined as the ratio of the amount of water required to refill the soil moisture profile to the total amount of water supplied to the field. The three types of common efficiencies of an irrigation system are: (i) application efficiency (AE), (ii) requirement efficiency (RE), and (iii) distribution uniformity (DU). AE is probably the most commonly applied measure in surface irrigation and may be defined as

$$AE = \frac{\text{volume of water added to the root zone}}{\text{total volume of water applied}} \tag{3.1}$$

The RE is an indication of how well the water requirements have been met and may be expressed as

$$RE = \frac{\text{volume added to the root zone}}{\text{water deficit prior to irrigation}} \tag{3.2}$$

WUE may also be characterized in terms of the DU or evenness of the applied water across the field. This may be defined as

$$DU = \frac{\text{average of the lowest 25\% of applied depths}}{\text{average applied depth in the whole field}} \tag{3.3}$$

WUE of typical surface irrigation systems under normal irrigation practices is low and is variable compared to the conventional pressurized systems. Labor requirements under pressurized system can also be as low as 10% of the labor required in typical surface irrigation systems.[38] In an evaluation of cotton under surface irrigation in Queensland, researchers obtained AE ranging from 17 to 100% and an average of 48%.[45] The low efficiencies are attributed to excessive deep drainage losses. The same study demonstrated the potential to increase the average efficiency to about 75% by increasing flow rates and reducing irrigation duration and to the range 85–95% by implementing some form of real-time optimization and control. The AE of 14–90% has been observed in furrow-irrigated sugarcane in the Burdekin region in Queensland.[37] An evaluation of bay irrigation in the Goulburn Murray Irrigation District (GMID) yielded[44] an average AE of 72%.[44]

Improvements have been made to the surface system in the recent past in order to reduce the labor requirement and to improve its WUE. Sufficient focus has gone to the latter especially because of the need to ensure sustainable use of water resources. The Federal and State governments in Australia have provided funding for improvements in the surface system with a condition that the water saved is surrendered and fed back to the environment.[36] These improvements have come in the form of upgrades in the physical irrigation on-farm infrastructure and changes in the irrigation management practices, loosely referred to as automation and control, respectively.

This chapter reviews efforts that have made toward to improve the WUE of the surface irrigation method. The design of an automated real-time optimization system for control of furrow irrigation is also discussed. Field tests demonstrated that the new system has the potential to significantly improve the WUE of the furrow system.

3.2 IRRIGATION AND WATER USE IN AUSTRALIA

Estimates from the Australian Bureau of Statistics (ABS) indicate that the agriculture industry consumed 6987 giga-liters of water during the period 2009–2010, 90% of which was used for the irrigation of crops[2,3] and pastures.[4] Australia is one of the driest continents in the world and therefore irrigation plays a significant role in the production of food and raw materials. Australia is also a vast country, but due to the scarcity of water resources, only a small proportion of the agricultural land is irrigated. For instance, in 2009–2010, less than 1% of the approximately 399 million hectares of agricultural land was irrigated.[4] Nevertheless, irrigation accounts for approximately 30% of food and non-food agricultural produce.[29]

Surface irrigation system has remained the main irrigation method in Australia, with 59% of the total irrigated land in 2013–2014.[1] However, this represents a decrease of 15% in the acreage under surface irrigation when compared to 1990. Koech et al.[25] indicated that there has been an uptake in the use of the sprinkler and the drip/trickle irrigation methods in the same period of time.

The decline in both the total irrigated land in Australia and the proportion of the acreage under irrigation was partly due to the severe drought experienced in the last decade, which also resulted in a reduction of the total land area under irrigation. On the other hand, some irrigators have converted from surface to the low-pressure systems (drip/trickle and sprinkler) which are regarded as more water efficient. Surface systems are comparatively simple, have lower energy, and initial capital requirements compared to conventional pressurized systems and are therefore expected to continue to dominate in the future.

3.3 IMPROVEMENTS OF EFFICIENCY OF SURFACE IRRIGATION SYSTEMS

Efforts to bring about WUE improvements in surface irrigation may be broadly categorized into: physical irrigation infrastructural and hardware development (commonly termed automation); and improved management practices and water metering regulatory requirements. However, it has been noted that certain irrigation management changes can only be implemented if the system is already automated.

3.3.1 AUTOMATION OF SURFACE IRRIGATION SYSTEMS

Automation and control engineering principles have traditionally been applied in industrial processes such as manufacturing and production. In the recent decades, however, these principles have also been applied in precision agriculture including surface irrigation systems. Published works suggest that improvements in the surface system and particularly automation dates back to the 1960s. It would appear that labor saving was the major driving factor behind these initial developments and hence automation was mainly focused on the water delivery system. The mode of water delivery system into a surface-irrigated field determines to a large extent the degree to which the irrigation system can be automated, controlled, and optimized. In the majority of surface irrigation systems, either a head ditch running along the edge of the paddock or a buried pipe equipped with riders are used as sources of water.

Earlier methods used to automate the water delivery system of surface systems included cut-back concept,[17] surge flow irrigation,[46] and cablegation.[21] The cut-back concept was applied in furrow systems in which water was distributed to individual furrows through metal or plastic tubes installed in the side of the concrete lined-ditch. In this system, the initially high furrow inflows were automatically "cut back" or reduced by lowering the depth of water in the supply ditch resulting in a more uniform water application. Surge flow irrigation is achieved by intermittent application of water to furrows as opposed to the conventional continuous flow. The concept was applied in furrow systems using gated pipe water application method and involved the use of an automatic valve to switch the flow between the two sides of the system.

Cablegation, on the other hand, uses a traveling plug inside a gated pipe system on a sloped headland. The slope causes the water application to be restricted to only those gates nearest to the plugs, and the flow gradually decreases as the pug moves further downstream. The above three automatic systems have demonstrated potential for reduced labor requirement, increased WUE, reduced run-off, and greater convenience for farmers. However, there is no indication that any of them has been adopted for commercial applications because they are mostly complex and expensive.

Over the years, significant amount of research has been directed toward investigating the mechanisms of controlling the flow of water into irrigation borders and basins. Metallic or concrete structures commonly referred to as "gates" have traditionally been used for the control of water into borders and basins as well as controlling the level of water in head ditches. Perhaps, the most notable research to date on the automation of gates was undertaken by Humpherys.[18–20] Gates with a single function (either open to admit water or shut off the flow) are described by Humpherys[18] while dual-function gates (open and close) are detailed in a report by Humpherys.[19] The control of these devices may be achieved by a mechanical timer or electronic solenoid, that is, they are time-based open-loop systems.[20] These two types of gates, however, require resetting prior to an irrigation event.

A variety of gates are widely used in the Australian irrigation industry, with many of them being manufactured locally (Table 3.1). A number of these gates can also measure the flow rate when fitted with sensors, thereby replacing Dethridge wheels, which have traditionally been used in the Australian irrigated agriculture.[43]

TABLE 3.1 Commercially Available Channel/Outlet Gates.

Type of gate	Manufacturer	Features
Combination gates	AWMA Pty Ltd.	Multi-leaf and multi-functional control gates
FlumeGate™	Rubicon	Combined flow measurement and control gate. Suitable for use in open canals
Padman's Box Culvert Stop	Padman Stops	Reinforced rubber flap door set in concrete structure. Used as bay outlets or channel check
SlipMeter™	Rubicon	Integrated control gate and flow meter. Can be controlled remotely

At the present stage, many irrigation districts especially in southern parts of Australia already have or are implementing some form of automation in their irrigation systems. It is likely that in the future, on-farm automation and irrigation channel control technologies will see more integration. Some of the automation projects in the irrigation industry in Australia have benefited from the Australian Federal and State governments financing with a precondition that the water saved be surrendered back to the environment.[36]

The use of automatic structures and devices in irrigation guarantees timely farm operations (such as opening and closing of inlet bay structures) and eliminates (or at least reduces) the element of human error. This leads to water savings, the magnitude of which depends in part on the robustness of the control strategy in place. The fact that the water-saving aspect of automation is somehow obscure is perhaps best illustrated by a survey undertaken by the team in Australia.[30] When asked about their perceptions of the benefits of automation, the percentage of farmers who considered labor saving and reduction of water usage as having the greatest benefits were 59 and 19.3%, respectively. The potential increase in the value of land as a result of automation was also widely recognized by the farmers.

Few researchers have attempted to quantify the benefits of automation in irrigation projects. There was estimated water saving of 5–9% in the Shepparton Irrigation Region.[28] Initial results from a bay irrigation project using an intelligent irrigation controller and wireless sensor network at Dookie, Northern Victoria, suggest that average water saving of 38% can be realized.[9]

3.3.2 SMART SURFACE IRRIGATION SYSTEMS

In the recent decades, great advancements have been realized in computing and communication technologies, including the Internet. Consequently, a variety of devices that can be used for purposes such as wireless data transmission or telemetry and feedback control are now commercially available. Various devices can also be controlled remotely using the Internet. An irrigation system with some or all of these devices and capabilities is generally referred to as a "smart irrigation." The majority of these improvements can only be feasible in surface irrigation systems that already have some form of automation.

In the context of surface irrigation, telemetry basically means accessing and/or transmitting data and controlling a system remotely. The use of telemetry system in surface irrigation is a fairly recent development. These systems are vital components of automatic surface irrigation methods for they allow measurement of various parameters (e.g., inflow, advance time, and soil moisture) from a remote location and the results conveyed to a central location mainly via some form of radio communication. Telemetry systems have been integrated with SCADA (supervisory control and data acquisition) in automated surface irrigation systems in Australia.[5,43]

Notable examples of smart surface irrigation systems that are commercially available in Australia include: Aquator,[13] FarmConnect™,[41] WiSA,[49] SamC-Gate Keeper,[35] and the variants of the observant systems.[33] The basic features of these systems are:

- Central control system located on a remote server, farm PC, or one of the radio nodes,
- Actuated bay gates or pipe and riser outlets and pump control units where required,
- Range of different in-field sensors including soil capacitance probes, water level sensors, flowmeters,
- Wireless radio telemetry for communication among the different system components, and
- The use of batteries and/or solar power where no power is available in the remote devices.[25]

Some of these systems allow an operator, for instance, to switch on a pump, open a bay gate, acquire soil moisture data, and automatically switch off the inflow from a remote location. And with the use of the Internet, the irrigation could potentially be controlled from anywhere in the world. FarmConnect™, for example, can graphically display the farm being irrigated by use of satellite mapping and GPS positioning.

Other recent concepts that are commonly associated with improved management of surface irrigation are feedback and real-time control systems. Feedback control refers to control and management decisions (e.g., time to cut off flow into an irrigation furrow or bay) made based on some form of measurement or feedback from the irrigation process. A common example is the placement of water sensors along the length of the furrow or bay, which are then triggered by the arrival of the water front. Feedback control may also be as simple as cutting off the flow when the water front reaches the end of the furrow. Real-time control, on the other hand, is similar in many aspects to feedback control; but as the name suggests the feedback obtained is processed in real-time and used for the management of the same irrigation. These systems also incorporate the measurement or estimation of the soil infiltration characteristics, or the rate of entry of water into the soil profile. This is especially important because in surface systems soil is used both to convey water from one end of the field to the other and the surface through which infiltration occurs. This thus makes the performance of surface systems dependent on the infiltration characteristics of the soil which vary temporally and spatially.[10,31,46]

Real-time control of furrow irrigation has the potential to overcome the effects of the spatial and temporal infiltration variability and proved a significant improvement in irrigation performance.[22,45] These systems inevitably involve a high level of automation and in many cases computer simulation modeling, which is discussed in greater detail in this chapter.

Examples of feedback control systems used in automated bay and basin systems have been described by several scientists.[7,16,33] These systems employ the use of water sensors that are triggered by the advancing water front to send a signal by telemetry to the gates to cut off flow. In the system described by BDA Group,[6,7] the feedback from sensors can also be used to continually adjust the flow rate. It is worth noting that in these systems the control is purely by time or distance to cut off flow, and no attempt was made to estimate the soil infiltration characteristics.[7]

Published works describing real-time control systems include a computerized furrow system,[15] advance rate feedback irrigation system,[27] and Programmable Logic Control or PLC.[42] In a computerized furrow irrigation system utilizing and adaptive algorithm,[15] water is delivered to a block of furrows and the flow rate is monitored while the outflow from selected furrows is monitored using a flume and a depth sensor installed near the downstream end of the field. Feedback obtained is used by a microcomputer to analyze the soil infiltration characteristics and then the inflow is adjusted accordingly. The advance rate feedback irrigation system[27] uses a similar approach but the system determines the appropriate time to cut off flow and the reduction in inflow necessary to achieve a desired net average infiltrated depth. Probably one of the most recent developments in real-time control surface systems is the use of an adaptive cablegation system using a PLC.[42] The system uses two advance points in a computer model to estimate in real time the application depth that will result in the optimal AE of the irrigation event.

Feedback and real-time control systems have generally performed well in research settings and demonstrated potential for labor savings and increased WUEs. However, none of the methods have been widely adopted by irrigators. This may have to do with the high initial costs of these systems and perceived complexities. A new real-time control furrow irrigation system is described later in this paper.

3.3.3 SURFACE IRRIGATION MODELING

Simulation modeling in surface irrigation is the process of mathematically describing the hydraulic characteristics of water as it flows from one end of the field to the other. Surface irrigation simulation models are useful tools both for the design and management stages of the surface systems. They are used to simulate irrigation events and determine efficiencies and uniformities. In the design stage, modeling can be used to optimize variables such as filed slope, length of the field, and design flow rate. Time to cut off flow, inflow rate, and desired depth of application are management decisions that can be optimized using simulation models. Simulation modeling provides an opportunity to identify and evaluate more efficient practices at a lower cost and shorter time compared to field trials.[40]

SIRMOD (surface irrigation simulation, evaluation, and design) developed by Utah State University[47] is perhaps the first surface irrigation model to be used commercially in Australia and is widely accepted as the standard for the evaluation and optimization of surface irrigation.[11] SIRMOD is commercially available through IRRIMATE™ (a suite of hardware and software tools developed by the National Centre for Engineering in Agriculture, NCEA, based at the University of Southern Queensland). New models that have since been developed include SISCO: surface irrigation simulation, calibration, and optimization.[12] This new software can use temporally variable inflow rates as well as spatially variable soil infiltration characteristics, surface roughness, slope, and furrow geometry. This makes it suitable for use in real-time controlled surface irrigation systems.

Within the cotton industry in Australia, the introduction of commercial Irrimate™ evaluation service in 2001 heralded a new era, whereby irrigators were presented with the opportunity to first evaluate the performance of their systems and then initiate improvements. Assessments made after the introduction of the service showed that most clients were able to save an average of 0.15 ML/ha/irrigation after undertaking the necessary design and management changes.[39]

There is abundant evidence showing that the Irrimate™ commercial service has resulted in significant increase in WUEs over time leading to large water savings. For instance, the BDA Group[7] estimated that the cotton industry has saved 400 GL over a 16-year period, corresponding to an increase in WUE of 10%. AE of individual irrigation events, which is the ratio of the volume of water added to the root zone to the total volume of water applied, has also improved considerably from the low (average) of 48%.[45] In an evaluation of 47 furrow irrigation events in the Gwydir and Namoi Valleys, Montgomery and Wigginton[32] found that almost half the events had an AE of over 90%, with only 35% of the events recording AE of less than 80%. They also reported average water saving of 0.18 ML/ha for each irrigation event that was optimized for water saving.

3.3.4 NON-URBAN WATER METERING

In Australia, regulatory requirements are now in place to ensure that non-urban water meters operate at an acceptable level of performance.[8] The requirements include pattern approval by the National Measurement Institute (NMI) and the ability of the meters to perform within maximum limits of error of ±5% in field conditions. This policy change came into effect after studies showed that because of inaccurate flow measurement techniques, the volume of water diverted for irrigation often exceeded entitlement volumes.[48] Studies that informed this change in policy include: Smith and Nayar[43] who showed that meter recording errors within the Coleambally Irrigation networks (prior to its meter upgrade) ranged from +20 to −30%; and Goulburn Murray Water[14] which showed that large Dethridge wheels operated with inaccuracies of between −18 to +3% in Goulburn–Murray irrigation areas.

This new policy came into effect in 2010 and it is expected that all non-urban water meters shall comply with the national meter standards by 2020. Therefore, the full impact of this policy change on the irrigation industry is still unknown, but it is likely that it will provide irrigators with more incentive to increase their WUEs. As already mentioned, some of the automation hardware used for channel flow control in surface irrigation systems are fitted with meters.

3.4 DESIGN OF A SMART FURROW IRRIGATION SYSTEM

As mentioned earlier in this paper, a number of automatic and real-time control systems have been developed for surface irrigation in the past. However, none has presented itself as viable commercially. A new automated real-time furrow irrigation system developed by the author jointly with others will now be described. The real-time concept employed implies that soil infiltration characteristics are estimated, analyzed, and used to optimize the same irrigation event to deliver optimum performance for the current soil conditions. The system has been kept simple, practical, and inexpensive in order to encourage adoptability. The inflow rate is preselected with the irrigation being controlled by varying the time to cut off to give the best performance for that particular furrow length and soil type.

The system (Fig. 3.1) is an integration of a computer hydraulic simulation model and the associated automation hardware and consists of:

- water delivery system,
- inflow measurement system,
- a water sensor to monitor the advance of water along the furrow,
- a computing system, and
- a radio telemetry system to facilitate communication among system components.

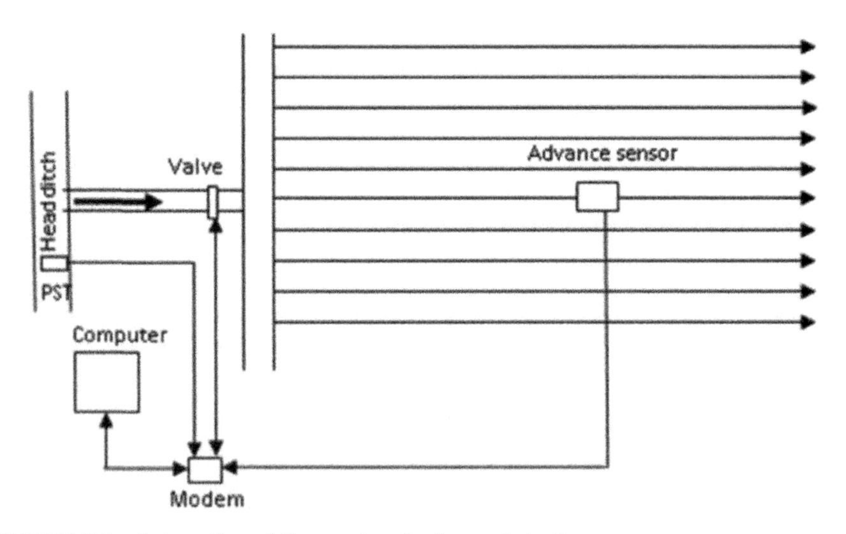

FIGURE 3.1 Automatic real-time system for furrow irrigation.

Water is supplied into a group of furrows from gated pipe, pipes-through-the-bank (PTB), or siphons, and the flow rate is determined through inference from pressure measurement using a pressure-sensitive transducer (PST). The signal output (current) from the PST is then converted into pressure head which is ultimately converted into flow rate using an appropriate calibration equation. Other flow measurement techniques can also be used, including flow meters that may be incompatible with the telemetry system. However, in this case, data would have to be manually entered into the computer model.

When the water front makes contact with the advance water sensor placed approximately midway down the furrow,[26] a signal is sent via radio telemetry to the computer which then calculates the current infiltration characteristics. Based on the user-defined optimization strategy, the hydraulic computer model developed determines the optimal time to cut off the inflow and displays the predicted performance measures and other variables on the main input screen of the model.

Trials for the system developed were undertaken on commercial furrow-irrigated cotton farm near St George and Dalby, Queensland Australia.[23,24] The system performed robustly in the field and demonstrated its potential for substantial water savings; however, the results suggested that there was further scope for improvement in performance.

The system generally predicted shorter irrigation times than those traditionally used by the farmer. This translates to water savings and higher application efficiencies. The field study concluded that the real-time control system is beneficial as it leads to water savings, reduced need for recycling, and potentially lower deep drainage losses. The automated inflow and advance measurement means reduction in the labor requirement. Complete automation of the system will lead to further labor savings. Further development and commercialization of the system, in conjunction with an industry private investor, is still underway (Fig. 3.2).

FIGURE 3.2 Automated furrow irrigation with a Rubicon BayDrive all-in-one in the open position viewed from upstream with the small diameter PTB's through the channel bank on the left.

3.5 SUMMARY

Surface irrigation systems are the most popular methods of irrigating crops and pastures in Australia and globally. However, these systems are often labor-intensive and exhibit low WUE. Rising labor costs especially in the developed countries and the increasing competition for the scarce water resources has put the system in to sharp focus. This paper reviews the measures that have been taken particularly in Australia in the recent decades to improve the WUE of surface irrigation. These measures range from design (including development of hardware and software) to changes in management practices of these systems. It is demonstrated that generally these measures have resulted in the optimization of WUE of the surface system. The paper also presents a conceptual design of a real-time optimization system for the control of furrow irrigation. Trials undertaken using the system demonstrate potential for water savings.

KEYWORDS

- **automation**
- **feedback control**
- **irrigation modeling**
- **real-time control**
- **smart irrigation**
- **surface irrigation**
- **water metering**
- **water-use efficiency**

REFERENCES

1. ABS. *Water Use on Australian Farms;* Australian Bureau of Statistics (ABS), Catalog No. 4618.0, Canberra, Australia, 2004, 2005, 2006, 2008, 2015.
2. ABS. *Water Use on Australian Farms*; Australian Bureau of Statistics (ABS), Catalog No. 4618.0, Canberra, Australia, 2010.
3. ABS. *Water Use on Australian Farms*; Australian Bureau of Statistics (ABS), Catalog No. 4618.0, Canberra, Australia, 2011.

4. ABS. *Year Book Australia*; Australian Bureau of Statistics (ABS), Catalog No. 1301.0, Canberra, Australia, 2012.
5. Armstrong, D. *Modelling Dairy Farming Systems*; Technical Bulletin, Department of Primary Industries, Victoria, Australia, 2009; pp 1–3.
6. BDA Group. *Cost Benefit Analyses of Research Funded by the Cotton Research and Development Corporation (CRDC)*; Report to Cotton Research and Development Corporation: Manuka, ACT, Australia, 2007, pp 1–41.
7. Clemmens, A. J. Feedback Control of Basin Irrigation System. *Irrig. Drainage Eng.* **1992,** *188*(3), 481–496.
8. Commonwealth of Australia. *National Framework for Non-urban Water Metering Policy Paper*. Australian Government: Canberra, December 2009, pp 1–40.
9. Dassanayake, D.; Dassanayake, K.; Malano, H.; Dunn, G. M.; Douglas, P.; Langford, L. In *Water Saving Through Smarter Irrigation in Australian Dairy Farming: Use of Intelligent Irrigation Controller and Wireless Sensor Network*, 18th World IMACS/MODSIM Congress, Cairns, Australia, July 13–17, 2009, pp 1–7.
10. Emilio, C.; Carlos, P. L.; Jose, R. C; Miguel, A. IPE: Model for Management and Control of Furrow Irrigation in Real Time. *Irrig. Drainage Eng.* **1997,** *123*(4), 264–269.
11. Gillies, M. H. Managing the Effect of Infiltration Variability on the Performance of Surface Irrigation. PhD Thesis, University of Southern Queensland, Toowoomba, 2009.
12. Gillies, M. H.; Smith, R. J. SISCO: Surface Irrigation Simulation, Calibration and Optimization. *Irrig. Sci.* **2015,** *33*, 339–355.
13. Poly, G. M. *Aquator*. 2009, http://gmpoly.com.au/product_range/aquator.(accessed March 3, 2016).
14. Goulburn Murray Water. *In-situ REVS Testing of Large Dethridge Meter outlets in the GMID*, 2008, http://www.gmwater.com.au/downloads/gmw/Current_Projects/Insitu_REVS_testing_of_LDMO_in_the_GMID.pdf.(accessed 4/3/2016).
15. Hibbs, R. A.; James, L. G.; Cavalieri, R. P. A Furrow Irrigation Automation System Utilizing Adaptive Control. *Trans. ASABE* **1992,** *35*(3), 1063–1067.
16. Humpherys, A. S.; Fisher, H. D. Water Sensor Feedback Control System for Surface Irrigation. *Trans. ASABE* **1995,** *2*(1), 61–65.
17. Humpherys, A. S. Automation of Furrow Irrigation Systems. *Trans. ASABE* **1971,** *14*(3), 460–470.
18. Humpherys, A. S. Semi-automation of Irrigated Basins and Borders: I. Single Function Turnout Gates. *Trans. ASABE* **1995,** *2*(1), 67–74.
19. Humpherys, A. S. Semi-automation of Irrigated Basins and Borders: II. Dual-function Turnout Gates. *Trans. ASABE* **1995,** *2*(1), 75–82.
20. Humpherys, A. S. Semi-automation of Irrigated Basins and Borders: III. Control Elements and System Operation. *Trans. ASABE* **1995,** *2*(1), 83–91.
21. Kemper, W. D.; Trout, T. J.; Kincaid, D. C. Cablegation: Automated Supply for Surface Irrigation. In *Advances in Irrigation*; Hillel, D., Ed.; Academic Press: London, 1986; pp 1–66.
22. Khatri, K. L.; Smith, R. J. Toward a Simple Real-time Control System for Efficient Management of Furrow Irrigation. *Irrig. Drain.* **2007,** *56*(4), 463–475.
23. Koech, R. K.; Smith, R. J.; Gillies, M. H. A Real-time Optimization System for Automation of Furrow Irrigation. *Irrig. Sci.* **2014,** *32*, 319–327.
24. Koech, R. K.; Smith, R. J.; Gillies, M. H. Evaluating the Performance of a Real-time Optimization System for Furrow Irrigation. *Agric. Water Manag.* **2014,** *142*, 77–87.

25. Koech, R. K.; Smith, R. J.; Gillies, M. H. Trends in the Use of Surface Irrigation in Australian Irrigated Agriculture. *Water J.* **2015,** *42*(5), 84–92.
26. Langat P. K.; Smith R. J.; Raine S. R. Estimating Furrow Irrigation Infiltration Function from Single Advance Point. *Irrig. Sci.* **2008,** *26*(5), 367–374.
27. Latimer, E. A.; Reddell, D. L. Components of an Advance Rate Feedback Irrigation System (ARFIS). *Trans. ASABE* **1990,** *33*(4), 1162–1170.
28. Lavis, A.; Maskey, R.; Lawler, D. *Quantification of Farm Water Savings with Automation*; Department of Primary Industries, Victoria, 2007; pp 1–9.
29. Leonardi, S.; Roth, S. *Irrigation in Australia: Facts and Figures*. National Program for Sustainable Irrigation (NPSI), 2008; 20 p.
30. Maskey, R.; Roberts, G.; Graetz, B. Farmers' Attitudes to the Benefits and Barriers of Adopting Automation for Surface Irrigation on Dairy Farms in Australia. *Irrig. Drain. Syst.* **2001,** *15*, 39–51.
31. McClymont, D. J.; Smith, R. J. Infiltration Parameters From Optimization on Furrow Irrigation Advance Data. *Irrig. Sci.* **1996,** *17*(1), 15–22.
32. Montgomery, J.; Wigginton, D. *Evaluating Furrow Irrigation Performance: Results from the 2006–2007 Season*; Surface Irrigation Cotton CRC Team Bulletin: 2008, 51 p.
33. Niblack, M.; Sanchez, C. A. Automation of Surface Irrigation by Cut-off Time or Cut-off Distance Control. *Trans. ASABE* **2008,** *24*(5), 611–614.
34. Observant. *One Farm, One Solution*. 2015, http://www.observant.net/.(accessed Mar 3, 2016).
35. PadMan Stops. *Innovative Irrigation Solutions*. 2015, http://www.padmanstops.com.au/?product=samc-gate-keeper.(accessed Mar3, 2016).
36. Plusquellec, H. Modernization of Large-scale Irrigation Systems: Is It an Achievable Objective or a Lost Cause? *Irrig. Drain.* **2009,** *58*, 104–120.
37. Raine, S. R.; Shannon, E. L. *Improving the Efficiency and Profitability of Furrow Irrigation for Sugarcane Production*; CSIRO Division of Tropical Crops and Pastures: Brisbane, 1996, 62 p.
38. Raine, S. R.; Foley, J. P. *Comparing Application Systems for Cotton Irrigation—What are the Pros and Cons*? Field to Fashion, 11th Australian Cotton Conference, August 13–15, Australian Cotton Growers Research Association Inc., Brisbane, 2002, 221 p.
39. Raine, S. R.; Montagu, K. *Improving Surface Irrigation Application Performance*; IREC Farmers' Newsletter, No. 172, Autumn 2006, 21 p.
40. Raine, S. R.; Walker, W. R. *A Decision Support Tool for the Design, Management and Evaluation of Surface Irrigation Systems*; Proc. National Conference, Irrigation Association of Australia, May 19–21, Brisbane, 1998, pp 117–123.
41. Rubicon Water. *FarmConnect Software*, 2014.: http://www.rubiconwater.com/catalogue/farmconnect-software-usa.(accessed Mar 3, 2016).
42. Shahidian, S.; Serralheiro, R. Development of Adaptive Surface Irrigation System. *Irrig. Sci.* **2012,** *30*, 69–81.
43. Smith, M.; Nayar, M. In *The First Stage of Coleambally Water Smart Australia Project: Building on CICL's Adoption of Total Channel Control (TCC)*, Irrigation Australia National Conference and Exhibition, May 20–22, Melbourne, 2008, 12 p.
44. Smith, R. J.; Gillies, M. H.; Shanahan, M.; Campbell, B.; Williamson, B. In *Evaluating the Performance of Bay Irrigation in the GMID*, Irrigation Association of Australia National Conference, October 18–21, Swan Hill: Victoria, 2009, 33 p.

45. Smith, R. J.; Raine, S. R.; Minkevich, J. Irrigation Application Efficiency and Deep Drainage Potential Under Surface Irrigated Cotton. *Agric. Water Manag.* **2005,** *71*, 117–130.
46. Walker, W. R. *Guidelines for Designing and Evaluating Surface Irrigation Systems*; FAO Irrigation Drainage Paper 45, 1989, 221 p.
47. Walker, W. R. *SIRMOD II—Irrigation Simulation Software*; Utah State University, Logan, 1997, 132 p.
48. Water Efficiency Division. *National Framework for Non-Urban Water Metering: Final Regulatory Impact Statement*; Department of Environment, Water, Heritage and Arts, Australian Government, June 2009, 33 p.
49. WISA. *Simple, Smart Irrigation Management*, 2015. http://www.wisagroup.com/. (accessed June,23 2015).

CHAPTER 4

HYDRAULIC DESIGN AND MANAGEMENT OF SURFACE IRRIGATION SYSTEM

PHILIP K. LANGAT[1] and RICHARD K. KOECH[2,*]

[1]*Land Resource Management & Agricultural Technology, University of Nairobi, Queensway House 7th Floor, Kaunda Street, P.O. Box 47309, Nairobi GPO 00100, Kenya*

[2]*Agricultural Water Management, School of Environmental and Rural Science, University of New England, Armidale NSW 2351, Australia*

[*]*Corresponding author. E-mail: richardkoech@hotmail.com*

CONTENTS

ABSTRACT

Factors controlling the surface irrigation hydraulic design and performance, and challenges faced by designers and managers have been presented in this chapter.

4.1 INTRODUCTION

Surface irrigation is the most commonly used method of irrigation worldwide and is likely to be same for the foreseeable future. The need for efficient and sustainable use of natural resources demands better management of the irrigation system to improve the current, highly variable (and often low) efficiency of surface irrigation systems, particularly in the developing world where food and fiber demand has also been increasing. On-farm water application methods include surface irrigation, pressurized systems (e.g., sprinklers), and micro-irrigation systems (e.g., trickle irrigation). The choice of the water application method depends on physical and socio-economic factors including product supply (i.e., cost, availability, and quality), soil conditions, field topography and geometry, type and value of the crop, cost and availability of labor, energy costs as well as the practicability and availability of various technologies. The efficiency of the irrigation method depends on irrigation conditions, management practices, and maintenance. Potential application efficiencies of the common irrigation systems still remain much high than attainable efficiencies. Low efficiency in surface irrigation is attributed to a lack of water control and soil spatial and temporal variability which make the irrigation system design and management. The potential efficiency of micro-irrigation systems is high but requires appropriate system design, maintenance, and operation, while pressurized systems can be efficient if designed appropriately and operated under low-wind conditions.

Surface irrigation is used in over 90% of world's total irrigated land despite its high labor and land leveling requirements, poor application efficiencies, and an inability to apply frequent and small applications. It is the most ancient and widespread method of water application. Pressurized and micro-irrigation systems, which are used in less than 10% of global irrigated land, offer the prospect of better irrigation performance than surface irrigation through increased farmer control of water application. With 70% of world's water use occurring in the agricultural sector, it is evident that a very substantial amount of water is being used by surface irrigation systems. While we are living in a high-tech world today, surface irrigation systems and their operation have changed little over the last decades. Siphon tubes and open ditches still account for approximately 75% of the surface irrigation systems. Because of its lack of complicated hardware, surface irrigation systems are often assumed to be the simplest of irrigation systems. In fact, to do an efficient and effective job of irrigating with a surface system requires substantially more knowledge and water management skill than is required for operation of pressurized systems. This requirement stems from the use of the soil as the transfer medium for moving water to various field locations. Pressurized systems use pipelines or drip tubing to facilitate this transfer. Using the soil to transport irrigation water, it results in different amounts of water infiltrating throughout the field (non-uniformity). This is due to both time differences for which water is in contact with the soil (intake opportunity time) and differences in the field's infiltration characteristics (spatial variability of infiltration). Unlike well-designed pressurized irrigation systems, surface systems apply water at a rate equal to or greater than the soil's intake rate, thereby making the spatial variability of infiltration characteristics a major source of application non-uniformity.

This chapter explores optimal design and hydraulic performance of surface irrigation.

4.2 SURFACE IRRIGATION METHODS

Surface irrigation application methods can be broadly classified into uncontrolled flooding and the controlled methods of:

- Border irrigation
- Basin irrigation
- Furrow irrigation

Border irrigation makes use of parallel ridges to guide a sheet of flowing water as it moves down the slope. The land is divided into a number of long parallel uniformly graded strips 10–100-m wide and 200–1000-m long, known as borders that are separated by low earth banks/ridges. It has no cross slope but a uniform gentle slope in the direction of irrigation. The essential feature is to provide a surface such that water can flow down with a uniform depth. Each strip is irrigated independently by a sheet of water confined by the border ridges. The precision of field topography is of critical consideration but the extended lengths permit better leveling through the use of farm machinery. Border irrigation is suitable for most soils where depth and topography permit the required land leveling at a reasonable cost and without permanent reduction in soil productivity, it is more suitable to soils having low-to-moderate infiltration rates such as loamy soils but unsuitable to coarse sandy soils having high infiltration rates. It is also not suited on soils having very low infiltration rates and is most suitable for irrigating all close growing crops like wheat, barley, fodder crops, and legumes.

Basin irrigation is a common form of surface irrigation, particularly in developing countries or regions with layouts of small fields. This is the simplest method of surface irrigation and is claimed to give higher application efficiencies. There are many variations in its use, but all involve dividing the field into smaller unit areas so that each has a nearly level surface. Bunds or ridges are constructed around the areas forming basins within which irrigation water can be controlled. The basins are filled to the desired depth, and the water is retained until it infiltrates into the soil. When irrigating paddy rice, or ponding water for leaching salts from the soil, the depth of water may be maintained for a considerable period of time by allowing water to continue to flow into the basins. This is similar to border irrigation except that here there is no longitudinal slope on the field and the length may be shorter. Basins may vary in size from 1 m^2 used for growing vegetables to as much as several hectares for the production of rice and other grain crops. Sandy soils require small basins, and clayey soils allow large basins. The objective in selecting the basin size is to enable flooding of the entire area in a reasonable length of time, so that the desired depth of water can be applied with a high degree of uniformity over the entire basin. Cotton, grain, maize, ground nuts, lucerne (alfalfa), pasture, and many other field crops are suited to this system of irrigation. It is seldom used for crops which are sensitive to wet soil conditions around the stems.

Furrow irrigation avoids flooding the entire field surface by channeling the flow along the primary direction of the field using furrows, creases, or corrugations. The size and slope of the furrow depends upon the crops grown, equipment used, and spacing between crop rows. The furrows run down the slope of the land, between individual rows of plants, at spacing typically 0.75–1.5 m. Water infiltrates into the soil and spreads laterally to irrigate the areas between the furrows. The length of time required for the water to flow in the furrows depends on the amount of water required to replenish the root zone and the infiltration rate of the soil and the rate of lateral spread of water in the soil. Both large and small irrigation streams can be used by adjusting the number of furrows irrigated at any one time to suit the available flow. The distinctive feature of furrow irrigation is that the flow into each furrow is set and controlled independently as opposed to borders and basins where the flow is set and controlled on a border-by-border or basin-by-basin basis. To supply water with borders, one is also limited by the available capacity and volume in the supply channel.

Of these methods, furrow irrigation is most commonly used worldwide. Furrow irrigation can be used to irrigate all crops planted in rows, including orchards. It is suitable for irrigating maize, sugarcane, tobacco, cotton, groundnut, potato, and other vegetables. It is suited to all soils except sandy soils due to high infiltration rates. Furrow irrigation is widely used as it provides better on-farm water management flexibility under many surface irrigation conditions and offers the irrigators more opportunity to manage irrigations toward higher efficiencies as field conditions may change for each irrigation across field and throughout a season.

4.3 FACTORS INFLUENCING HYDRAULCS OF SURFACE IRRIGATION

The performance of on-farm irrigation management is a function of the farmer's management skills and irrigation system design.[5,6,9] However, design and management of surface irrigation is a complex process owing to the interaction of several hydraulic and environmental variables.[3] The factors controlling surface irrigation performance are largely those governing the unsteady flow hydraulics in open channels as well as those that are specific to the irrigation function.[10] It is necessary to understand the role of these factors if irrigation performance is to be improved. In general, the main factors influencing irrigation performances include fixed (soil infiltration characteristic and surface roughness), system design factors (inflow rate, field slope, and furrow length), and management factors (inflow rate, time to cut-off, and depth of application).

4.3.1 SOIL INFILTRATION CHARACTERISTICS

The infiltration characteristic of the soil is an explicit component in the continuity equation otherwise known as volume balance equation[14] and is a major determinant of the efficiency and uniformity of surface irrigation application.[10] Infiltration affects the advance, recession, run-off, and volume of infiltration during irrigation.[2] Its temporal and spatial variability makes surface irrigation management a complex process. Design, evaluation, and simulation of surface irrigation systems rely on the knowledge of the infiltration characteristic. The infiltration characteristic may be influenced by agronomic management practices (e.g., cultivation, compaction) but is generally regarded as a fixed parameter in relation to a single irrigation event.

4.3.2 INFLOW (DISCHARGE) RATE

Inflow rate has a large effect on the advance of irrigation water along the furrow, but it has little effect on the rate at which the water recedes off the field. Larger inflow rates cause more rapid advance.[14] Inflow rate also affects irrigation performance of furrows via changes in the depth of water in the furrow during the event and the impacts on wetted perimeter and infiltration.

4.3.3 TARGET DEPTH OF APPLICATION

The target depth of application (Z_{target}) is defined as the amount of water required to be applied to completely fill the root zone soil water deficit.[8] It has historically been regarded as a fixed factor but it is determined by soil water deficit at the time of irrigation and hence is more appropriately regarded as a management variable. Z_{target} does not affect the advance and recession of irrigation water flow but has an impact on the performance parameters (e.g., application efficiency) of the irrigation.

4.3.4 TIME TO CUT-OFF

The time at which the irrigation application is cut-off is an important management variable that is closely associated with inflow rate. Early inflow cut-off may result in inadequate depths of application, poor uniformity, and the water advance not reaching the tail end of the field. Similarly, late cut-off commonly results in low application efficiencies due to percolation losses below the root zone and run-off from the end of the field. Cut-off time has no influence on the rate of advance or recession, but does strongly influence the application efficiency of the irrigation event.

4.3.5 SURFACE ROUGHNESS

The roughness of the irrigated surface impacts on resistance to flow and affects both the depth (area) and velocity of flow. A rough surface has a high resistance (i.e., high Manning n) and, therefore, a greater depth or area of flow and low flow velocity for a given inflow rate (Q). Cultivation, crop growth, soil dispersion, and erosion can all cause changes in the surface roughness from one irrigation event to the next. The relationship of the discharge, cross-section area of flow (A), furrow slope (S_o), hydraulic radius (R), and hydraulic resistance (n) is given by the Manning equation:

$$Q = \frac{AR^{2/3} S_O^{\ 1/2}}{n} \quad (4.1)$$

4.3.6 FIELD SLOPE

Field slope affects the water advance and recession via its effect on the volume of water in the furrow during the irrigation. For a given inflow rate, increasing the furrow slope increases the rates of advance and recession. Field slope is usually fixed at the design stage for any irrigated field.

4.3.7 FIELD LENGTH

The length of the field affecting the performance of surface irrigation may affect the rate of advance but not recession. Long fields commonly have low uniformity of application due to the time required for the advance over the whole field. Table 4.1 provides a summary of how the variables can impact irrigation advance and performance and their importance in terms of irrigation system design and management.[10]

4.4 FARM FIELD CONTROL AND MANAGEMENT CHALLENGES

4.4.1 HYDRAULIC EVALUATION OF SURFACE IRRIGATION

Surface irrigation is often perceived to be an inexpensive, inefficient method of irrigating crops, bound by inherent characteristics, traditional practices, and most wasteful water application. The efficiency of surface irrigation is a function of the field design, infiltration characteristic of the soil, and the

TABLE 4.1 Effect of Surface Irrigation Parameters on Advance and Performance.[10]

Factor	Impact on advance	Impact on performance	Factor type	Comment
Depth of application	–	*	Management	Higher depth of applications—more efficient
Field slope	*	*	Design	Steep slope—rapid advance and recession
Inflow rate	***	***	Design and management	High inflow rate—fast advance, increased tail-water runoff
Length of field	–	**	Design	High efficiency and uniformity difficult on long fields
Soil infiltration characteristic	***	***	Fixed	High infiltration rate—slow advance and rapid recession
Surface roughness	*	*	Fixed	Rough surface—slower advance
Time to cut-off	–	***	Management	Determines opportunity time and deep percolation loss

***High, **medium, *low.

irrigation management practice. The complexity of the interactions of these factors makes it difficult for irrigators to identify optimal design or management practices. A well-designed and managed surface irrigation system may have application efficiencies of up to 90%. However, in practice, many irrigation systems worldwide operate with significantly lower and highly variable efficiencies. Application efficiencies for individual irrigations ranging from 14 to 90% and with seasonal efficiencies commonly between 31 and 62% in Australian sugar and cotton industry.[10]

The soil infiltration characteristic is one of the most important determinants of surface irrigation performance. Infiltration often varies temporally and spatially and, thus, makes the management of surface irrigation a complex process.[1] The spatial and temporal variations commonly found in infiltration characteristics for a particular field also raise concerns regarding the adequacy of generalized design and management guidelines for surface irrigation.

Field evaluation of irrigation systems is used to measure irrigation performance and plays a fundamental role in improving surface irrigation. It provides data for modeling the complex surface irrigations systems processes, model validation, and updating and optimization programming. Data from the water advance and recession are used to derive infiltration parameters,[14] which may be then applied to design, management, and control of the irrigation system in a simulation or optimization model. This information may be used to develop real-time irrigation management decisions.[7] Information obtained may be used to identify the effective and feasible alternatives to improve the system performance including adjustment of management factors (e.g., flow rate, time of irrigation, depth of application) or modification of design factors (e.g., field slope or length) or a combination of both to improve the system.[3] The problems of over-irrigation or under-irrigation, poor distribution of irrigated water over the field, excessive tail water run-off, and deep percolation losses can be easily detected with a field evaluation of surface irrigation systems.[13]

In surface irrigation system, soil surface is used to convey water across the field and thus low capital cost, although design and management problems are often experienced. Both design and management depend, to a high degree, on the soil hydraulic properties such as infiltration rate and surface roughness which can be difficult to measure or predict accurately. Hence, it is common to use a trial-and-error approach to develop proper design and management strategies. A surface irrigation event consists of four phases (Fig. 4.1), namely: advance phase, wetting or ponding phase, depletion phase (vertical recession), and recession phase (horizontal recession).

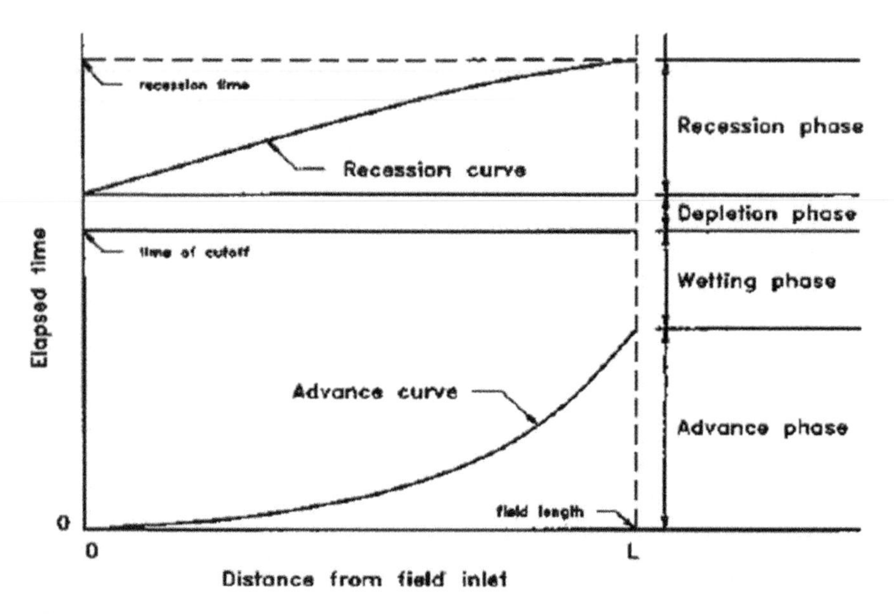

FIGURE 4.1 Time-space trajectory of water during a surface irrigation showing its advance, wetting, depletion, and recession phases.[13]

When water is applied to the field, it advances across the surface until it reaches the end of the field and it is called advance phase. Then the irrigation water either runs off the field or begins to pond on its surface. The time interval between the end of the advance and when the inflow is cut-off is called the wetting or ponding phase. The volume of water on the surface begins to recede after the water is no longer being applied. It either drains from the surface (runoff) or infiltrates into the soil. In describing the hydraulics of the surface flows, the drainage period is subdivided into the depletion phase (vertical recession) and the recession phase (horizontal recession). Depletion is the interval between cut off and the appearance of the first bare soil under the water. Recession begins at that point and continues until the surface is drained.

4.4.2 MANAGEMENT METHODS OF SPATIAL AND TEMPORAL VARIABILITIES

Advance phase has an important role in furrow design procedure and is commonly computed repeatedly when a significant number of alternative irrigation scenarios are examined during the optimization process[14] in such of optimum hydraulic performance of the system. Accurate models for predicting advance time based on computer simulation (kinematic wave, zero inertia, full dynamic, and volume balance) are available. The applicability of the simulation models depends on the sophistication and complexity level associated with the various optimization techniques and thus such techniques are not easily adopted for routine furrow design applications and hence generic parameters are often applied in design and management factors (e.g., wetted perimeter, inflow) and soil intake characteristics across the field.[12] Management methods of spatial and temporal variabilities require intensive data and are often based on volume balance model which assumes depth of flow.

Spatial and temporal variabilities in infiltration have a major impact on field-scale irrigation performance. While temporal infiltration variability may be managed or controlled by infield data collection and real-time optimization,[1] spatial variability requires that the field representative infiltration function is modified to reflect the variations in hydraulic factors (e.g., wetted perimeter, inflow) and soil intake characteristics across the field.[12]

4.4.3 INFILTRATION

Soil infiltration is the most important factor in the surface irrigation process and often the volume of infiltration per unit stream length is difficult to quantify. Theoretically, it is possible to solve equations of unsaturated flow in the soil medium, but this requires empirical soil data relating moisture content to piezometric head and hydraulic conductivity.[12,14] Moreover, in natural soils, surface conditions such as cracking, sealing, and consolidating when wetted are often time-dependent[11] and can be of major significance, as can other heterogeneity and anisotropy factors. The major issues are the infiltrated-depth function of time (volume per unit infiltrating area) and the inclusion of furrow wetted perimeter in calculations of the infiltrated volume per unit length of furrow. How these affect performance prediction, and hence design and also appropriate functional form, their spatial and temporal variabilities, and estimation of the parameters are not yet clear.

4.4.4 INFLUENCE OF WETTED PERIMETER

Furrow cross-section characteristics play a role in irrigation performance and are commonly used as input to simulation or design programs.[14] It is customary in intensive field research to measure furrow cross-sections and derive relationships for cross-sectional area and wetted perimeter as functions of flow depth. The challenge is the accuracy with which such measurements must be made, nor the extent and effect of their spatial and temporal variations (e.g., through erosion), nor the number of cross-sections that should be defined is known.

Wetted perimeter plays a very direct role in infiltration but whether the effect is exactly proportional is not clear. There is some experimental evidence supporting both that wetted perimeter be raised to a power somewhat greater than unity in its relationship with infiltration, and less.[12] The common practice of expressing furrow-infiltration data (collected by inflow/outflow measurement in furrow sections) and entering it into simulation models directly and how well matched are the wetted perimeters in test and simulation still pose a challenge. In real-time control and management of surface irrigation, the quest to use minimum data is enormous[4] because intensive field data is time-consuming and costly.

4.5 SUMMARY

Surface irrigation remains the most commonly used method of water application and will remain so for the foreseeable future. Optimal design and management of this method for efficient water use and sustainability therefore requires more attention from researchers, design engineers, and the land users. Factors controlling the surface irrigation hydraulic design and performance, and challenges faced by designers and managers have been presented in this chapter.

KEYWORDS

- **irrigation efficiency**
- **irrigation management**
- **surface irrigation**
- **surface roughness**
- **wetted perimeter**

REFERENCES

1. Camacho, E.; Perez-Lucena, C.; Roldan-Canas, J.; Alcaide, M. IPE: Model for Management and Control of Furrow Irrigation in Real Time. *J. Irrig. Drain. Eng.* **1997,** *123*(4), 264–269.
2. Fonteh, M. F.; Podmore, T. Physically Based Infiltration Model for Furrow Irrigation. *Agric. Water Manag.* **1993,** *23*, 271–278.
3. Horst, L. *The Dilemmas of Water Division: Considerations and Criteria for Irrigation System Design*; International Water Management Institute (IWMI): Colombo, 1998; 110 p.
4. Khatri, K. L.; Smith, R. J. Real Time Prediction of Soil Infiltration Characteristics for the Management of Furrow Irrigation. *Irrig. Sci.* **2006,** *25*(1), 33–43.
5. Kay, M. Recent Development for Improving Water Management in Surface and Overhead Irrigation. *Agric. Water Manag.* **1990,** *17*, 7–23.
6. Kay, M. *Smallholder Irrigation Technology: Prospects for Sub-Saharan Africa*; Food and Agriculture Organization of the United Nations International Programme for Technology Research in Irrigation and Drainage (IPRID): Rome, 2001; 120 p.
7. Koech, R. K.; Smith R. J.; Gillies, M. H. A Real Time Optimization System for Furrow Irrigation. *Irrig. Sci.* **2014,** *32*(4), 319–327.
8. Pereira, L. S.; Trout, T. J. Irrigation Methods. In *CIGR Handbook of Agricultural Engineering*; Van Lier, H. N., Pereira, L. S., Steiner, F. R., Eds.; Kyoto, Japan; 2007; Vol. 1, pp 297–379.
9. Raine, S. R.; McClymont, D. J.; Smith, R. J. *The Effect of Variable Infiltration on Design and Management Guidelines for Surface Irrigation*, ASSSI National Soil Conference, Brisbane, 1998, 130 p.
10. Raine, S. R.; Smith, R. J. *Simulation Modelling for Surface Irrigation Evaluation. Training in Using SIRMOD and Infilt.V5;* National Centre for Engineering in Agriculture Publication 1000008/4, USQ: Toowoomba, 2005; 121 p.
11. Schwankl, L.; Raghuvanshi, N. S.; Wallender, W. W. Furrow Irrigation Performance Under Spatially Varying Conditions. *J. Irrig. Drain. Eng.* **2000,** *126*(6), 355–361.
12. Strelkoff, T. S.; Clemmens, A. J.; Bautista, E. Field-parameter Estimation for Surface Irrigation Management and Design. *Watershed Manag. Oper. Manag.* **2000,** *105*(142), 1–10.
13. Walker, W. R. *Guidelines for Designing and Evaluating Surface Irrigation Systems*; FAO Irrigation and Drainage Paper No. 45, Rome, 1989, 145 p.
14. Walker, W. R.; Skogerboe, G. V. *Surface Irrigation: Theory and Practice*; Prentice-Hall: Englewood Cliffs, NJ, USA, 1987; 235 p.

PART II

Technological Interventions in Management of Soil Properties

CHAPTER 5

SPATIAL AND TEMPORAL VARIABILITY OF SOIL WATER STORAGE IN A FARMER'S FIELD†

SUSMITHA S. NAMBUTHIRI*

Washington State Department of Health, WA 98512, USA

*Corresponding author. *E-mail: susmitha.agri@gmail.com*

CONTENTS

†This chapter is adapted from Susmitha Nambuthiri, "Soil Water and Crop Growth Processes in a Farmer's Field." PhD Dissertation, submitted to The Graduate School, University of Kentucky, Lexington, Kentucky, USA, 2010.

ABSTRACT

Spatial and temporal variability in soil water storage (SWS) was determined by the interaction of different soil, land, and climatic factors. Absence of temporal stability in the spatial pattern of SWS for the entire profile and different soil depths was confirmed by cumulative probability analysis, relative deviation of mean SWS, MABE analysis spearman correlation, and coherency.

5.1 INTRODUCTION

Farmers experience spatial and temporal variability in grain yield even though they manage field crops spatially homogeneously. Spatial and temporal variability in crop production systems is due to the interaction among various factors such as soil, crop, weather, topography, and other factors in the field. A better understanding of the factors and processes causing spatial variability is important in developing site-specific management strategies.

Accurate assessment of soil water content across the landscape is required to understand spatially varying factors particularly those that affect crop production in a field. Soil water capacitance sensors provide an indirect, non-destructive, and rapid way of estimating volumetric soil water content compared to conventional gravimetric soil sampling. The sensors function based on responses to soil electromagnetic properties and measurement of the dielectric constant (*K*) or relative permittivity of the soil–water–air mixture to estimate soil water content. The *K* of water (78.54 at 22°C) is large compared with that of the soil matrix (<10) and air (1). Therefore, soil water content strongly influences the *K* of the soil–water–air mixture.[29]

Portable and commercially available soil moisture sensing capacitance probe (Diviner 2000 developed by Sentek, Australia) was used in the study. In their study, Evett et al.[6] and Geesing et al.[10] recommended the Diviner probe for routine field soil water studies compared to many other commercially available electromagnetic sensors due to its temperature insensitivity and accuracy of soil water measurements under field soil conditions. The zone of major influence of these sensors represents a cylinder of soil, 10 cm along the axis of the probe, and a circle with 10 cm diameter around the wall of the PVC access tube as reported by Paltineanu and Starr.[28] Thus, the water content may be expressed either as volumetric percentage or as depth of water (millimeters per 10 cm soil depth increment) based on an in-built calibration equation of the probe.

The universal calibration equation supplied by the manufacturer is based on a variety of different soils. The great variability of the K of soil minerals (4–9) and organic matter (1–4) makes it necessary to calibrate soil moisture sensors for a particular soil and for each soil horizon and location according to Baumhardt et al.[1] Accuracy of the moisture estimates from the capacitance probe is affected both by sensor bias and precision according to Evett et al.[6] Thus, a highly precise reading does not always assure high accuracy of measurement. Evett et al.[6] noticed that precision of the probe is affected by soil type, temperature fluctuations, soil wetness, and the calibration equation used. Higher precision is reflected by a smaller SD of soil water content measurements.

High spatio-temporal variability of vadose zone soil moisture has been observed at small spatial scale by many researchers, Brocca et al.[2] and Hupet and Vanclooster.[16] Investigators Gomez-Plaza et al.[11] and Van Pelt and Wierenga[33] have observed that soil moisture played a critical role in the growth and development of crops and their grain yield spatial patterns. Spatial and temporal analysis of soil moisture is helpful in characterizing its inherent spatial variability under field conditions. The near-surface soil moisture content is determined by the interaction of different factors such as topography, vegetation, and soil properties with incoming rainfall as reported by Jacques et al.[18] Analysis of temporal dynamics of soil moisture gained attention with the pioneering work of Vachaud et al.[32] who introduced the temporal stability concept of soil moisture that refers to the ability of some locations to represent the mean, standard deviation, and extreme values of soil moisture at any time of the year. It may also be used to optimize the sampling scheme under field conditions and to better understand processes affecting the soil moisture spatial pattern. Temporal stability of soil moisture reflects the temporal persistence of spatial structure of soil moisture under field conditions. Van Pelt and Wierenga[33] analyzed the temporal stability of the soil water matric potential with a view of optimizing sampling strategy. Temporal persistence of the spatial pattern of soil water content of an area was observed by Gomez-Plaza et al.[12] and Grayson et al.[13] Grayson et al.[13] studied three catchment areas having significant relief and observed that although the overall spatial soil moisture patterns were not time stable, the measurements in a specific subset of the locations could represent mean soil moisture over their areas of interest. Temporal persistence of soil water storage (SWS) at different depths was observed by Pachepsky et al.[27] Time instability of spatial patterns was also found in agricultural fields by many researchers Comegna and Basile[4] and Grayson et al.[13]

The objectives of the current study were to analyze the spatial relationship and persistence of spatial patterns of SWS in heterogeneous field soil conditions and to evaluate the sensitivity of DSSAT crop simulation model to reflect site-specific inputs on soil variability in crop production in a nitrogen fertilizer treatment study.

5.2 MATERIALS AND METHODS

Field experiments were carried out for corn in a farmer's field during 2007 and in Winter Wheat in 2008. The field is located near Princeton (37.045°N, 87.862°W), Caldwell County, Western Kentucky. The field has been under no-till cultivation. The farmer follows corn and double crop soybean after winter wheat. Crider silt loam (Mesic Typic Paleudalfs), a deep, well-drained, moderately permeable soil is the major soil series of the field (Soil Survey Geographic Database, 2008). An ET106 weather station (Campbell Scientific, Inc.) was installed within the field in May 2007 to automatically monitor weather conditions of the field. The weather station could provide data on air temperature (°C), precipitation (mm), relative humidity (%), solar radiation (W m^{-2}), total solar radiation (J m^{-2}), soil temperature (°C) for the upper 10 cm, wind speed (m s^{-1}), wind direction (°) at 10-min intervals throughout the study. The data were downloaded from the station on biweekly basis.

5.2.1 SOIL WATER CONTENT SAMPLING

In order to study the spatial and temporal properties of soil water content in the field, a transect considering the varying landscape of the field was selected. An initial soil sampling was conducted in December 2006 to collect preliminary information on the spatial structure of soil water content in the field. The total length of the transect selected was 465 m. Soil samples were collected from 0 to 10 cm depth. Soil sampling points were located at 5-m intervals, and at each sampling point, samples were taken in duplicate and were separated by about 10 cm. Nested sampling was carried out at 1-m interval for every 25 m to determine small-scale spatial variability of soil water content distribution. Soil samples were thus taken at every 24, 25, and 26 m along the transect.

The soil samples were stored in airtight plastic containers; fresh weight was obtained and samples were oven dried at 105°C for 24 h to obtain a constant dry weight. Gravimetric water content was calculated as follows:

$$\theta_g = \frac{W_w - W_d}{W_d}, \tag{5.1}$$

where W_w is the wet weight of soil sample in grams, W_d the dry weight of the soil sample in grams, and θ_g the gravimetric soil water content.

The field average gravimetric soil water content was 0.273 g g^{-1} of soil. The field was fairly wet since the sampling was conducted following a rainy day. The two replicates were spatially separated by approximately 10 cm only. The average variance between replicates along the transect was 0.0000851 g g^{-2}. The spatial structure of the gravimetric soil water content was characterized using a geostatistical tool, semivariogram suggested by Nielsen and Wendroth.[26] Semivariogram quantifies the spatial continuity of the gravimetric soil water content. In this spatial analysis, gravimetric water content A_i is taken at a lag distance of h at location x_i and was compared to another observation taken at $A_{i+h}(x_{i+h})$. For N pairs of values of A_i separated by a lag distance of h, the average difference for each lag class can be obtained from the semivariogram as below:

$$\gamma(h) = \frac{1}{2N(h)}\sum_{i=1}^{N(h)}\left(A_i(x_i) - A_i\ (x_i + h)\right)^2. \tag{5.2}$$

Based on the semivariogram results, the average standard deviation (SD) between two replications, when lag was 5 m, was 0.0131816 g g^{-1}. The nugget variance (variance at zero lag) was 0.016125 g g^{-2}. Average SD between two replications, when the lag is non-uniform, was 0.0133721 g s^{-1}. The nugget variance was 0.016733 g g^{-2}. From the semivariogram for uniform lag intervals, the nugget variance was 0.00013 g g^{-2}. The spatial dependence of soil water content existed to a range of 38 m. The differences in soil water content between the pairs of all values separated by distances greater than 38 m were not spatially dependent.

The semivariogram for non-uniform lag intervals provided the nugget variance of 0.00014 g g^{-2}. The spatial dependence of soil water content exists to a range of 40 m. The differences in soil water content between the pairs of all values separated by distances greater than 40 m are not spatially dependent. Average of total variance of soil water content along the transect was 0.000274 g of water per gram of soil.

Therefore, it was clear that one cannot avoid the nugget variance by decreasing the sampling interval. This variance may be due to error associated with the sampling device, auger, or any other analytical error. Thus, sampling at distances smaller than 1 m is not going to give additional information about the spatial pattern of soil water content distribution along the transect. The information gained about the spatial dependence of soil water content distribution was used while installing soil water access tubes.

By nested sampling at every 25-m interval, nugget variance was slightly increased. Average variance between replications and field average of variance in gravimetric water content along the transect were also slightly increased with nested sampling. The spatial dependence of gravimetric water content values also increased along the transect.

5.2.2 INSTALLATION OF SOIL MOISTURE SENSING CAPACITANCE PROBE ACCESS TUBES

Based on the semivariogram results of gravimetric soil water content obtained from the initial field study, 45 soil water content sensor access tubes (made from PVC) were installed at 10-m intervals along the transect. The tubes were installed at 44 locations at a depth of 0–80 cm and in one location at a depth of 0–60 cm due to the presence of rock beyond 60 cm soil depth. The tubes were 1-m long with an inside diameter of 5.10 cm and an outside diameter of 5.65 cm. After installation, each tube was left with an extended section of about 5 cm above the soil surface to prevent water entry into the tube. A plastic cap was firmly fitted to the upper end of each tube. A compression rubber plug was used to seal the bottom of the pipe against water and vapor.

5.2.3 SOIL WATER-SENSING CAPACITANCE PROBE

The hand-held Diviner probe measures soil moisture content at regular intervals of 10 cm down through the soil profile. After drying the condensation water from the access tube walls with a dry cloth, readings were taken by lowering the probe down the PVC access tube while point measurements of soil water content for every 10-cm soil depth were recorded.

Three replications were taken at each time of soil water content measurement. The measurements were taken within a time interval of approximately 20 s for which no variation of soil temperature or soil water content was expected. Precision of the probe was assessed by calculating the standard deviation (SD) of soil water content estimates from the three repeated measurements of soil water content taken with the insertion and removal of the probe in a single orientation at each location. The effect of orientation on precision of the probe was evaluated by taking three replicate measurements for each of four different directions by rotating the probe by 90° steps at individual locations in a few times of the study.

5.2.4 STATISTICAL DATA ANALYSIS

Descriptive statistics analysis of each variable considered in the study was performed using MS Excel. Regression analysis was performed to determine the relationship between crop reflectance and crop growth parameters using MS Excel. Linear regression models were fitted for each data set. Pearson correlation coefficients between crop growth parameters and reflectance at individual wavelength bands or indices were calculated at different growth stages using Proc Corr [SAS Institute, 2001].

5.2.5 SPATIAL STATISTICAL DATA ANALYSIS

The spatial structure of various soil and crop properties was characterized using variograms to understand the spatial continuity of the measured parameters. The autocorrelation length was used to analyze the spatial relation between one variable measured at different locations in the field. It is considered as a diagnostic measure to analyze spatial process in the field.

Autocorrelation function $r(h)$ was calculated considering n measurements of a soil property A_i measured at locations x and $x + h$ separated by a specified distance h using the relation, where cov and var are covariance and variance, respectively, as Nielsen and Wendroth[26] calculated:

$$r(h) = \frac{\mathrm{cov}\left[\left(A_i(x),\ A_i(x+h)\right]\right.}{\sqrt{\mathrm{var}\left[A_i(x)\right]}\sqrt{\mathrm{var}\left[A_i(x+h)\right]}}. \tag{5.3}$$

Spatial association between various crop growth parameters and canopy reflectance was analyzed using the cross-correlation function. The cross-correlation function $r_c(h)$ between two soil properties A_i and B_i observed at locations x_i and x_{i+h} is calculated with cov denoting the covariance, var the variance, and h the lag distance as Nielsen and Wendroth[26] calculated:

$$r_c(h) = \frac{\mathrm{cov}\left[\left(A_i(x_i),\ B_i(x_i+h)\right]\right.}{\sqrt{\mathrm{var}\left[A_i(x_i)\right]}\sqrt{\mathrm{var}\left[B_i(x_i+h)\right]}}. \tag{5.4}$$

5.2.6 SEMIVARIOGRAM

The spatial structure of variables was characterized using a geostatistical tool, semivariogram, as suggested by Nielsen and Wendroth.[26] Semivariogram quantifies the spatial continuity of the variable. In this spatial analysis, gravimetric water content A_i taken at a lag distance of h at location x_i was compared to another observation taken at $A_{i+h}(x_{i+h})$. For N pairs of values of A_i separated by a lag distance of h, the average difference for each lag class can be obtained from the semivariogram:

$$\gamma(h) = \frac{1}{2N(h)} \sum_{i=1}^{N(h)} \left(A_i(x_i) - A_i(x_i + h) \right) \quad (5.5)$$

5.2.7 SPECTRAL ANALYSIS

The periodic behavior of soil and crop properties is analyzed with spectral analysis. A spectrum identifies the periodically repeating variance components. The analysis involves calculation of autocorrelation function $r(h)$ of a property and its substitution into the following equation:

$$S(f) = 2\int_0^{\infty} r_c(h)\cos(2\pi fh)\,dh, \quad (5.6)$$

where S is the spectrum and f the frequency equal to p^{-1} where p is the period.[26]

Co-spectral analysis was carried out to identify the spatial frequencies for which two sets of observations are correlated with each other. In the analysis, the cross-correlation coefficient $r_c(h)$ of the two properties under study is calculated to partition the total covariance of them. The co-spectrum was calculated as follows:

$$\mathrm{Co}(f) = 2\int_0^{\infty} r_c(h)\cos(2\pi fh)\,dh, \quad (5.7)$$

where Co is the co-spectrum and f the frequency equal to p^{-1} where p is the period.[26]

Quadrature spectrum was used to identify the lag between two sets of observations, which are correlated at the same frequency:

$$Q(f) = 2\int_0^{\infty} r_c'(h)\cos(2\pi fh)\,dh, \quad (5.8)$$

where $r_c' = 0.5\{r_c(h > 0) - r_c(h < 0)\}$. This subtracting procedure is used to reinforce cyclic variations described by a sine function and eliminates that described by a cosine function as Nielsen and Wendroth[26] calculated.

5.2.8 COHERENCY ANALYSIS

Coherency analysis was carried out to measure the significance of the correlation between two sets of observations $A_i(x_i)$ and $B_i(x_i)$ for various frequencies f. The coherency is calculated from

$$\mathrm{Coh}(f) = \frac{\mathrm{Co}^2(f) + Q^2(f)}{S_A(f)\,S_B(f)}, \tag{5.9}$$

where $Q(f)$ is the quad spectrum, $Co(f)$ the cospectrum, and $S_A(f)$ and $S_B(f)$ are the spectra of the two sets of observations $A_i(x_i)$ and $B_i(x_i)$, respectively, as suggested by Nielsen and Wendroth.[26] Coherency values ranged from 0 to 1 and is analogous to the coefficient of determination of a simple linear regression between the two variables.

5.2.9 STATISTICAL METHODS

Linear regression equations were generated from the sensor output and field-measured volumetric soil water content. The coefficient of determination (r^2) provided the degree of linear association between factory (default)-calibrated water content to field-measured volumetric water content. To compare the field- and factory-calibrated Diviner estimates of soil water contents with gravimetrically derived volumetric water content, statistical tests such as RMSE (root mean square error) and Md (mean difference) were used as suggested by Nielsen and Wendroth.[26] The RMSE and Md were calculated as

$$\mathrm{RMSE} = \sqrt{\frac{1}{4}\sum_{i=1}^{i=4}\left(E_i - M_i\right)^2}, \tag{5.10}$$

$$\mathrm{Md} = \frac{\sum_{i=1}^{i=4}\left(E_i - M_i\right)}{45}, \tag{5.11}$$

where E is the value of the soil moisture content estimated by the Diviner (either factory-calibrated or field-calibrated), M the corresponding gravimetrically derived soil volumetric water content, i corresponds to the number of calibration days, and n is the number of measurements which is the total number of locations in the field.

The coefficient of determination (r^2) provided the degree of linear association of field-calibrated and factory (default)-calibrated water content to gravimetrically derived volumetric water content. The Md measured the average difference of factory- and field-calibrated, Diviner-estimated water content from gravimetrically derived volumetric water content measurements. The sign and value of Md indicate the degree of coincidence between the factory- and field-calibrated water content and actual volumetric water content measurement.

5.2.9.1 FREQUENCY DISTRIBUTION

The frequency distribution under the driest and wettest time periods was computed to investigate whether or not one location keeps its rank in the frequency distribution. If this occurs and assuming the probability function as normal, we can select the particular location with a probability of 50% to characterize the field-mean soil moisture. Similarly, other particular locations, associated with cumulative probabilities of 17 or 83% by considering one SD from field mean, SWS $(\bar{x} \pm 1\text{SD})$, can be selected.

5.2.9.2 RELATIVE DIFFERENCES APPROACH

This technique is based on the parametric test of the relative differences introduced by Vachaud et al.[32] The difference (Δ_{ij}) between an individual measurement of SWS, S_{ij} at location i and time j and the daily spatial mean of SWS, $\bar{S}_j$ at the same time from all locations is calculated. Specifically, the relative difference, δ_{ij}, is defined as below. Temporal mean relative difference of location i, $\bar{\delta}_i$, and its standard deviation σ (δ_i) are determined for location I, calculated as below:

$$\Delta_{ij} = S_{ij} - \bar{S}_j, \tag{5.12}$$

$$\bar{S}_j = \frac{1}{n} \sum_{i=1}^{i=n} S_{ij}, \tag{5.13}$$

$$\delta_{ij} = \frac{\Delta_{ij}}{\overline{S}_j}, \tag{5.14}$$

$$\overline{\delta}_i = \sum_{j=1}^{j=m} \delta_{ij}, \tag{5.15}$$

$$\sigma(\delta_i) = \sqrt{\frac{1}{m-1}\sum_{j=1}^{j=m}\left(\delta_{ij} - \overline{\delta}_i\right)^2}, \tag{5.16}$$

where n is the number of sampling locations and m the number of sampling times ($m = 36$ in the present study).

$\delta_{ij} = 0$ implies that moisture content of location i on jth day is equal to field mean on day j. Thus, for any location i, the temporal average δ_i and the temporal standard deviation σ (δ_{ij}) can be calculated for the $m = 36$ days of measurements. $\overline{\delta}_i$ and σ (δ_i) are used to rank the locations (from lowest to highest relative difference from the mean) and to assess temporal stability of spatial variable pattern. This approach helps to identify locations which represent field mean, $\overline{\delta} = 0$, and locations which systematically overestimate field mean $\delta_{ij} > 0$, also locations which systematically under estimate field mean $\delta_{ij} > 0$. A temporally stable location is characterized by a low value of σ (δ_{ij}). Gomez-Plaza et al.[12] and Greyson et al.[13] used the value of σ (δ_i) as a major criterion in selecting temporally stable sites. Locations can be identified for continuous monitoring of field mean soil moisture as well as extremely wet and dry soil moisture conditions as noticed by Mohanty and Skaggs[24] and Vachaud et al.[32] The extent of temporal variability of soil moisture in each location relative to field mean is also obtained from this approach.

5.2.9.3 SPEARMAN RANK CORRELATION COEFFICIENT

Vachaud et al.[32] suggested the application of the nonparametric Spearman rank correlation coefficient test to evaluate the persistence of soil moisture spatial patterns at each observation time. The test was carried out using Proc Corr of SAS. Spearman rank correlation coefficient, r_s, is defined as

$$r_s = 1 - \frac{6\sum_{i=1}^{i=n}\left(R_{ij} - R_{ij}\right)^2}{n\left(n^2 - 1\right)} \tag{5.17}$$

where R_{ij} is the rank of the SWS S_{ij} for the measurement campaign j of location i and $R_{ij'}$ is the rank of the SWS at the same location, but for the measurement campaign j'. n is the number of sampling locations and is equal to 45 in the present study. An r_s equal to 1 corresponds to perfect time persistency between sampling dates j and j'.

5.2.9.4 MEAN ABSOLUTE VALUE OF BIAS ERROR (MABE)

Mean absolute value of bias error (MABE) is a tool introduced to identify temporally stable sites representing field mean SWS. A critical value of 5–10% of MABE was used to select representative stable sites. MABE and the associated temporal standard deviation were calculated as

$$\text{ABE}_j(i) = \frac{\left|\delta_j(i) - \overline{\delta(i)}\right|}{1 + \overline{\delta(i)}}, \tag{5.18}$$

$$\text{MABE}(i) = \frac{\sum_{j=1}^{m} \text{ABE}_j(i)}{m}, \tag{5.19}$$

$$\sigma_{\text{ABE}}(i) = \frac{\sqrt{\sum_{j=1}^{m}\left(\text{ABE}_j(i) - \text{MABE}(i)\right)^2}}{m-1}, \tag{5.20}$$

where ABE is the absolute value of bias error for a given relative difference, $\overline{\delta(i)}$; MABE (i) is the time-averaged ABEj (i); σ_{ABE} (i) is the temporal standard deviation of ABE for location i. A value of σ_{ABE} (i) close to zero indicates high temporal stability.

5.3 RESULTS AND DISCUSSION

5.3.1 SPATIAL AND TEMPORAL DISTRIBUTION OF SWS

5.3.1.1 SOIL TEXTURE

Variability in soil texture existing in the field within each soil layer and also between adjacent soil layers is evident from Table 5.1. Silt content was the largest in all the soil layers followed by clay and sand contents. The sand and clay contents varied more than silt content both along the transect and across soil depths in the field as evident from the coefficient of variation percentage. The variability in sand content was the highest along the transect and across the soil depths followed by clay content. Variability in silt content slightly increased with soil depth. Textural classes varied from silt loam, silty clay loam, and silty clay within the same soil layer along the transect for all the soil depths.

TABLE 5.1 Soil Texture of the Experimental Field with Percentage of Soil Particles (mean ± SD) in Each Depth. Coefficient of Variation (%) is Given in Parentheses.

Depth (cm)	Sand (>0.05 mm)	Silt (0.002–0.05 mm)	Clay (<0.002 mm)	Major texture class
0–15	4.95 ± 1.21 (24.50%)	73.14 ± 5.10 (6.97%)	21.92 ± 5.13 (23.42%)	Silt loam
15–30	4.08 ± 1.55 (37.89%)	68.23 ± 5.10 (7.48%)	27.69 ± 4.61 (16.64%)	Silty clay loam
30–60	4.74 ± 1.71 (36.13%)	65.70 ± 5.45 (8.30%)	29.56 ± 4.44 (15.02%)	Silty clay loam
60–90	5.75 ± 1.82 (31.60%)	67.60 ± 8.86 (13.11%)	26.66 ± 8.32 (31.20%)	Silty clay loam

Spatial distribution of depth-averaged silt and clay content is presented in Figure 5.1. The sand and silt contents showed an opposite trend to each other. The silt content ranged from 60.22 to 78.46% and the clay content varied from 17.86 to 32.93%. The sand content varied from 2.6 to 7.19% along the transect.

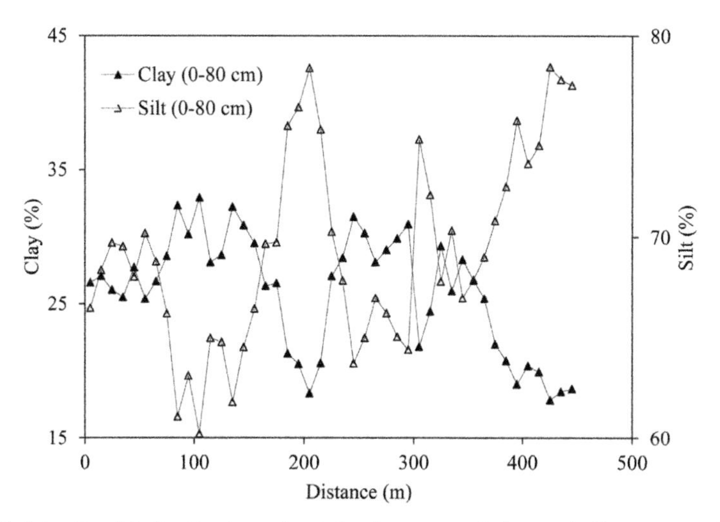

FIGURE 5.1 Spatial distribution of weighted average of clay and silt contents along the transect.

5.3.1.2 ELEVATION

Figure 5.2 shows the distribution and spectrum of elevation along the transect. Spatial variability is evident from Figure 5.2(a). Elevation ranges from 469 to 485 m with a standard deviation of 4.55 m. Spectrum showed a peak at a frequency of 0.031 m^{-1} which was close to a distance of 300 m along the transect.

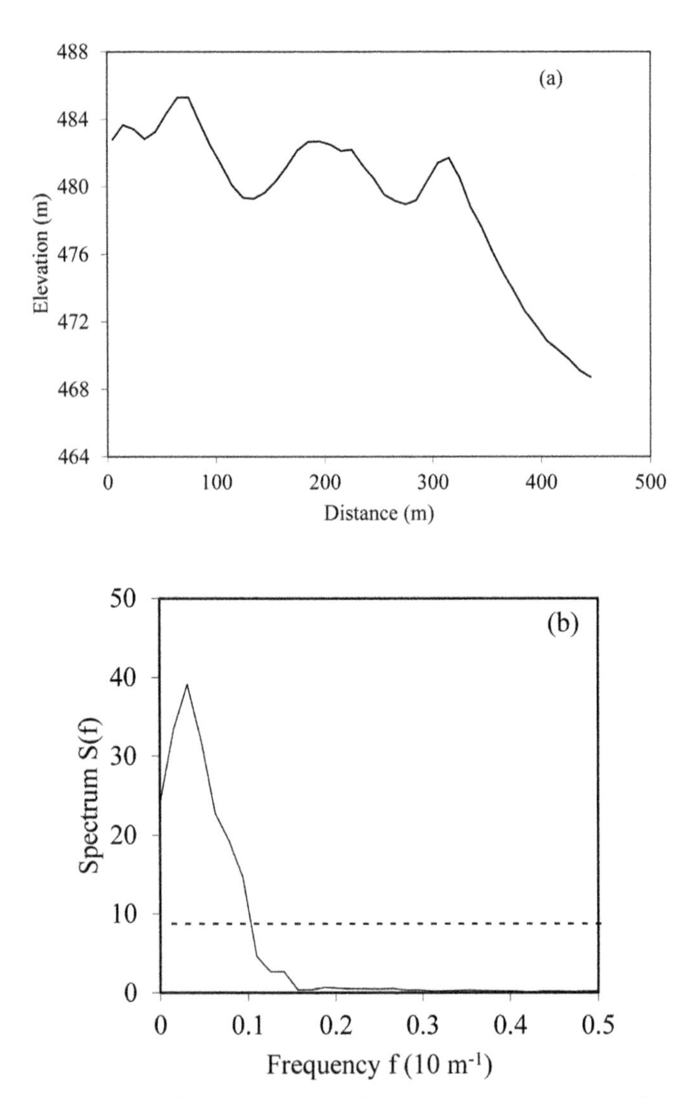

FIGURE 5.2 (a) Spatial distribution of elevation and (b) spectrum of elevation.

SWS of the 80-cm depth was obtained by summing the SWS of every 10-cm layer. The spatial and temporal variability in SWS in the field is evident in Figures 5.3 and 5.4. Figure 5.3 presents spatial distribution of SWS from May 11, 2007 to April 2, 2008. Corn was grown from April 20 to August 15, 2007, and winter wheat was grown from October 17 to June 25, 2008. The wettest day, December 10, showed 306.98 mm and SD of 10.47 mm and September 29 was the driest day with mean SWS of 165.11 mm and SD of 19.42 mm. Figures 5.3 and 5.4 showed similar spatial patterns of SWS especially when the time interval was very close

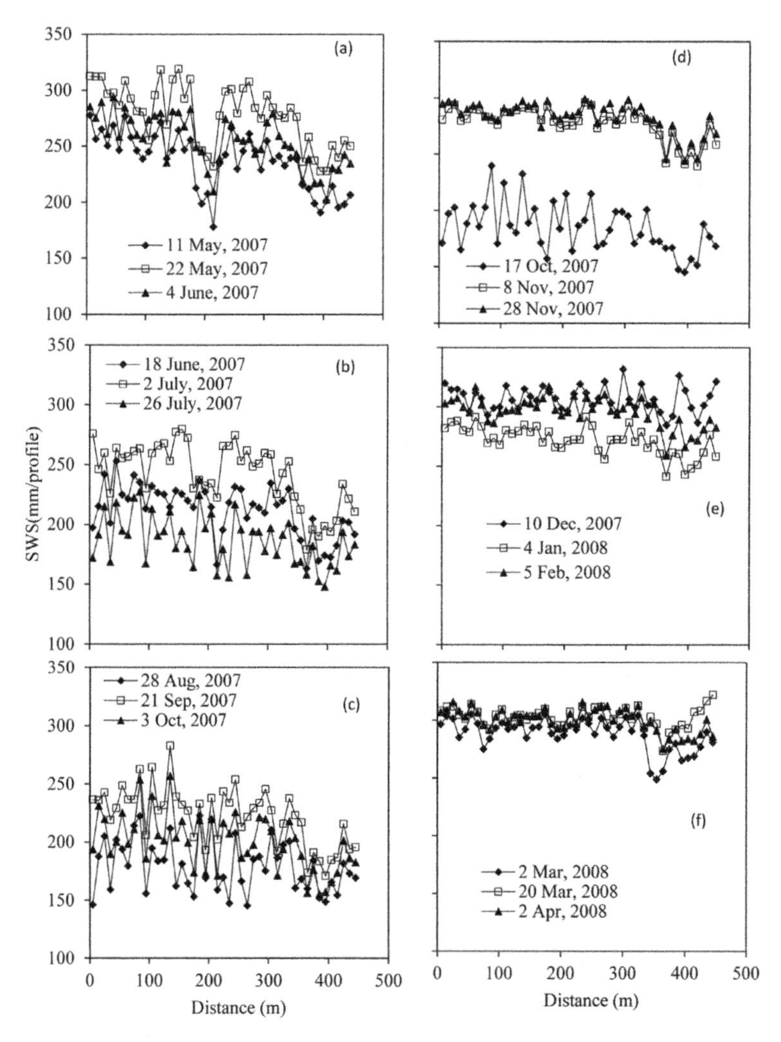

FIGURE 5.3 Spatial distribution of SWS at different soil water content measuring dates.

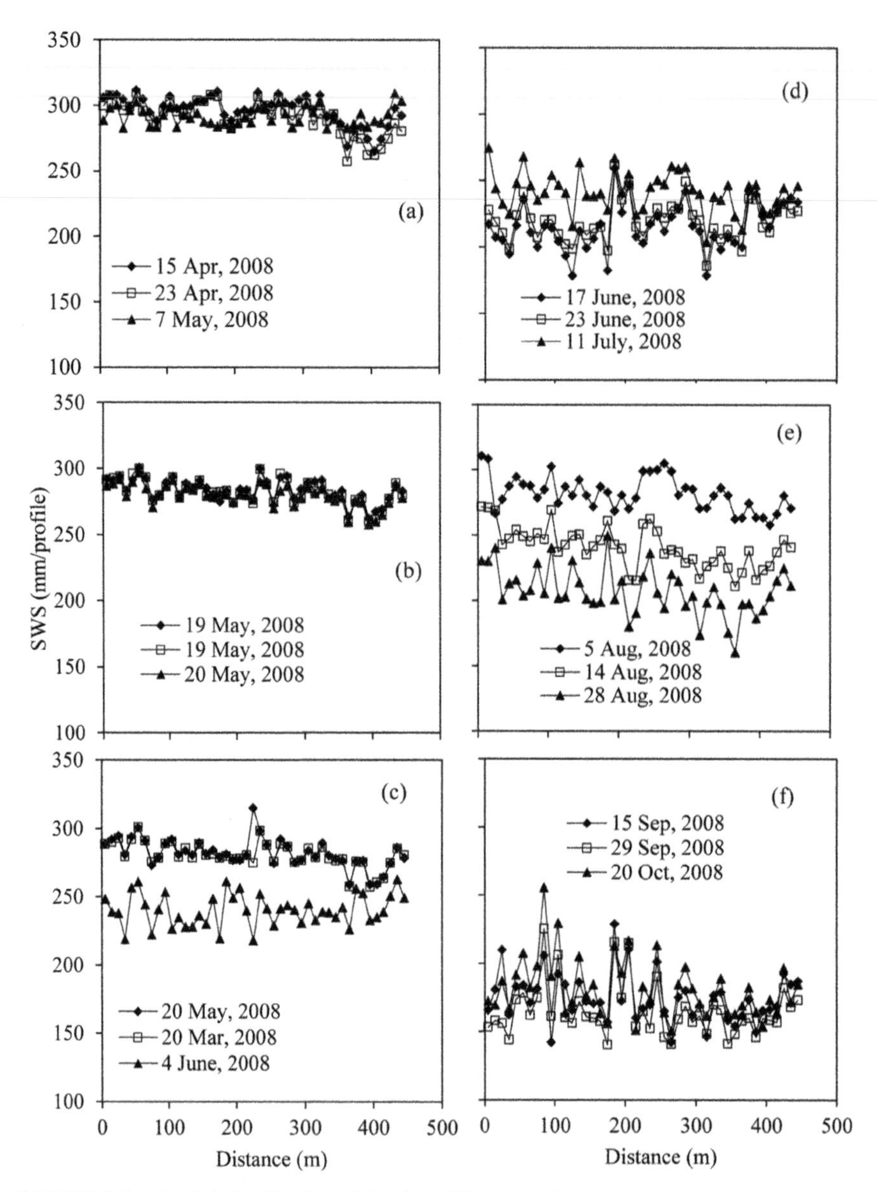

FIGURE 5.4 Spatial distribution of SWS at different soil water content measuring dates.

between successive measurements. The similarity was very evident as, for example, in Figure 5.5(a) and (b), where the time interval between measurements was 1 week or less. Effect of precipitation was also very evident as can be seen in Figure 5.5(e). The field received 10.54 cm precipitation between the soil water content measurement on July 11 and August 5, 2008. The average SWS increased by about 3.9 cm from July 11 to August 5. The precipitation received during August 6 to August 14 was 1.78 cm and the SWS decreased by about 4 cm until August 14. The precipitation received was 0.51 cm from August 14 to August 28 and the average SWS decreased again by about 3.3 cm on August 28th measurement. Among the three measurement days, August 5 showed the largest SWS as the precipitation received by that time was the largest followed by August 14 and August 28. Corn was grown during these measuring days. Brocca et al.[2] also reported strong dependency of the soil moisture on precipitation with a steep raise in soil moisture after a storm event and a slow recession in periods without rainfall. They attributed the decline in soil moisture to evapotranspiration.

Figure 5.5 shows the similar nature of spatial patterns in SWS among different days with soils containing high clay content in the transect being wetter due to their high water-holding capacity and silt loam-textured soils toward the end of the transect being drier due to their fast water releasing nature. Locations on the hilltop soils were drier than locations in the foot slope. Spatial variation in the soil moisture distribution pattern over seasons was reported by Zhou et al.[35] Lateral flow and redistribution were considered as responsible for the temporal instability of soil moisture spatial pattern as reported by authors Lin[20] and Grayson and Western.[13] Gish et al.[11] attributed the uncertainty in determining soil hydrological properties at scales larger than small plots to a high degree of spatial and temporal variability in subsurface flow pathways.

Mean SWS ranged from 221 to 265 mm across the transect and the mean SWS was 249 mm. Location 6 showed the highest mean SWS and location 37 the lowest during the study.

Location 30 showed the highest maximum SWS and is a silty clay loam location storing large amounts soil water and location 37 showed the lowest maximum SWS among the 45 locations during the study. Location 37 was with silt loam texture showing good drainage. Location 19 had the largest minimum SWS and location 18 showed the lowest minimum SWS. A spatial SD of 10.57 mm for the time-averaged SWS shows the existence of high spatial variability. SD of SWS ranged from 26.09 to 57.93 mm, which was negatively correlated with mean SWS, with a Pearson correlation coefficient of 0.88 ($P < 0.0$ 1). This means that SWS was more heterogeneously distributed in the field under drier conditions.

Location 9 showed smallest SD and its clay content was the highest especially in the lower layers. Largest SD was found in location 18 and it showed high silt content in all the layers. SD was on an average 45 mm and CV around 18%. Location 18 showed the largest SD and CV% of SWS and location 9 showed the lowest SD and CV of SWS.

While considering the temporal mean of total space average SWS during the entire study period, December 10, 2007, was found to be the wettest day with the largest mean SWS (306.98 mm) and the largest maximum SWS (331.80 mm) of the study. It is clear that the temporal variability is larger than the spatial variability of SWS. This is mainly due to crop cultivation throughout the study. Kachanoski and de Jong[19] observed that root water uptake and canopy cover vary with crop growth stages. Soil water input in the form of rainfall also varies with time.

It is evident that the range in the spatial mean SWS (43.5 mm) was smaller than the range of temporal mean SWS (57.67 mm). The value of SD and CV% for the mean, maximum, and minimum SWS was also 3–4 times larger for the temporal SWS values than the spatial SWS values. Significant variability in soil moisture content exists along the length of the transect, and the variability increases with decreasing transect–mean moisture content. The topographic and soil attributes may operate jointly to redistribute soil water under wet conditions, whereas under dry conditions, relative elevation, aspect, and soil texture are important in contributing to the variability as reported by Famiglietti et al.[9] In the present study also, variability in elevation and soil texture was evident in contributing to large spatial variability in SWS.

Figure 5.5(a) shows the spatial variability of the mean, minimum, and maximum total SWS observed during the study and we can see similar spatial pattern for these parameters for some of the locations. SD between different locations for the mean SWS was 10.57%, SD for the maximum SWS was 9.7%, and SD for the minimum SWS was 17.45%. The high variability in SD for the minimum SWS could be due to variability in soil texture in the horizontal and vertical directions.

Figure 5.5(b) presents weekly cumulative precipitation for a better understanding of the temporal dynamics of SWS. From Figure 5.5(b), the temporal variability in rainfall is evident. The field experienced drought during middle of July to middle of September 2007 causing a huge decline in SWS resulting in decreased grain yield. After each precipitation, an increase in SWS was observed. December 10, 2007, was a winter day and was the wettest day of the study. December to March were wet months due to high precipitation. The driest day of measurement of the study was September 29, 2008, and was not preceded by any precipitation for a few weeks.

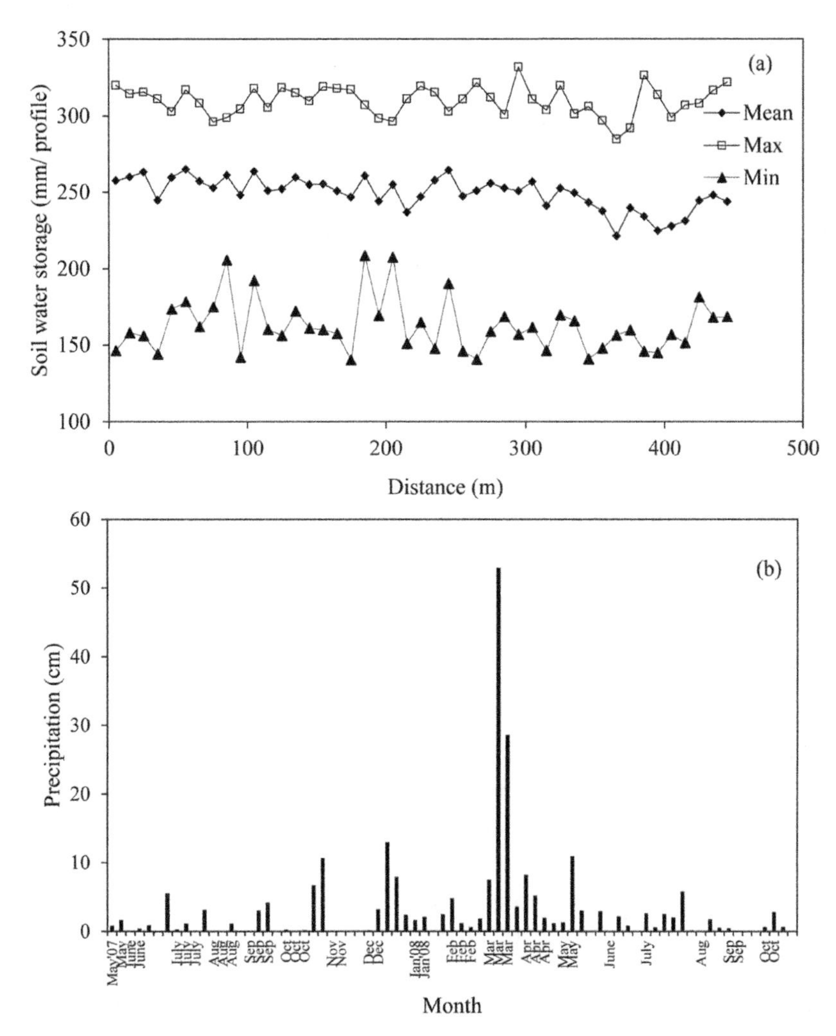

FIGURE 5.5 (a) Spatial distribution of mean, maximum, and minimum of total SWS during the study; (b) cumulative rainfall received on a weekly interval during the study.

5.3.1.3 VARIABILITY IN SPATIAL AND TEMPORAL DISTRIBUTION OF SWS WITH SOIL DEPTH

Figure 5.6(a)–(c) shows the spatial distribution of mean, maximum, and minimum SWS of each profile during the study. Spatial variability in soil moisture within the field was clear across depths. Similar spatial patterns can be seen among the mean, maximum, and minimum SWS. The clayey textured soils toward the middle of the transect were more wet due to their high water-holding capacity, and silt loam textured soils toward the end of the transect were drier due to their faster drainage and release of water for plant uptake than the clayey textured soils. The differential behavior of soil texture was more evident from the minimum SWS values. The deeper layers showed high SWS compared to the upper layers as also reported by Lin.[20] The time dependency of the depth was most pronounced for the deeper layer and could be related to temporal variability in crop growth and root water uptake. The temporal patterns of mean, minimum, and maximum SWS were very similar among the four groups of layers. The deeper three layers showed very similar magnitude and spatial pattern of mean and maximum SWS throughout the study as also observed by Hupet and Vanclooster.[16] The minimum SWS (Fig. 5.6(c)) showed an overlapping among the four soil layers.

From Figure 5.6, an increase in the mean and maximum SWS with depth is clear. This could be due to the increased clay content in deeper layers and also could be due to the flux coming to these layers.

Figure 5.7(a)–(c) reveals that the temporal dynamics of soil moisture was depth-dependent. Temporal variations of the mean, maximum, and minimum SWS were depth-dependent. The observed range of mean SWS (Fig. 5.7(a)) for each 0–20 cm depth was 31.56–70.81 mm with an average of 52.31 mm. SD of mean SWS was 12.69 mm for the 0–20 cm layer. The maximum SWS in this layer ranged from 91.18 to 41.29 mm with an SD of 11.81 mm. The range for minimum SWS was from 21.06 to 63.60 mm with an SD of 14.20 mm. The high variability in SWS of 0–20 cm layer could be because this part of the profile was subjected to dynamic environmental conditions.

For the 20–40 cm layer, mean SWS ranged from 41.76 to 78.24 mm with an average of 63.68 mm. The SD of mean SWS was 11.75 mm for the 20–40 cm layer. The maximum SWS varied from 55.66 to 87.93 mm with an SD of 9.43 mm. The range for minimum SWS was larger than that of mean and maximum SWS as it varied from 22.82 to 70.71 mm. The SD was also larger than that of mean and maximum SWS and its value was 15.94 mm. The larger SD for the minimum SWS could be attributed to soil textural variability in the field.

The 40–60 cm layer showed mean SWS range from 42.04 to 79.36 mm with an average of 66.01 mm. The SD of mean SWS was 12.42 mm for the 40–60 cm layer. The maximum SWS in this layer ranged from 64.83 to 90.60 mm with an SD of 7.71 mm. The minimum SWS varied from 24.78 to 67.40 mm with an SD of 14.33 mm.

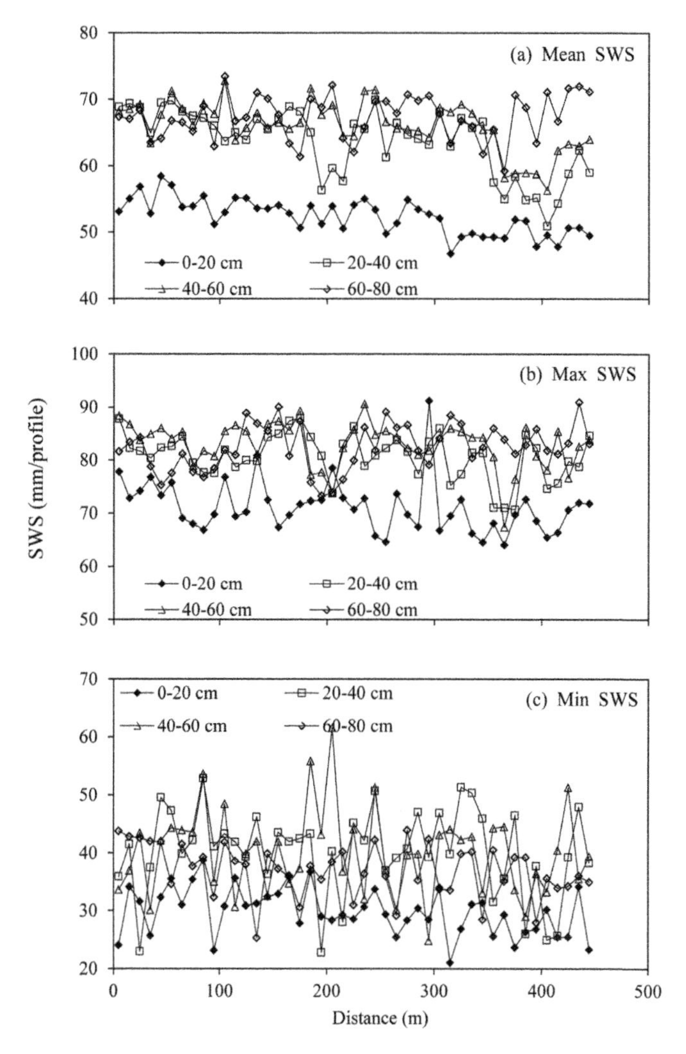

FIGURE 5.6 Spatial distribution of: (a) mean, (b) maximum, and (c) minimum SWS at different soil depths.

The maximum SWS varied from 44.61 to 90.91 mm with an SD of 7.75 mm. The minimum SWS showed very similar range in SWS as that of 40–60 cm depth. Thus, the mean, maximum, and minimum of field mean SWS increased with soil depth. The temporal dynamics of soil moisture in the intermediate layers (40–60 cm) and deeper layers (60–80 cm) could be mainly influenced by root water uptake and upward flux as observed by Hupet and Vanclooster[16] in corn.

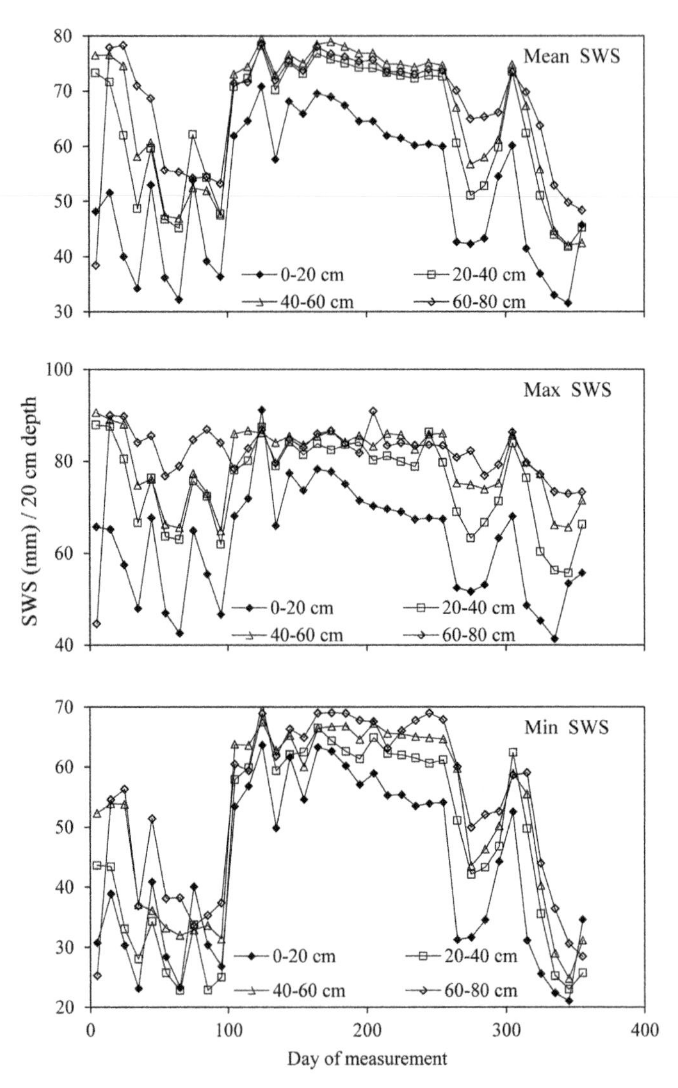

FIGURE 5.7 Temporal distribution of: (a) mean, (b) maximum, and (c) minimum SWS at different soil depths.

Figure 5.8(a) presents the spatial distribution of standard deviation of SWS at different soil depths. The mean SD was very similar among the four layers ranging from 11.38 mm in the 60–80 cm layer to 13.12 mm in the 0–20 cm layer (13.12 mm). The high SD in the 0–20 cm layer could be due to its exposure to dynamic atmosphere. The increased mean SD of deeper layers could be due to spatial variability in crop growth.

Figure 5.8(b) shows the temporal distribution of standard deviation of SWS for different soil depths. The SD of 0–20 cm depth varied from 2.90 to 8.12 mm with an average value of 4.45 mm. The range of SD was from 3.40 to 11.82 mm in the 20–40 cm depth with 6.49 mm as mean SD. SD varied from 2.92 to 10.06 mm with a mean value of 6.01 mm in the 40–60 cm depth. The SD of mean SWS for the 60–80 cm depth varied from 11.24 to 2.97 mm with an average of 5.97 mm. SD decreased under wet conditions and increased under dry conditions. The large SD of mean SWS of deeper layers could be due to variability in root water uptake and soil texture along the transect.

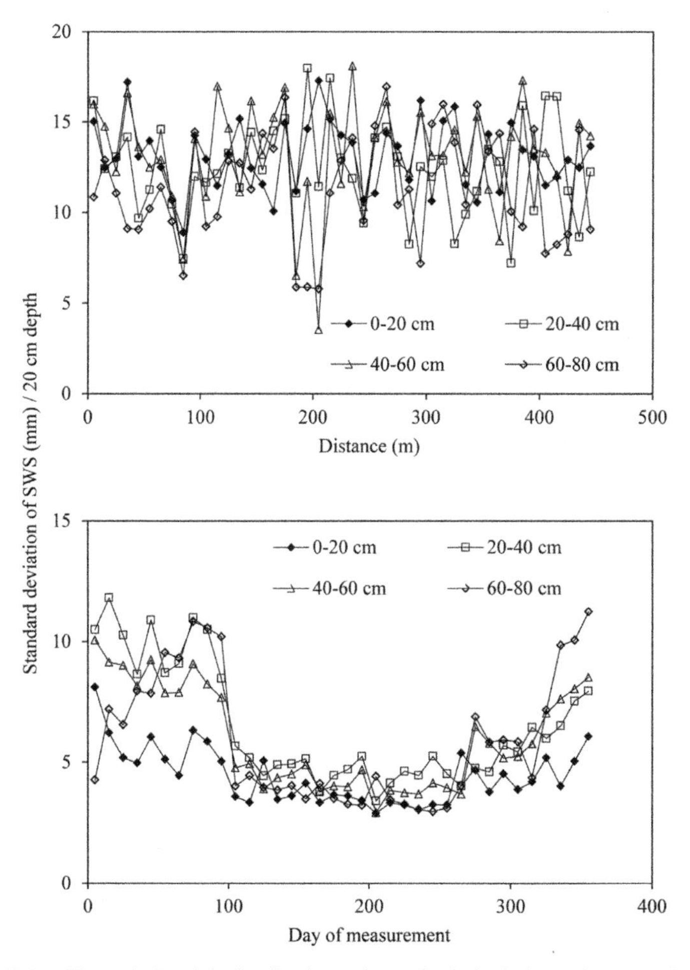

FIGURE 5.8 (Top, a) Spatial distribution of standard deviation of SWS; (bottom, b) temporal distribution of standard deviation of SWS at different soil depths.

5.3.1.4 SPATIAL STRUCTURE ANALYSIS OF SWS

In Figure 5.9(a), the distribution of profile mean SWS and its variance is presented. Crop cultivation existed in the field during the study. From Figure 5.9(a), it can be seen that there exists a positive relationship between mean SWS and its variance when the mean SWS was below about 200 mm. When the mean SWS is above 200 mm, there exists a negative relationship between mean SWS and its variance. A rain after a long dry period showed a steep decrease in variance. Jacques et al.[18] have noticed a threshold value of soil moisture content (around 0.36 cm^3 cm^{-3}), above which variance decreases and below the threshold value, variance increased or decreased with rain events and they attributed the trend to soil water redistribution processes influenced by interacting factors such as soil and topography.

Hupet and Vanclooster[16] observed an increasing standard deviation with decreasing mean soil moisture. This negative correlation is consistent with the previous findings of Famiglietti et al.,[8] who analyzed the surface soil moisture of different cultivated, relatively less sloppy remote sensing pixels. Hupet and Vanclooster[16] attributed the negative correlation in their study to less pronounced topographic features existing in their field. Different drainage rates among the textural groups were considered an important source of high soil moisture variability in the mid-range of mean soil moisture content.[7]

The relationship between mean SWS and its SD for different soil depths is presented in Figure 5.9(b). The trend was not clear from the data. For the 0–20 cm depth, the range of SWS and SD were narrow compared to deeper layers. In deeper layers, the relationship is more scattered. The three deeper layers exhibit wide range of SWS and SD of SWS. This could be because the deeper layers are more influenced by variations in water redistribution affected by topography and variability in crop water uptake.

Contradicting results on the behavior of mean and variance of SWS was reported by researchers such as Famiglietti et al.[9] and Mohanty et al.[23] They attributed the behavior to location-specific interaction between climate, soil, vegetation, and mean soil water condition affecting the spatial structure of soil moisture in the near-surface horizons. In the present study, even when mean soil moisture content was almost constant between two sampling days, change in field variance existed and that could be due to water distribution across the landscape as suggested by Mohanty et al.[23]

The spatial pattern of soil moisture in a landscape was most influenced by topography during wet periods, but soil moisture patterns depended mainly on soil properties during dry periods. During the fall and winter seasons, soils remained wet and the variance in SWS was smaller than that observed during drier times. Zhou et al.35 observed little variation in soil moisture during wet periods even though topographic redistribution of soil water existed in the field. Pachepsky et al.27 suggested that differences in soil structure can be responsible for the variations in soil water content. Guber et al.14 reported the dependence of water retention on the size of soil aggregates. Vachaud et al.32 observed that the variability of soil water contents was related to the variations in soil texture. Van Wesenbeeck and Kachanoski34 also obtained a negative correlation between soil water content and variance for the soil water content measured under a corn crop.

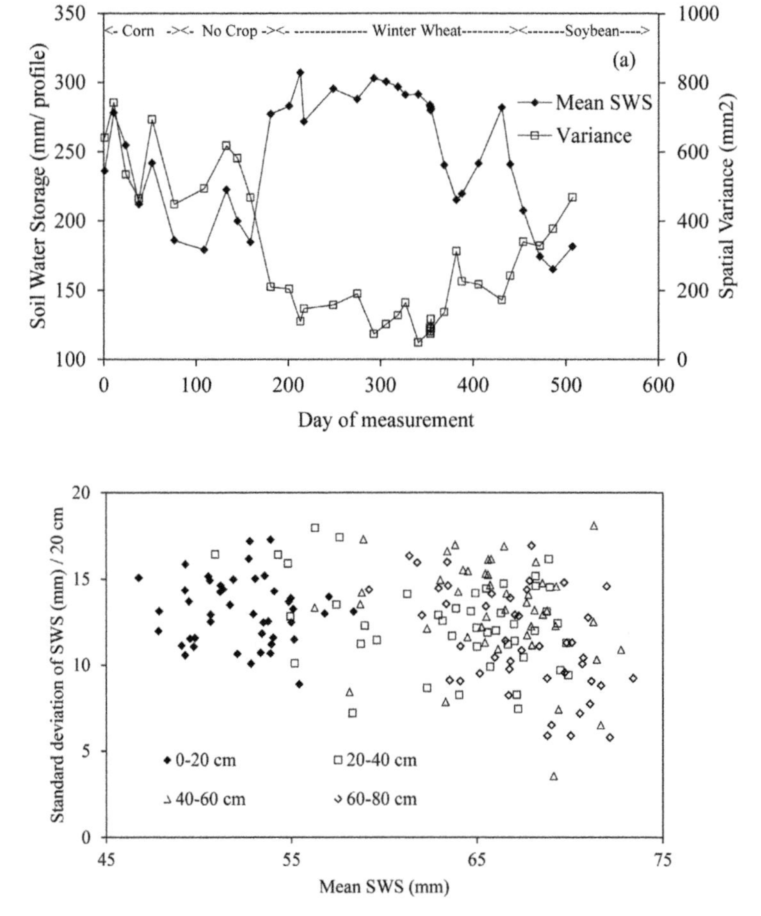

FIGURE 5.9 (a) Temporal distribution of total mean SWS and corresponding variance and (b) relationship between depth-wise mean SWS and corresponding SD along the transect for each day of measurement in the study.

Autocorrelogram of total SWS at different soil water contents measuring dates indicated that among the 36 times of SWS measurements, 22 times showed significant autocorrelation. During wet periods, the autocorrelation distance ranged from 40 to 50 m and during intermediate moisture conditions autocorrelation distance ranged from 10 to 20 m. During all these significantly autocorrelated measuring days, crops were growing in the field. As suggested by different authors, the spatial structure of soil moisture was affected by location-specific interaction between climate, soil, vegetation, and mean soil water condition.

Figure 5.10(a) shows the temporal distribution of autocorrelation length of SWS and its corresponding variance for each measurement. The autocorrelation length showed a wide range of values for the same value of SWS and variance. From the figure, it is evident that the relationship between variance and autocorrelation length is not clear. Hupet and Vanclooster[17] and Van Wesenbeeck and Kachanoski[34] have observed lack of spatial structure during crop growth and considered vegetation as the main factor controlling the development of the soil moisture patterns. But in Figure 5.10(a), spatial structure existed considering the value of autocorrelation length during the cultivation of corn and wheat. But during the period without any crop growth and also during soybean growth, spatial structure of SWS is lacking by considering the zero value of autocorrelation length. Hupet and Vanclooster[17] attributed small-scale variability in the crop growth as the cause for lack of spatial structure. Figure 5.10(b) shows the temporal distribution of precipitation and autocorrelation length of SWS. During both wet and dry periods, significant autocorrelation either existed or not. So the relationship is not clear. But in general, during wet times, significant autocorrelation exist compared to dry times where a range of autocorrelation exist.

Figure 5.11(a) shows the relationship between autocorrelation length and mean SWS. The autocorrelation length shows a wide range of values for the same value of mean SWS and also a wide range of values of mean SWS exists for the same autocorrelation length. For a wide range of mean SWS, the spatial structure was either lacking or shows a wide range of values of autocorrelation length. So from the figure, it is obvious that the relationship between mean SWS and autocorrelation length is not clear.

Figure 5.11(b) presents the relationship between the autocorrelation length and the range of SWS for each day of measurement in the study. The range of SWS was calculated by subtracting the minimum SWS from the maximum SWS for each day. Here also, the relationship between the mean SWS and the autocorrelation length is not clear. This could be due to the spatial variability in soil texture and root water uptake.

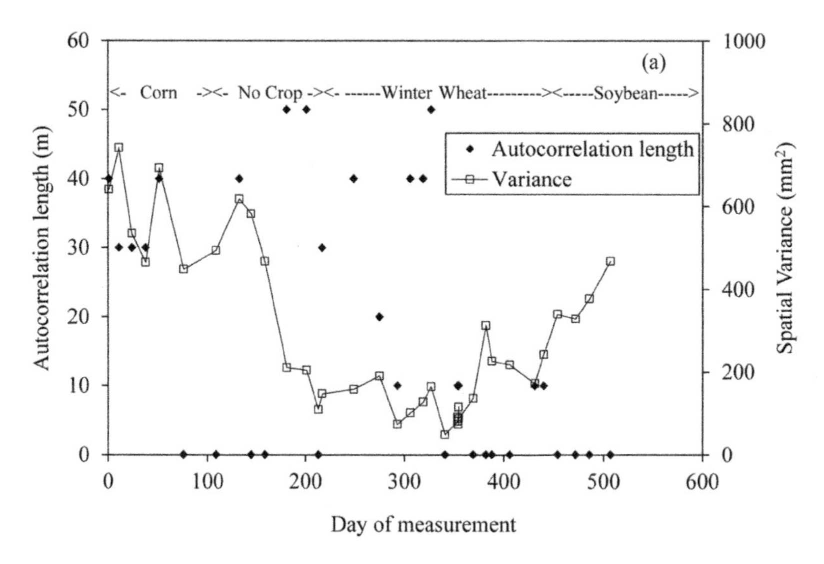

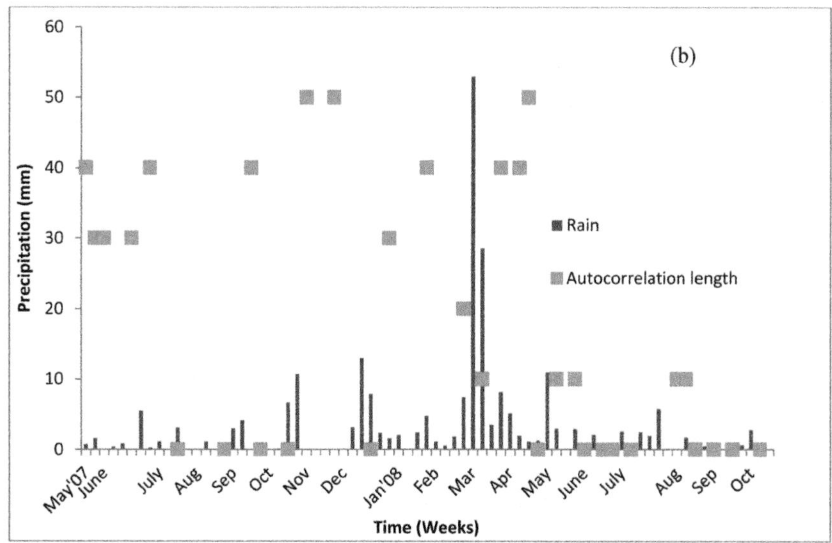

FIGURE 5.10 (a) Temporal distribution of autocorrelation length and variance of total SWS and (b) temporal distribution of autocorrelation length and precipitation.

When the time interval between soil water measuring day and that of rainfall was within 3–4 days, either small or no autocorrelation length was observed in the field. In order to get a good spatial structure of SWS, a waiting period of at least 1 week is required after a rainfall to measure soil water content using Diviner probe.

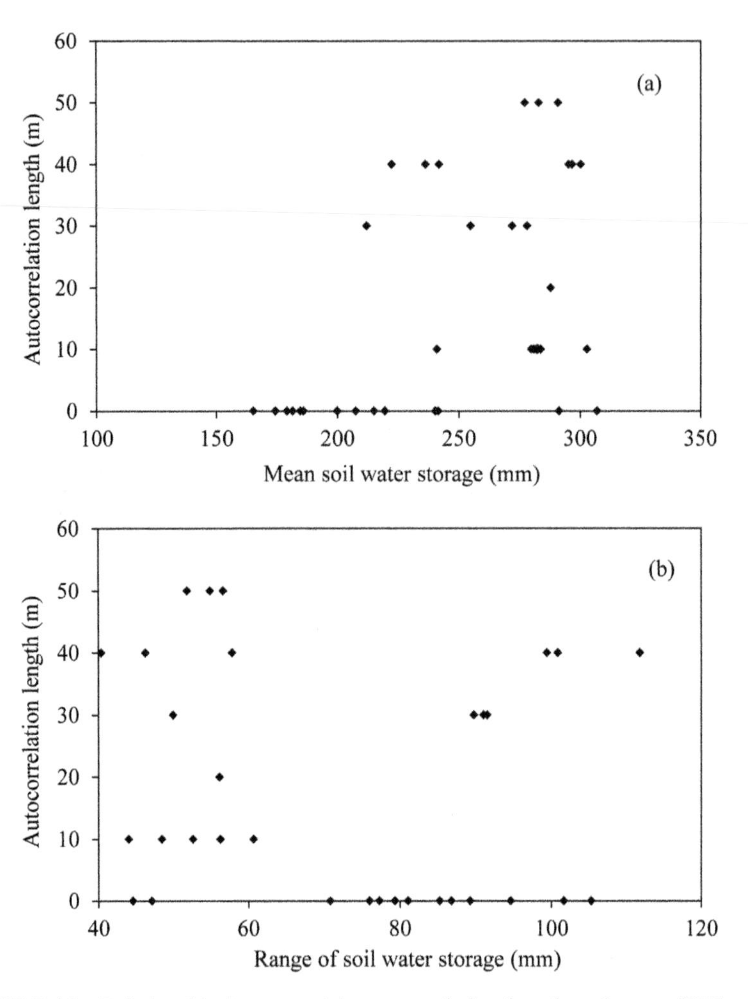

FIGURE 5.11 Relationship between: (a) autocorrelation length and mean SWS and (b) autocorrelation length and range of SWS for each day of measurement in the study.

5.3.1.5 ANALYSIS OF TEMPORAL PERSISTENCE OF SWS

5.3.1.5.1 Frequency Distribution

Ranking stability of locations in the cumulative probability function for the dates on which the lowest (September 29, 2008, 165.11 ± 19.42 mm) and the highest (December 10, 2007, 271.86 ± 12.14 mm) mean SWS values were recorded is shown in Figure 5.12(a) for the entire profile. The rank correlation coefficient between these 2 days was -0.116. From the figure, it can be noticed that only a few locations on the transect kept the same rank in the two extreme conditions. This indicates that the wettest locations and the driest ones were not located in the same spot. There does exist the possibility of some positions changing ranks appreciably from the dry to the wet condition; this is due to site-specific variability in vegetation growth, moisture redistribution processes due to spatial variability in elevation, and so on. Identification of sites that represent the field average SWS (cumulative frequency = 0.5) was also not time stable as location 19 represented field average under driest conditions, whereas location 38 was representing field average under wettest conditions for the entire profile. Both the locations were silt loam with silty clay loam in one of the deep layers. Location 19 was located on hill top with more sand and silt resulting in low water-holding capacity and faster drainage due to its position. Location 38 was located in back slope with large clay content resulting in high water-holding capacity. Locations 37, 42, and 38 were ranked driest on September 29 and locations 27, 39, and 30 were ranked wettest on September 29. Locations 27, 39, and 30 were ranked driest on December 10, 2007, and locations 9, 22, and 26 were ranked wettest on December 10 for the 0–80 cm depth.

For the 0–20 cm layer, different sites represented the field average SWS (Fig. 5.12(b)) under extreme moisture conditions, as location 32 represented field average under driest conditions, whereas location 29 represented field average under wettest conditions. Location 32 was located on a hill top with faster drainage due to its position. Location 29 was located in a valley with large clay deposits resulting in high water-holding capacity. Locations 42, 35, and 38 were ranked driest on September 29 and these profiles located on hill top and the locations ranked as wettest on September 29; locations 9, 22, and 26 were located on foot slopes having large deep silt content. Locations 32, 38, and 43 were ranked driest on December 10 and were located on slopes leading to their faster drainage. Locations 12, 6, and 9 were ranked wettest on December 10 and were located on foot slopes leading to deep silt content.

Based on the cumulative probability function, temporal stability in SWS for the layer 20–50 cm depth was not observed. Location 26 was representing field average under driest conditions, whereas location 15 was representing field average under wettest conditions. Locations 37, 38, and 42 were ranked driest on September 29 and locations 23, 39, and 17 were ranked wettest on September 29. Locations 41, 20, and 3 were ranked driest on December 10 and locations 33, 9, and 19 were ranked wettest on December 10 for the 20–50 cm depth.

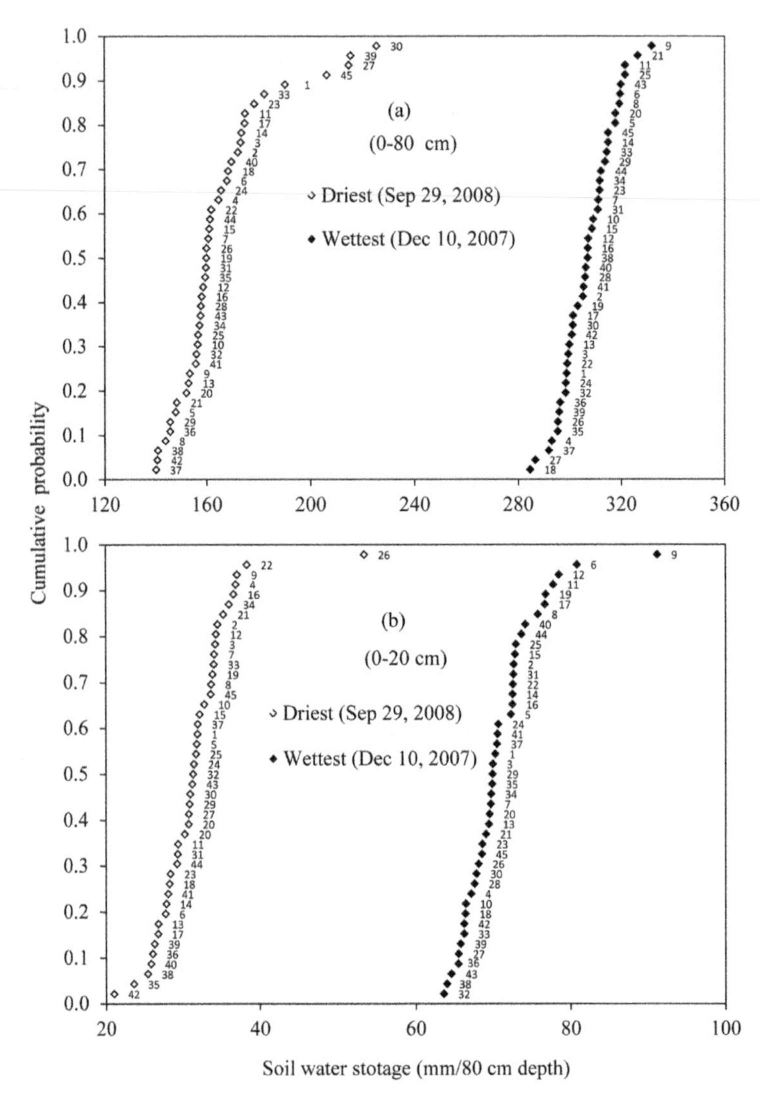

FIGURE 5.12 Cumulative probability functions of SWS in the driest and wettest days for: (a) 0–80 cm and (b) 0–20 cm soil depths.

The cumulative probability function for the 50–80 cm depth interval also did not show temporal stability in SWS. Under driest conditions, location 21 represented field average, whereas location 36 represented field average under wettest conditions. Locations 20, 14, and 35 were ranked driest on September 29 and locations 43, 18, and 38 were ranked wettest on September 29. Locations 28, 40, and 34 were ranked driest on December 10 and locations 11, 38, and 43 were ranked wettest on December 10 for the 50–80 cm depth. Location 9 was representing wet conditions of the deeper depths. This location also was high in clay content throughout the profile and was located in the back slope area. Field mean SWS was represented by different locations for different soil depths. This shows the poor temporal stability in the spatial pattern of SWS of the field.

5.3.1.5.2 Relative Differences Approach

Figure 5.13(a) presents the ranked inter-temporal relative deviation from field mean total SWS. Total SWS for the 36 measurements in time pooled together for each location, with the $\overline{\delta}_i$ values ranked in ascending order from left to right. It is clear that certain locations systematically either overestimate or underestimate the field average soil water content, irrespective of the observation date. $\overline{\delta}_i$ values ranged from 6.92 to −11.41%. The $\sigma(\delta_i)$ values ranged from 2.99 to 12.20% with an average value of 5.95%. Locations with mean relative difference close to zero and low standard deviation $\sigma(\delta_i)$ were identified to represent field mean SWS and to reduce the number of measurements needed to characterize the SWS of field.[4,13,32]

The standard deviation of the relative difference was also widely used to judge time stability of soil water contents as observed by Brocca et al.[2] and Starks et al.[30] Van Pelt and Wierenga[33] defined representative location as to estimate field mean SWS within 5% and have low variances. But here all locations are showing $\overline{\delta}_i$ values more than 5% and also high variance values. The high values of both $\overline{\delta}_i$ and $\sigma(\delta_i)$ for most of the locations indicates that the temporal persistence in spatial patterns of SWS does not exist in the field. The $\sigma(\delta_i)$ values were high for both wet and dry locations. It indicates the influence of dynamic variables such as vegetation and precipitation. Martinez-Fernandez and Ceballos[22] noticed high temporal stability in SWS under dry conditions and lowest temporal stability in SWS during transition from dry to wet conditions. However, Hupet and Vanclooster[16] reported weaker temporal stability in SWS during dry periods and they explained that with temporal variability in evapotranspiration at different locations in their field.

Field mean SWS was represented by location 17 with $\overline{\delta}_i$ value of 0.20% and $\sigma(\delta_i)$ value of 4.32%. Even though the $\sigma(\delta_i)$ value is high, location 17 is mostly representative of field SWS. Location 17 was located in the slope and the soil textural class was silty clay loam. Locations 27, 39, and 30 systematically overestimated the mean soil moisture value and represent wet conditions in this field. Locations 37, 42, and 38 systematically underestimate the mean soil moisture value and are representative of dry conditions. The cumulative probability function for the entire profile and upper 0–20 and 20–50 cm sections of depth also showed these locations as representative of wet and dry conditions in the field, respectively, when the entire profile is considered. As observed in the present study, Brocca et al.[2] attributed the increased variability of both $\overline{\delta}_i$ and $\sigma(\delta_i)$ to both the decreasing clay content and the increasing terrain slope. Low temporal stability in SWS of silt loam soils was reported earlier by Mohanty and Skaggs.[24] The wider variability of $\sigma(\delta_i)$ was linked to topography effects as Lin et al.[21] noticed. In dry locations, the predominance of the sand fraction was reported by Martinez-Fernandez and Ceballos[22] to reduce the retention of water and so temporal stability is relatively high for these locations.

Figure 5.13(b) presents the results on relative differences (RD) for the 0–20 cm depth. $\overline{\delta}_i$ values ranged from 12.40 to −12.28% and $\sigma(\delta_i)$ values ranged from 4.33 to 15.21% and with 8.32% as average $\sigma(\delta_i)$ in the 0–20 cm layer. Field mean SWS was represented by location 30 with $\overline{\delta}_i$ value of −0.65% and $\sigma(\delta_i)$ value of 8.27%. Location 30 was located in the back slope and the soil textural class was silty clay loam. At locations 6, 3, and 5, the mean soil moisture value was systematically overestimated and represented wet conditions of field. At locations 32, 40, and 42, the mean soil moisture value was systematically underestimated and was representative of dry conditions. The cumulative probability function for the entire profile also showed location 42 (silt loam) as representative of dry conditions in the field.

Lowest time stability was observed for the uppermost 0–20 cm layer due to its direct exposure to surrounding dynamic atmospheric conditions. Pachepsky et al.[27] also reported the weakest time stability at the shallowest observation depth among five depths measured up to 0.95 m.

Relative differences (RD) for the 20–50 cm depth. $\overline{\delta}_i$ values ranged from 10.75 to −20.96% and $\sigma(\delta_i)$ values ranged from 3.87 to 16.29% with 8.01% as average $\sigma(\delta_i)$ in the 20–50 cm layer. Location 4 with $\overline{\delta}_i$ value of 0.24% and $\sigma(\delta_i)$ value of 5.71% was located in the foot slope and the soil textural class was silty clay loam. Locations 9, 5, and 25 systematically overestimate the mean soil moisture value and represent wet conditions of field. Locations 41, 37, and 39 systematically underestimate the mean soil moisture value and are representative of dry conditions. The cumulative probability function for the entire profile also showed location 37 (silt loam) as representative of dry conditions in the field and location 9 (silt loam) as representative of wet conditions.

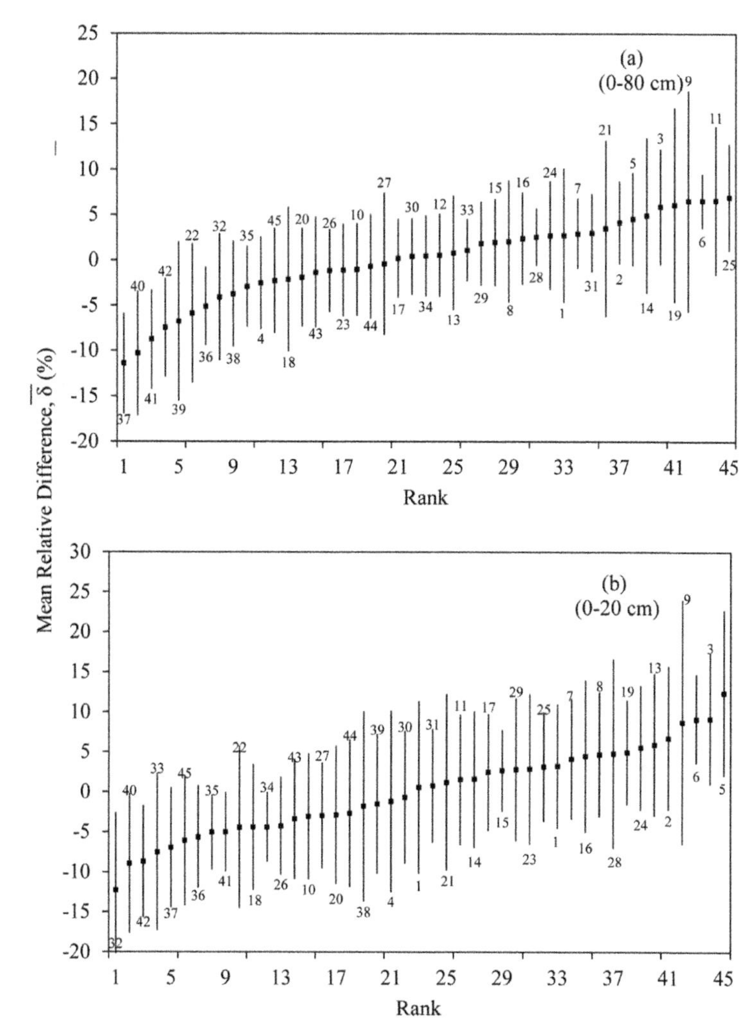

FIGURE 5.13 Ranked inter-temporal relative deviation from field mean SWS: (a) for 0–80 cm and (b) 0–20 cm. Vertical bars correspond to associated temporal standard deviation. Numbers refer to measuring locations.

Relative differences (RD) for the 50–80 cm depth range from 12.40 to −12.28% and $\sigma(\delta_i)$ values range from 4.33 to 15.21% with 8.32% as average. Location 30 represented field mean SWS with $\bar{\delta}_i$ value of −0.65% and $\sigma(\delta_i)$ value of 8.27%. Location 30 was located in the back slope and the soil texture was silty clay loam. Locations 6, 3, and 5 were the wettest locations and represent wet conditions of field. Locations 32, 40, and 42 were maintained driest and are representative of dry conditions. Location 42 (silt loam) was found as representative of dry conditions in the field in the cumulative probability function for the entire profile. The higher variability in SWS in the 50–80 cm layer could be attributed mainly to the impact of variable root water uptake[16] and water redistribution due to soil textural differences.

From the RD analysis, it was clear that mean, wet, and dry SWS at different depths were represented by different locations. Lin et al.[21] also reported a similar result and attributed that to variability in soil type and topography of the field. It makes selection of field mean SWS monitoring locations less reliable. Martinez-Fernandez and Ceballos[22] did not observe any specific pattern of stability with respect to depth, whereas Comegna and Basile[4] found a time-stable spatial structure for the water content in the upper 90 cm of the soil profile. Ranking stability of lower layers was expected to be greater than the upper layers of the profile due to the reduced dependence on the climatic, biological, and hydrological factors that determine the soil moisture dynamics as observed by Pachepsky et al.[27] Cassel et al.[3] observed greater temporal persistence of water content in deep soil layers than in shallow layers under a wheat crop which was attributed the persistence to the impact of crop root water uptake. The large variability in both $\bar{\delta}_i$ and $\sigma(\delta_i)$ shows temporal instability. Temporal stability during dry periods was attributed to stable property like soil texture as suggested by Brocca et al.[2] and Martinez-Fernandez and Ceballos.[22] It has been observed that sampling sites with moderate to moderately high clay content have shown more pronounced time stability. The less stability (a higher $\sigma(\delta_i)$) observed during dry period could be attributed to the dynamic nature of vegetation involved in the present study. Hupet and Vanclooster[15] have shown the role of vegetation in profile soil water distributions resulting in temporally unstable spatial structure.

As observed in the present study, locations along the foot slope were reported by Li et al.[21] and Zhou et al.[35] to be wetter than the field average throughout the soil profiles due to their high water-holding capacity resulting from high clay content, while the locations at the convex hill slope were much drier than the field average due to their high sand content.

5.3.1.5.3 Mean Absolute Value of Bias Error (MABE)

MABE is a new concept used to represent the temporal stability of locations for mean SWS. A critical value of 5% is used and locations having mean MABE below 5% are considered time stable as Hu et al.[15] identified. Figure 5.14 shows the rank-ordered MABE for the 0–80 cm (a) and 0–20 cm

(b) depths. The value of mean MABE ranges from 2.16 to 9.12% with an average value of 4.58% for the 0–80 cm layer. Thirty-three locations were showing mean MABE < 5%. So, the representative nature of these locations in estimating mean SWS is high. The σ MABE ranges from 1.74 to 6.77% with an average value of 3.68%. Locations 6, 28, and 7 were showing the lowest mean MABE (all <2.5%). These locations are positioned on slope, hilltop, and back slope areas, respectively, and these locations have silty clay loam soil texture. So the soil texture is playing an important role in keeping MABE low due to its high clay content. Locations 39, 19, and 9 were showing the highest mean MABE (all >8%) and were located on slope, hilltop, and slope positions, respectively. These locations have silt loam texture and its temporal stability is low. Locations 6, 33, 28, and 7 were showing the low σ MABE, whereas locations 14, 21, and 9 were showing high σ MABE. So, the mean MABE and σ MABE are positively correlated ($r = 0.82$).

Figure 5.14(b) shows the rank-ordered MABE for the 0–20 cm depth. The value of MABE ranges from 3.10 to 10.28% with an average value of 6.25% for the 0–20 cm layer. Only nine locations showed mean MABE <5%. The σ MABE ranges from 2.78 to 9.21% with an average value of 5.27%. Locations 15, 34, and 6 were showing the lowest mean MABE (all <2.5%) and they have silty clay loam texture. These locations were positioned on slopes as well. Locations 38, 4, and 9 were showing the highest mean MABE (all >9.5%) and have silt loam texture. These locations were positioned along slopes. Locations 35, 34, and 17 were showing the low σ MABE, whereas locations 19, 38, and 9 were showing high σ MABE. So, the mean MABE and σ MABE are positively correlated ($r = 0.70$).

The value of mean MABE for the 20–50 cm depth ranges from 6.07 to 13.93% with an average value of 2.75% for the 20–50 cm layer. Only 18 locations showed mean MABE < 5%. The σ MABE ranges from 2.17 to 11.32% with an average value of 5.19%. Locations 6, 28, and 7 were showing the lowest mean MABE (all <3.2%), whereas locations 41, 39, and 20 were showing the highest mean MABE (all >13%). These locations are positioned along slopes and have silt loam soil texture. So, the mean MABE and σ MABE are positively correlated ($r = 0.89$).

The rank-ordered mean MABE for the 50–80 cm depth ranges from 2.21 to 14.48% with an average value of 6.12% for the 20–50 cm layer. Only 17 locations showed mean MABE < 5%. The σ MABE ranges from 1.88 to 11.12% with an average value of 5.01%. Locations 7, 29, and 5 were showing the lowest mean MABE (all <2.7%). These locations have silty clay loam soil texture and were placed on hilltop, slope, and slope, respectively. Locations

14, 20, and 21 were showing the highest mean MABE (all >10.84%) and were located on valley floor, hilltop, and hilltop, respectively. Locations 29, 7, and 25 were also showing the low σ MABE (<2.18%), whereas locations 11, 21, and 14 were showing high σ MABE (>9.7%). So, the mean MABE and σ MABE are positively correlated ($r = 0.89$).

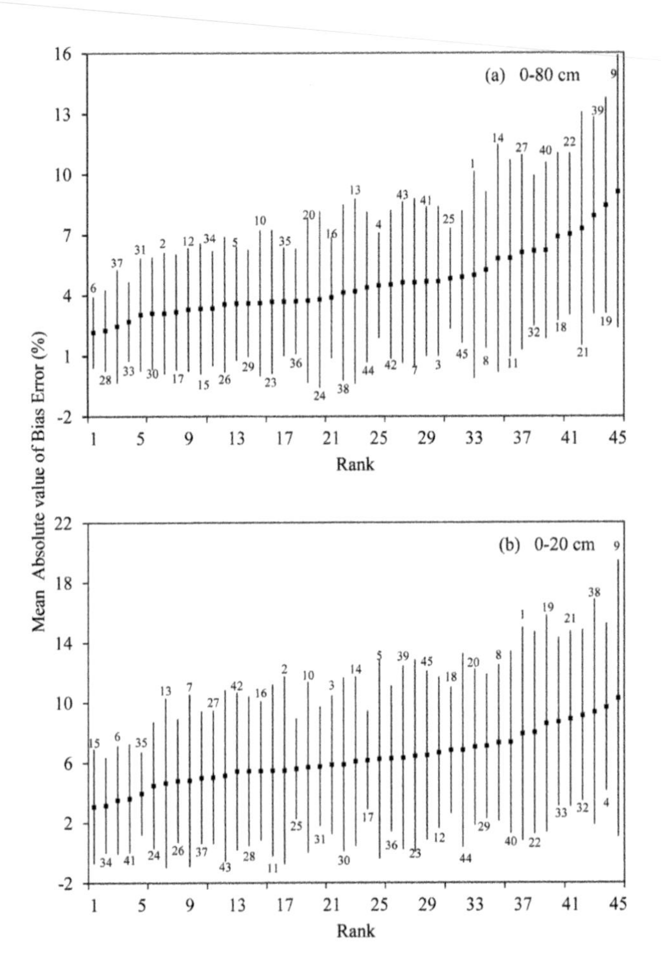

FIGURE 5.14 Rank-ordered MABE for (top: a) 0–80 cm and (bottom: b) 0–20 cm depths.

The mean MABE and σ MABE did not change with depth. But many researchers[3,14] have reported increased temporal stability of soil water content with increased soil depth. The representative nature of these locations in estimating mean SWS is low because different locations are representing low mean MABE at different soil depths. Soil texture is playing an important role as silty clay loams are mostly recorded with low MABE and silt loams with high MABE. Vegetation also is important and was a major dynamic variable in the field. Comegna and Basile[4] also reported the high uncertainty involved in predicting soil moisture from previous measurements at the transect scale due to seasonal changes in vegetation altering the spatial patterns of soil moisture between measurements.

5.3.1.5.4 Spearman Rank Correlation Coefficient

The correlation coefficient between SWS measured on different dates ranges between 0.12 and 0.95 between consecutive days of measurement. The correlation between successive pairs of measurement dates is an indication of the temporal persistence of the spatial pattern.[19] A decrease in the coefficient between successive pairs of measurement dates is an indication that there has been a change in the spatial pattern of soil water content and therefore the spatial pattern is not stable over time. The coefficient of 0.12 was obtained between December 10, 2007, the wettest day of the study, and January 4, 2008, even though January 4, 2008, was a wet day. December 10 has a coefficient of <0.45 with all the other dates. So, even between wet days, for example, December 10, 2007, and March 20, 2008, the coefficient was low (0.44). But the coefficient between dry days was always >0.76, for example, between July 26, 2007, and September 29, 2008, the coefficient was 0.76 even though the time interval was more than 1 year. High coefficients (>0.90) were observed only between successive measurement days. The correlation between dry and wet days was poor, for example, between August 28, 2007, and November 28, 2007, the coefficient was 0.40 even though the time interval was only 3 months. September 29, 2008, was the driest day of the study and had high correlation only with measuring days during the dry period in 2007 (July–October). It showed poor correlation with all other days in the study. Thus, the patterns of temporal stability do not persist across the whole time period analyzed. Kachanoski and de Jong[19] observed the lowest values of the coefficient during periods of transition from dry to wet. But the coefficients were not poor between August 28, 2007 (dry day) and December 10, 2007 (wettest day). So, the correlation did not improve after recharge as observed by Kachanoski and de Jong.[19] But the coefficients were low when the time interval between the days was large (June 4, 2007, and June 4, 2008) even though the mean SWS was very close between those 2 days. The low value of coefficient could be due to differences in spatial variance between these two times.

5.3.1.5.5 *Squared Coherency Analysis*

Squared coherency analysis was used to measure the consistency of a significant linear relationship between SWS at different spatial scales as Kachanoski and de Jong[19] reported. Figure 5.15 shows the coherency spectra obtained between different days of measurement. The days were with or without crop, wet or dry days. The coherency (Fig. 5.15a) was significant only in the medium scale even though the spatial variability was low on wet days. Winter wheat was present in the field on both the days, but the root water uptake would be more evident in March 20, 2008, than December 20, 2007.

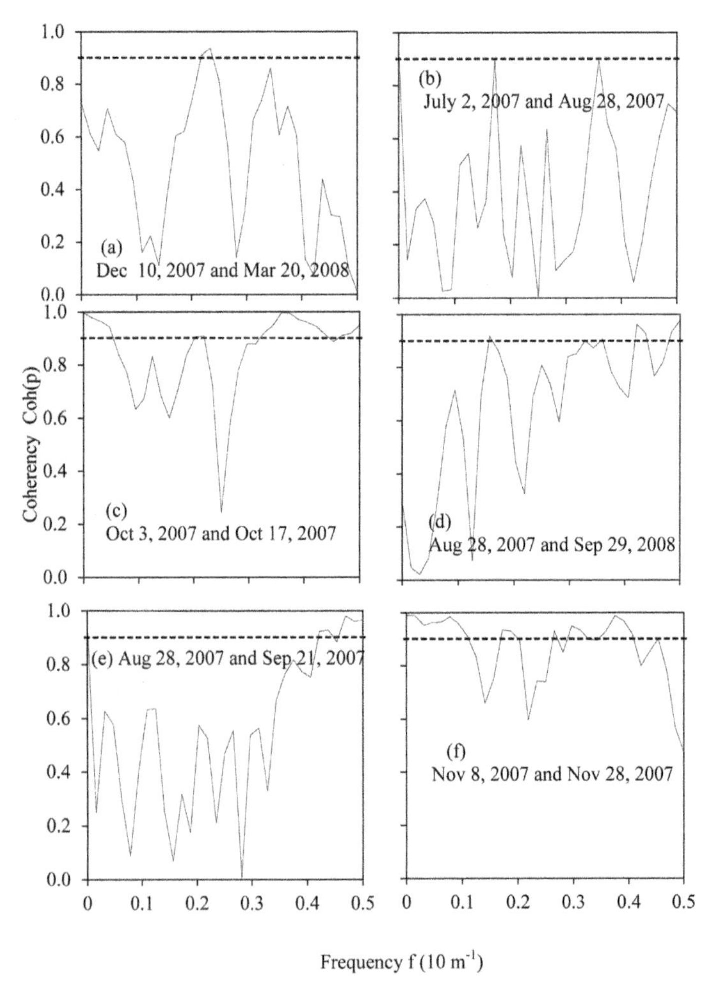

FIGURE 5.15 Coherency functions for SWS under different soil moisture conditions.

The coherency (Fig. 5.15b) between a wet (crop) and a dry day (no crop) was significant only at large scale similar to the length of transect. The coherency (Fig. 5.15c) between October 3 and 17 of 2007 was significant at medium scales and there was no crop during these two times, and the time interval between measurements was only 2 weeks. The mean, maximum, minimum, SD, and CV% of soil moisture were very close between these 2 days. The coherency was significant only at small scales in Figure 5.15(d) when 2 dry days were correlated. Both days had a crop growing, and the time interval was more than 1 year. The significance at smaller scales could be due to small scale variability in soil texture. Figure 5.15(e) shows significance between a dry day (August 28) and a day after recharge (September 21) at small scales. The coherency (Fig. 5.15f) between two adjacent days on November 8 and 28, 2007, was significant at small, medium, and large scales. Winter wheat was dormant during this period and might not have influenced the SWS much. The coherency results thus indicated poor temporal persistence of SWS in the field.

5.3.1.6 FACTORS INFLUENCING SPATIO-TEMPORAL DYNAMICS OF SWS

A better understanding of the spatio-temporal dynamics of SWS can be achieved by analyzing various interacting factors. Many investigators[9,24] have studied the influence of different factors on soil moisture content such as the texture of soil, which determines the water-holding capacity, the precipitation history, the slope and aspect of the land surface, which affect the water-redistribution processes, runoff, and infiltration, and the heterogeneity in vegetation and land cover, which influences evapotranspiration and deep percolation. Tomer and Anderson[31] found that 51–77% of spatial variability in soil water content could be explained by a combination of elevation, slope, and curvature in a sandy hill slope. Da Silva et al.[5] found that clay content and organic matter can serve as good explanatory variables, whereas topographic variables cannot. However, Gomez-Plaza et al.[12] found topography and vegetation rather than soil properties to be important factors in deciding time-stable locations. Mohanty and Skaggs[24] found better time stability on a sandy loam soil than a silt loam and reported that time-stable sites showed poor relationships to soil and topographic properties, thus suggesting the absence of a single dominant control.

5.3.1.6.1 *Soil Texture*

There was significant spatial variability of soil texture by depths (Table 5.2). Silt was the most common followed by clay and sand. CV% of soil texture was highest for sand, followed by clay and then silt at various depths. The high spatial variability in sand and clay are influencing the soil water dynamics.

TABLE 5.2 Descriptive Statistics of Soil Texture at Different Soil Depths in Percentages.

Soil Texture	Max.	Min.	Mean	CV%	SD
0–15 cm					
Sand	7.32	2.14	4.95	24.48	1.21
Silt	82.90	61.91	73.08	7.00	5.11
Clay	32.29	11.97	21.97	23.43	5.15
		15–30 cm			
Sand	9.21	1.57	4.08	37.89	1.55
Silt	80.57	60.39	68.23	7.48	5.10
Clay	33.88	16.31	27.69	16.64	4.61
		30–60 cm			
Sand	9.31	1.28	4.74	36.13	1.71
Silt	75.66	51.14	65.70	8.30	5.45
Clay	40.78	21.31	29.56	15.02	4.44
60–90 cm					
Sand	8.75	1.75	5.74	31.26	1.79
Silt	80.46	47.94	67.53	12.99	8.77
Clay	44.22	13.59	26.73	30.81	8.23

Spatial variability in the distribution of each soil texture is clear from the figures. Sand showed variability in autocorrelation length with depth. Silt and clay showed two lags of significant autocorrelation for the upper two layers. Highest autocorrelation length of all the soil texture observed at the deepest depth could be due to the highest SD value noticed at the deepest depth. Spatial variability in SWS was attributed to internal drainage processes influenced mainly by the local characteristics of porous medium as Comegna and Basile[4] reported.

It was observed that the zones with coarser soil particle size classes, silt loams, were generally drier than adjacent finer particle size class soil. The drier areas were located at the upper part of the slope and wetter areas at the lower valleys. Elevation also played a major role in the redistribution of water.

Soil texture affects the distribution, the vertical and lateral water transmission, and retention properties, and thus influences the soil moisture dynamics. Significant soil moisture variations were observed even over very small distances due to variations in soil particle and pore sizes as Mohanty and Skaggs[24] noticed. Time stability of the spatial distribution of moisture was explained by the deterministic impact of soil texture in the uppermost 1 m of soil profiles as reported by Vachaud et al.[32] Positive correlation between water content and clay content up to 70 m was observed by Munoz-Pardo et al.[25]

Significant relationship exists between soil texture and soil hydraulic properties. The increased variability observed in the mid-range of SWS could be attributed to soil texture, as it influences soil hydraulic properties mainly in the mid- to low-soil water pressure range as Mohanty and Skaggs[24] noticed. Significant soil moisture variations existed even over very small distances due to variations in soil particle and pore sizes. Da Silva et al.[5] found that the influence of landscape attributes on the redistribution of soil water was reflected, in part, in the spatial distribution of soil water as affected by soil texture.

Figure 5.16 shows the cross-correlogram of total SWS at different soil water content measuring dates versus various soil texture of 0–15 cm depth. From Figure 5.16, it is clear that correlation between sand and SWS is negative and was significant for four lag distances. The negative value was due to the drainage property of sand resulting in an inverse relationship with SWS. The silt was also showing a negative correlation with SWS due to its drainage property.

The clay content was showing positive correlation for four lag distances and could be attributed to the high water-holding capacity of clay. A similar relationship was found for deeper soil depths. Figure 5.17 shows the spectra of depth-averaged soil texture particles such as sand (a), silt (b), and clay (c) particles. Spectral peak was observed at every 160 m along the transect for the three soil particles. Small-scale variability in soil texture existed in the field but was not pronounced for any of the soil particles.

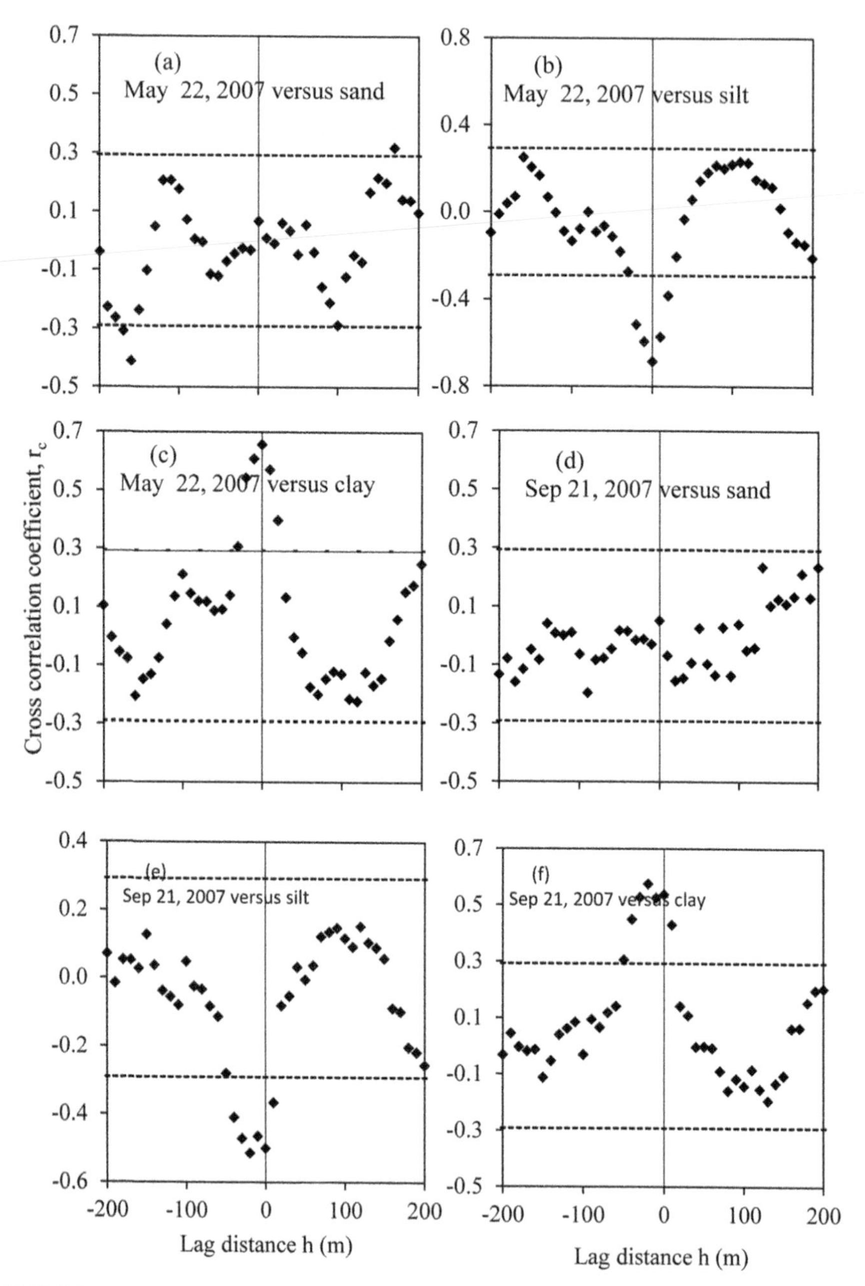

FIGURE 16 Cross correlogram of total SWS at different soil water content measuring dates versus various soil texture of 0-15 cm depth. The dotted lines represent 95% significance level.

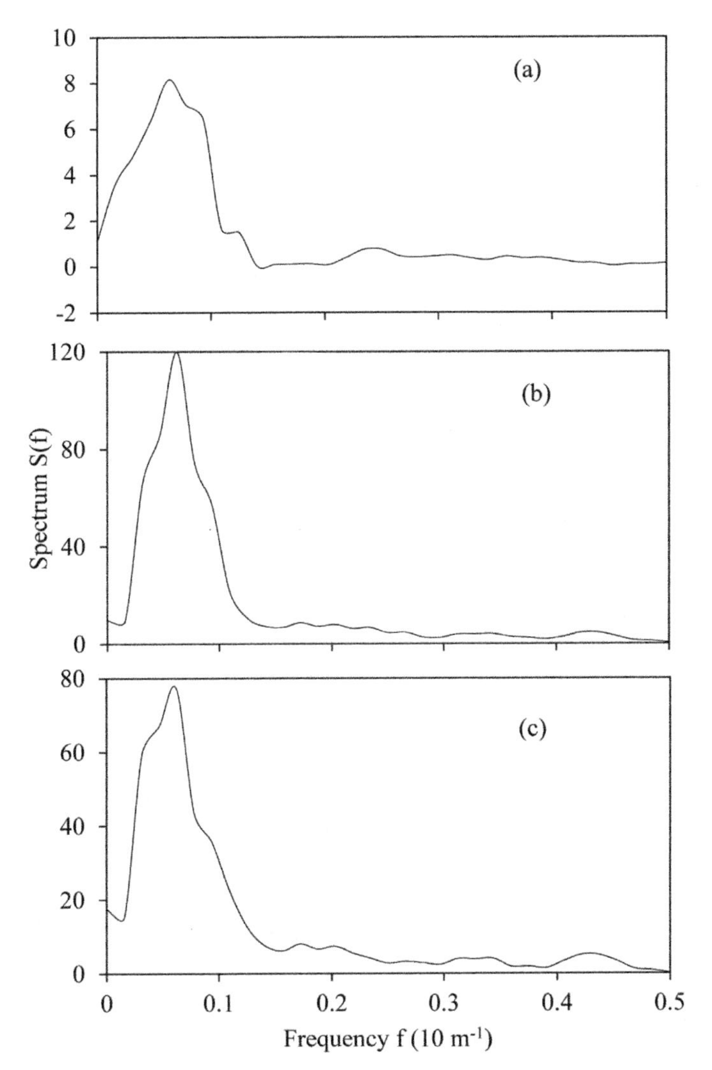

FIGURE 5.17 Spectra of depth-averaged (a) sand, (b) silt, and (c) clay particles.

5.3.1.6.2 Elevation

Topography plays an important role in the spatial distribution of soil moisture at different scales. Variations in slope, aspect, curvature, upslope contributing area, and elevation all affect the distribution of soil moisture near the land surface. Soil moisture varies as a result of water-routing processes. Topography was important in the lateral redistribution of soil moisture and created soil moisture variability. Mohanty et al.[23] observed time instability of soil moisture patterns and attributed them to the impact of subsurface flows in the redistribution of soil moisture especially for areas of relatively shallow soils and suggested that the local topography exerts a dominant control on these subsurface flows. Figure 5.18 presents cospectrum and quadrature spectrum of elevation and SWS at different soil moisture conditions. In Figure 5.18(a) and (b), peaks for both the spectra occurred at the same frequency at which the peak of elevation was observed. Corn was in the initial growing stages on these 2 days. So, the impact of vegetation on SWS in the form of evapotranspiration was less. Figure 5.18(c)–(f) shows a shift in the cyclic patterns of SWS. Active crop growth existed in the field during these days. The impact of soil textural variability was also on dry days (Fig. 5.18c and d). The shift could be thus attributed to the variability in vegetation and soil texture. Cross-correlogram of total SWS at different soil water content measuring dates versus elevation shows that the correlation ranged from 120 to 130 m with elevation leading the SWS.

5.3.1.6.3 Precipitation

Precipitation is the most important climatic variable influencing soil moisture content. Studies have observed that as the soil becomes wet, variability in soil moisture increases, and decreases with increasing time since the last rainfall as Mohanty and Skaggs[24] reported. But in the present study, an opposite trend was observed as also reported by others, for example, Famiglietti et al.[7] The precipitation data (Fig. 5.5b) indicated obvious temporal variability of rainfall in the field. The weekly cumulative precipitation ranged from 42.96 to 0.00 cm. The largest precipitations occurred during winter months of December to March and the lowest during July to August of both the years. Da Silva et al.[5] observed that during snow melt, soils approach saturation, spatial variability in soil water content begins to develop as drainage begins, evolves through the growing season, and then diminishes as evapotranspiration declines and late rainfalls become increasingly prevalent.

Hupet and Vanclooster[17] reported that the rainfall reaching the ground under the corn canopy was very variable with CV ranging between 77.87 and 188.7%. Some locations in their study received up to 4.47 times more water than the incident rainfall. Van Wesenbeeck and Kachanoski[34] made the same observation and attributed that to the rainfall distribution in the form of stem flow or through fall under the canopy, while inter row locations received much less water.

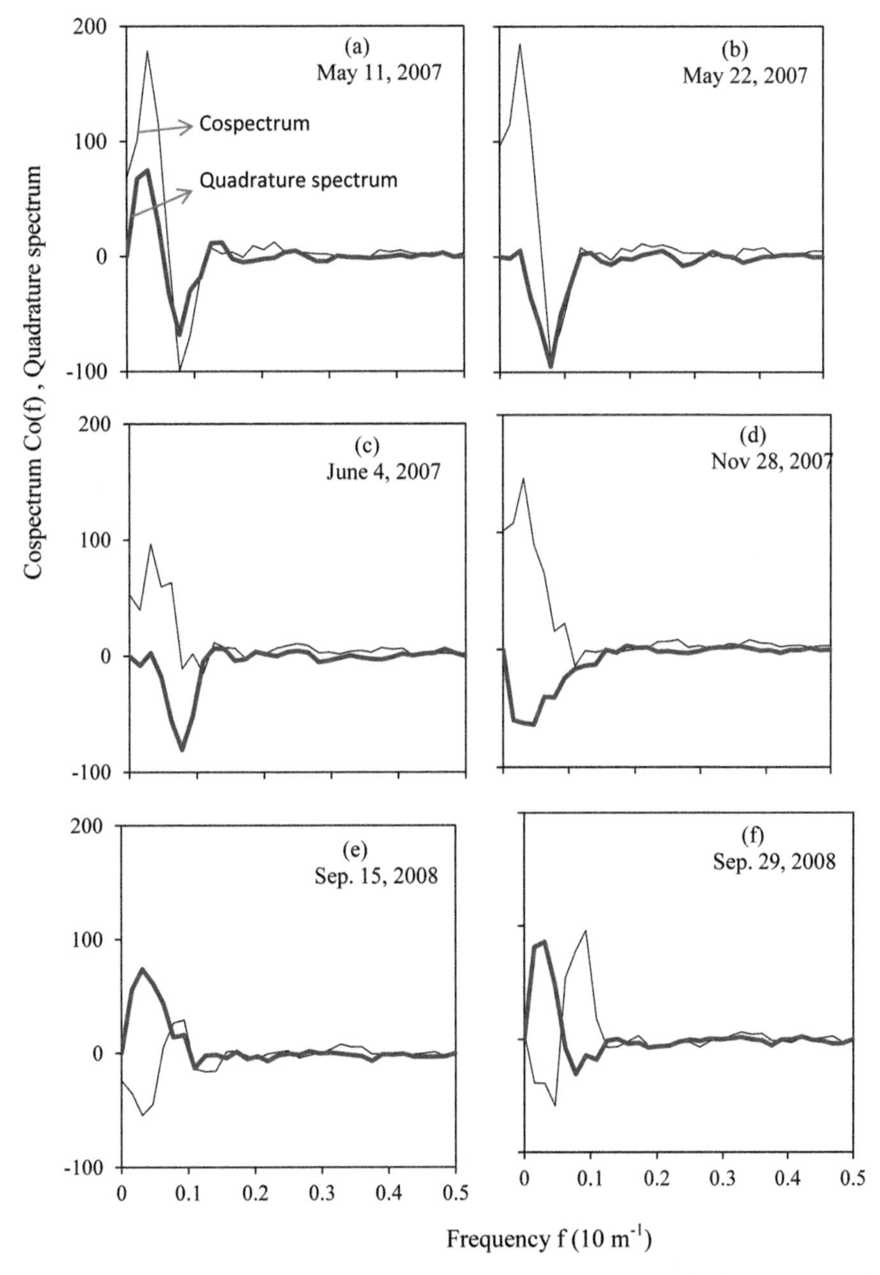

FIGURE 5.18 Cospectrum and quadrature spectrum of elevation and SWS at different soil moisture conditions.

5.3.1.6.4 Vegetation

Crop growth in the field influences the soil moisture dynamics as it affects infiltration, runoff, and evapotranspiration as reported by many authors.[16,24] The effect of a corn crop that changes continually over the growing season is an additional factor causing a systematic change in soil water content over space and time as Van Wesenbeeck and Kachanoski[34] observed. Soil water dynamics is one of the major causes of yield variability. Spatial and temporal analysis of crop growth parameters of both the crops showed the existence of variability in the field. Thus, the variability in crop growth is resulting in variability in root water uptake and evapotranspiration. Higher SWS observed during May 2007 and December 2007 to February 2008 could be attributed to low root water uptake as the crop growth was only initiated during these times. The spatial and temporal variability in crop growth contributed to spatial and temporal variability in evapotranspiration and was a major cause for lack of temporal stability in SWS. Vegetation is a more dynamic variable compared to other variables such as soil texture and elevation of the field. Van Wesenbeeck and Kachanoski[34] found that the variability of soil moisture is lowest with full canopy cover and highest with partial coverage. But in the present study, variability was obtained throughout the crop growth. The variability in temporal variance between row positions and inter-row positions in a corn field was attributed to the preferential drying and recharge in the row area of corn as Van Wesenbeeck and Kachanoski[34] reported. Hupet and Vanclooster[17] reported that the root water uptake was extremely variable within corn rows with CV values ranging between 14 and 137%. Rainfall reaching the ground under the corn canopy was very variable with CV ranging between 78 and 189%.

5.4 SUMMARY

Spatial and temporal analysis of soil moisture is helpful in characterizing its inherent spatial variability under field conditions. Spatial and temporal variability in SWS was observed in the field. The variability was determined by the interaction of different factors such as soil texture, topography, vegetation, and precipitation as evident in the study.

Absence of temporal stability in the spatial pattern of SWS for the entire profile and different soil depths was confirmed by cumulative probability analysis, relative deviation of mean SWS, MABE analysis spearman correlation, and coherency. Locations representing field mean SWS varied with soil depth, so no location could be found to represent field mean and extreme SWS conditions. Temporal persistence was high between successive measuring days at small, medium, and large scales. The temporal stability of SWS was low (r <0.40) between wet days when the time interval was more than 2 weeks. But coherency and the spearman coefficient were significant between two dry days even when the time interval was more than 1 year. Large variability in SWS under dry soil conditions than wet soil was attributed to small-scale variability in soil texture, vegetation, and topography.

KEYWORDS

- **soil moisture dynamics**
- **spatial continuity**
- **spatial statistics**
- **spatial structure**
- **spatial variability**
- **temporal persistence**
- **temporal stability**
- **temporal variability**

REFERENCES

1. Baumhardt, R. L.; Lascano, R. J.; Evett, S. R. Soil Material, Temperature, and Salinity Effects on Calibration of Multisensor Capacitance Probes. *Soil Sci. Soc. Am. J.* **2000,** *64,* 1940–1946.
2. Brocca, L.; Morbidelli, R.; Melone, F.; Moramarco, T. Soil Moisture Spatial Variability in Experimental Areas of Central Italy. *J. Hydrol.* **2007,** *333,* 356–373.
3. Cassel, D. K.; Wendroth, O; Nielsen, D. R. Assessing Spatial Variability in an Agricultural Experiment Station Field: Opportunities Arising from Spatial Dependence. *Agron. J.* **2000,** *92,* 706–714.
4. Comegna, V.; Basile, A. Temporal Stability of Spatial Patterns of Soil Water Storage in a Cultivated Vesuvian Soil. *Geoderma* **1994,** *62,* 299–310.
5. Da Silva, C. D.; Andrade, A. S.; Alves, J.; de Souza, A. B.; de Brito Melo, F.; Filho, M. A. Calibration of a Capacitance Probe in a Paleudult. *Sci. Agric.* (Piracicaba, Braz.) **2007,** *64*(4), 1–8.
6. Evett, S. R.; Tolk, J. A.; Howell, T. A. Soil Profile Water Content Determination: Sensor Accuracy, Axial Response, Calibration, Temperature Dependence, and Precision. *Vadose Zone J.* **2006,** *5,* 894–907.

7. Famiglietti, J. S.; Ryu, D.; Berg, A. A.; Rodell, M.; Jackson, T. J. Field Observations of Soil Moisture Variability Across Scales. *Water Resour. Res.* **2008,** *44,* W01423.
8. Famiglietti, J. S.; Devereaux, J. A.; Laymon, C. A.; Tsegaye, T.; Houser, P. R.; Jackson, T. J. Ground-based Investigation of Soil Moisture Variability Within Remote Sensing Footprints During the Southern Great Plains Hydrology Experiment. *Water Resour. Res.* **1999,** *35,* 1839–1851.
9. Famiglietti, J. S.; Rudnicki, J. W.; Rodell, M. Variability in Surface Soil Moisture Content Along a Hillslope Transect: Rattlesnake Hill, Texas. *J. Hydrol.* **1998,** *210,* 259–281.
10. Geesing, D.; Bachmaier, M.; Schmidhalter, U. Field Calibration of a Capacitance Soil Water Probe in Heterogeneous Fields. *Aust. J. Soil Res.* **2004,** *42*, 289–299.
11. Gish, T. J.; Walthall, C. L.; Daughtry, C. S. T; Kung, K. J. Using Soil Moisture and Spatial Yield Patterns to Identify Subsurface Flow Pathways. *J. Environ. Qual.* **2005,** *34*, 274–286.
12. Gomez-Plaza, A.; Alvarez-Rogel, J.; Albaladejo, J.; Castillo, V. M. Spatial Patterns and Temporal Stability of Soil Moisture Across a Range of Scales in a Semi-arid Environment. *Hydrol. Process.* **2004,** *14,* 1261–1277.
13. Grayson, R. B.; Western, A. W. Towards a Real Estimation of Soil Water Content from Point Measurements: Time and Space Stability of Mean Response. *J. Hydrol.* **1998,** *207*(1–2), 68–82.
14. Guber, A. K.; Gish, T. J.; Pachepsky, Y. A.; Van Genuchten, M. T.; Daughtry, C. S. T.; Nicholson, T. J.; Cady, R. E. Temporal Stability in Soil Water Content Patterns Across Agricultural Fields. *Catena* **2008,** *73*, 125–133.
15. Hu, W.; Shao, M. A.; Reichardt, K. Using the Mean Absolute Bias Error (MABE) as a Criterion to Identify Sites for Mean Soil Water Storage Evaluation. *Soil Sci. Soc. Am. J.* **2010,** *74*, 110–114.
16. Hupet, F.; Vanclooster, M. Intraseasonal Dynamics of Soil Moisture Variability Within a Small Agricultural Maize Cropped Field. *J. Hydrol.* **2002,** *261*, 86–101.
17. Hupet, F.; Vanclooster, M. Micro-variability of Hydrological Processes at the Maize Row Scale: Implications for Soil Water Content Measurements and Evapotranspiration Estimates. *J. Hydrol.* **2005,** *303*, 247–270.
18. Jacques, D.; Mohanty, B.; Timmerman, A.; Feyen, J. Study of Time Dependency of Factors Affecting the Spatial Distribution of Soil Water Content in a Field Plot. *Phys. Chem. Earth B* **2001,** *26*, 629–634.
19. Kachanoski, R. J.; de Jong, E. Scale Dependence and the Temporal Persistence of Spatial Patterns of Soil Water Storage. *Water Resour. Res.* **1988,** *24*, 85–91.
20. Lin, H. Temporal Stability of Soil Moisture Spatial Patterns and Subsurface Preferential Flow Pathways in the Shale Hills Catchment. *Vadose Zone J.* **2006,** *5*, 317–340.
21. Lin, H. S.; Kogelmann, W.; Walker, C.; Bruns, M. A. Soil Moisture Patterns in a Forested Catchment: A Hydropedological Perspective. *Geoderma* **2006,** *131,* 345–368.
22. Martinez-Fernandez, J.; Ceballos, A. Temporal Stability of Soil Moisture in a Large-field Experiment in Spain. *Soil Sci. Soc. Am. J.* **2003,** *67*, 1647–1656.
23. Mohanty, B. P.; Skaggs, T. H.; Famiglietti, J. S. Analysis and Mapping of Field-scale Soil Moisture Variability Using High-resolution, Ground-based Data During the Southern Great Plains 1997 (SGP97) Hydrology Experiment. *Water Resour. Res.* **2000,** 36(4), 1023–1031.
24. Mohanty, B. P.; Skaggs, T. H. Spatio-temporal and Time-stable Characteristics of Soil Moisture Within Remote Sensing Footprints with Varying Soil, Slope, and Vegetation. *Adv. Water Resour.* **2001,** *24,* 1051–1067.

25. Munoz-Pardo, J.; Ruelle, P.; Vauclin, M. Spatial Variability of an Agricultural Field: Geostatistical Analysis of Soil Texture, Soil Moisture and Yield Component of Two Rainfed Crops. *Catena* **1990,** *17,* 369–381.
26. Nielsen, D. R.; Wendroth, O. *Spatial and Temporal Statistics: Sampling Field Soils and Their Vegetation*; Catena Verlag GMBH: Reiskirchen, Germany, 2003; 110 p.
27. Pachepsky, Y.; Guber, A.; Jacques, D. Temporal Persistence in Vertical Distributions of Soil Moisture Contents. *Soil Sci. Soc. Am. J.* **2005,** *69,* 347–352.
28. Paltineanu, I. C.; Starr, J. L. Real-time Soil Water Dynamics Using Multisensor Capacitance Probes: Laboratory Calibration. *Soil Sci. Soc. Am. J.* **1997,** *61*, 1576–1585.
29. Robinson, D. A.; Jones, S. B.; Wraith, J. M.; Or, D.; Friedman, S. P. A Review of Advances in Dielectric and Electrical Conductivity Measurement in Soils Using Time Domain Reflectometry. *Vadose Zone J.* **2003,** *2,* 444–475.
30. Starks, P.; Heathman, G.; Jackson, T. J.; Cosh, M. H. Temporal Stability of Soil Moisture Profile. *J. Hydrol.* **2006,** *324,* 400–411.
31. Tomer, M. D.; Anderson, J. L. Variation of Soil Water Storage Across a Sand Plain Hill Slope. *Soil Sci. Soc. Am. J.* **1995,** *59,* 1091–1100.
32. Vachaud, G.; Passerat de Silans, A.; Balabanis, P.; Vauclin, M. Temporal Stability of Spatially Measured Soil Water Probability Density Function. *Soil Sci. Soc. Am. J.* **1985,** *49*, 822–828.
33. Van Pelt, R. S.; Wierenga, P. J. Temporal Stability of Spatially Measured Soil Matric Potential Probability Density Function. *Soil Sci. Soc. Am. J.* **2001,** *65*, 668–677.
34. Van Wesenbeeck, I. J.; Kachanoski, R. G. Spatial and Temporal Distribution of Soil Water in the Tilled Layer Under a Corn Crop. *Soil Sci. Soc. Am. J.* **1988,** *52*, 363–368.
35. Zhou, X.; Lin, H.; Zhu, Q. Temporal Stability of Soil Moisture Spatial Variability at Two Scales and Its Implication for Optimal Field Monitoring. *Hydrol. Earth Syst. Sci. Discuss.* **2007,** *4*, 1185–1214.

CHAPTER 6

USE OF SPECTRAL DATA FROM ON-THE-GO MULTISPECTRAL CAMERAS TO MONITOR SOIL SURFACE MOISTURE: THE PARTIAL LEAST-SQUARE REGRESSION FOR DATA MINING, ANALYSIS, AND PREDICTION

EDUARDO A. RIENZI[1,*], BLAZAN MIJATOVIC[1], CHRIS J. MATOCHA[1], FRANK J. SIKORA[2], and TOM MUELLER[3]

[1]Department of Plant and Soil Sciences, University of Kentucky, Lexington, KY 40546-0091, USA

[2]Regulatory Services Soil Testing Laboratory, 103 Bruce Poundstone Regulatory Service, Lexington, KY 40546-0275, USA

[3]Decision Science and Modeling, John Deere & Co., Urbandale, IA 50263, USA

[*]Corresponding author. E-mail: eduardo.reinzi@uky.edu

CONTENTS

ABSTRACT

This study highlighted several important issues that illustrate the potential of the spectra to hold information useful for the management. In addition, these findings demonstrate that by using partial least-square regression for data mining, it is possible to isolate regions of the NIR spectra that pose relevant information for prediction of soil moisture. Several regions of wavelengths were found that allow developing a simple index to assess the soil water content status. Therefore, several important and useful tasks could be performed with spectral reflectance that could contribute substantially with the irrigation management in daily operations.

6.1 INTRODUCTION

Visible-near-infrared (vis-NIR) diffuse reflectance spectroscopy is one important source of information to infer or quantify soil properties status. The measurements are rapid and cheap, and a spectrum can be used for various soil properties.[6,8,11] Inference is possible because vis-NIR spectra provide an integrative measure of soil that accounts for the soil's color, iron oxides, clay and carbonate mineralogy, organic matter content and composition, the amount of water present, and the particle size.[11] Since early 1970s, it has been reported that soil may be identified by their reflectance characteristics.[4] Some specific wavelengths can describe the entire spectral curve by unique correlation with "soil energies" that represent the soil chromophore. A chromophore is a substance (chemical or physical) that significantly affects the shape and nature of a material spectrum. A given soil sample consists of a variety of chromophores, which can vary with the environmental conditions and the status of the five soil formation factors (climate, topography, parent material, organic matter, and time).[3] Because the spectral reflectance of a soil sample is the result of the entire chromophore interaction with the incident electromagnetic energy, the resulting spectral curve can serve as a footprint to the chromophore existence in the sample.[3] However, NIR spectra are difficult to interpret directly because of the overlap of weak overtones and combinations of fundamental vibrational bands. As a result, multivariate calibration is required for quantitative analysis of sample constituents by NIR spectra. Among various calibration methods, partial least-squares (PLS) regression was used with substantial success to predict several soil characteristic and properties. The techniques involved in the PLS regression procedure works by extracting linear combinations of predictors in successive waves, which are selected to accomplish the objective of explaining response variation and predictor variation. Compared to a multiple lineal regressions, PLS regression seeks factors (named latent variables) that explain both dependent and independent variables' behavior. By accounting for predictor variation, it was assumed that the new observations should be better predicted when the predictors are highly correlated.

To empower the use of proximal sensors, like those carried with unmanned aerial vehicles, a procedure was developed for taking advantage of the variables important for predictions selected with PLS. The procedure to create the index does not require specific technical background and can be solved with any spreadsheet calculator and displayed in a map with QGIS, the open source software.

6.2 MATERIALS AND METHODS

Air-dried surface soil samples (0–5 cm depth) from the North American Proficiency Testing (NAPT) soil library (n = 149) were obtained for this research. The samples included Alfisols (n = 18), Andisols (n = 1), Aridisols (n = 18), Entisols (n = 30), Inceptisols (n = 1), Mollisols (n = 54), Ultisols (n = 16), Vertisols (n = 2), Spodosols (n = 5), and unknown soil orders (n = 4). Soil organic carbon (OC) data were obtained from the NAPT database measured with the Walkley–Black method,[12] and soil texture with a micropipette method.[5] Soil OC in the samples ranged from 2 to 60 g kg^{-1}. A summary of the variation of these selected soil parameters are shown in Table 6.1. To assess the effect of soil moisture on spectral reflectance response, several gravimetric water content levels (15, 20, and 25%) were imposed on the samples by adding deionized water by weight and labeled as WC0, WC15, WC20, and WC25.

TABLE 6.1 Mean Values of Selected Soil Parameters Used in This Study.

Soil	SOC (g kg^{-1})	Sand (%)	Silt (%)	Clay (%)
Alfisol	15.25 (8.20)	30.11(11.09)	51.06 (10.30)	19.59 (4.95)
Aridisol	5.46 (2.89)	65.68 (18.64)	22.06 (16.10)	12.28 (6.15)
Entisol	12.37 (12.51)	53.80 (24.60)	27.87 (15.68)	18.30 (12.59)
Molisol	15.30 (7.30)	35.20 (18.50)	42.41 (18.50)	22.06 (7.31)
Spodosol	16.51 (6.66)	71.52 (27.85)	20.87 (21.54)	7.20 (7.28)
Ultisol	7.26 (5.02)	57.71 (19.79)	28.55 (17.86)	13.37 (9.02)

The air-dry samples that were referred as the 0% soil moisture actually present water contents that ranged from 0.2 to 0.7g per 100 gon an oven dry basis. This discrepancy was considered negligible for the considerations of this study. Before running the PLS regression, the first derivative was taken.[9] Vis-NIR reflectance spectra were measured with a laboratory spectrometer Shimadzu UV 3101 PC. Spectra were collected over a wavelength range of 400–2220 nm at 2-nm increments, using an integrating sphere attachment.

Soil samples were located in a rectangular cell holder and referenced to barium sulfate as a 100% reflectance standard. GRAMS/32 AI (version 6.00) software from Galactic Industries Corporation (Salem, NH) was used for spectral processing. The data were analyzed with the SAS PLS procedure (SAS 9.3 for Windows, SAS Institute Inc., Cary, NC), using the random option for cross-validation. The number of iterations was set to 100, and 30% of observations were excluded in each random subset (e.g., NTEST = 48). The Rannar, Lindgren, Geladi, and Wold (RLGW) factor extraction algorithm was implemented because it is an iterative approach and the most efficient procedure to use when there are many predictors and few variables.[7] The number of factors to extract was selected by minimizing the predicted residual sum of squares (PRESS). To select the variables important for prediction (VIP), the procedure of Wold[13] was used. From the VIP selected, a simple index for predicting and assessing the soil moisture condition was designed by using a common spreadsheet and the open code geographic information system, QGIS.

6.3 RESULTS AND DISCUSSION

Figure 6.1 shows an example of two selected soils with contrasting soil texture (Molisol and Entisol) from the database used in this study and the effects on the reflectance produced under different soil water contents.

The curves downscale progressively as the water content increases in both soils, independent of its sand or OC content (Fig. 6.1(a) and (b)). However, it is also evident that when these soils reach 20% of soil moisture, the decreasing effect of moisture on reflectance is not as large as occurred when moving from WC0 to WC15.

Figure 6.2 displays the wavelength selected through the process that were identified as VIP. An arbitrary limit of one (1) was imposed on the results to help in the discussion of results. It can be observed that some of the wavelength present large VIP values with increasing soil water content. Additionally, one can observe that despite the differences in soil texture and soil OC content (Table 6.1), these soil variables were not always most important as could be expected. The corresponding values of VIP for sand, silt, and clay soil content and OC were labeled for easy identification (Fig. 6.2).

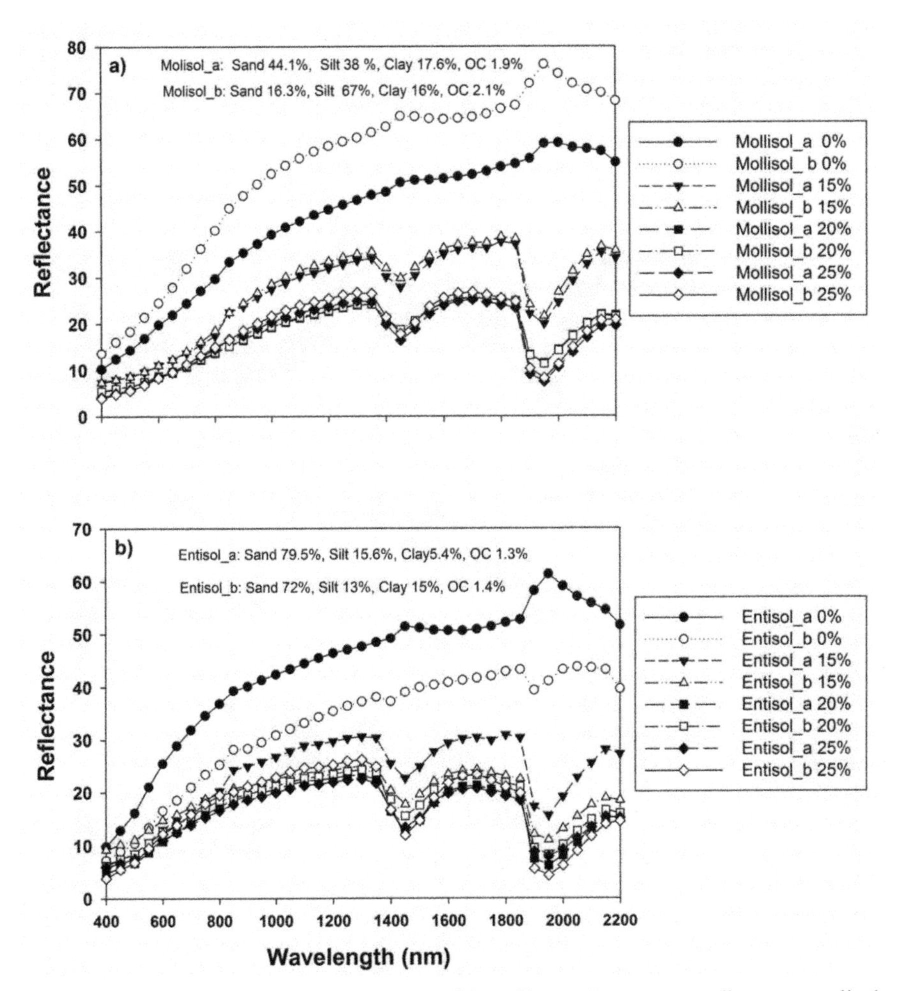

FIGURE 6.1 Values of reflectance measured in soil samples corresponding to a molisol (top: a) and an entisol (bottom: b) with different water content. The measurement was taken every 2 nm, but expressed every 50 nm for easy display.

The results show that OC and clay content always are moving as pairs in the ranking of VIP and they are showing a clear displacement from low to high positions as the soil is increasing in water content (Fig. 6.2(a)–(d)). Moreover, OC and clay content were at the highest position for 20–25%

of soil moisture (Fig. 6.2(c) and (d)), confirming that their influence was increased with the increment in soil water content.[2] This could be considered irrelevant when the same soil is assessed, but must be considered when an area with different soils is under study. Notice that the sand and silt content were always in low positions compared to the OC and clay content. The low relevance of sand and silt as predictors could be a consequence of the low influence on color when the soil become wet. In addition, the results

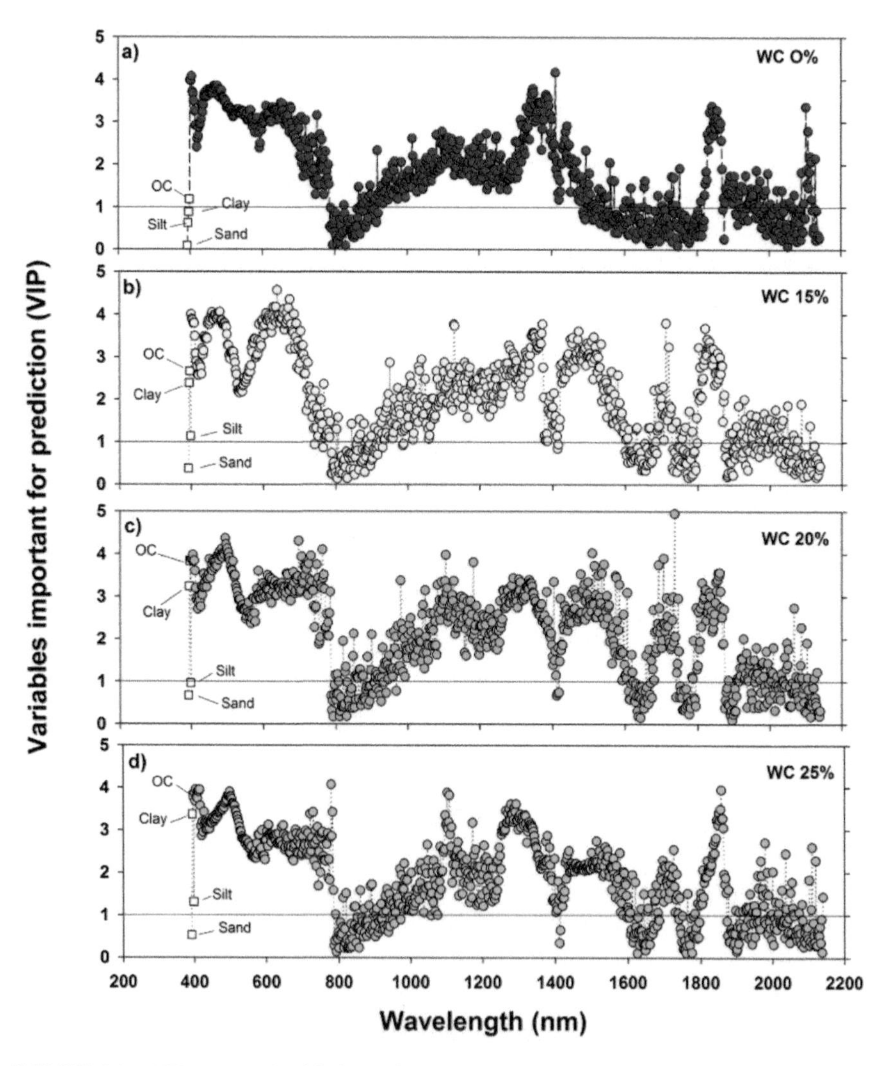

FIGURE 6.2 VIP extracted with the PLS regression analysis. A line was set at a value of 1 (one) as an arbitrary limit to compare the relevance of the variables.

expressed as VIP show that a few groups of wavelengths seem to hold information useful for the predictive purposes. Even though the change in soil moisture from WC0 to WC25 is too large, the PLS results illustrate this behavior. If we divided the VIP in sections of 400 nm approximately, we can observe that the first region of variables higher than 1 are very useful to identify differences in soil moisture from WC0 to WC20 (Fig. 6.2(a)–(d)). However, WC20 and WC25 seem very similar to be isolated with those wavelengths. The second region from 800 to 1300 nm approximately seems inadequate to detect differences in soil moisture because they have several wavelengths with almost the same values of importance. However, this region cannot be ignored because it has been mentioned to be successful to assess soil moisture by other investigators.[10]

From 1300 to 1700 nm approximately, these groups of wavelength give valuable information to separate WC0 from WC15 and WC20 (Fig. 6.2(a)–(c)). It can be observed that this section has several wavelengths that modify its relevance with the increase in soil moisture. From 1700 to 2100 nm approximately, this region can help to isolate WC0 from the rest of soil water content but seems to confuse the identity of the other soil moistures. In general, the regions we found are consistent with the findings others.[3]

An interesting consequence of this finding from a practical point of view is that we can use these wavelengths to create an index for classifying water content in soils. To illustrate how we can take advantage of this information, a VIP-reflectance curves relationship is displayed in Figure 6.3. As an example, only the VIP-0 and VIP-15 with the corresponding WC0 to WC15 spectra curves are shown. The red squares on the VIP graph scatter show the region where the wavelengths were selected based on the importance for the water content prediction. The arrows indicate how these regions look like on the spectra reflectance measured in a Molisol and an Entisol soils. These reflectance curves suffered when the soils became wet (Fig. 6.3).

The numerical differences developed between the reflectance values at specific wavelengths are correlated with the amount of water content in the soil. Thus, the combination of these ranges of wavelength would be useful to detect the stage of soil moisture in the field. It was mentioned that it is feasible to get a good prediction ability using discrete wavelengths instead of the full spectrum.[1] These results support this statement, and the authors believe that the development of wavelength-based sensors to measure soil properties will contribute to produce a drastic reduction in the measurement costs.

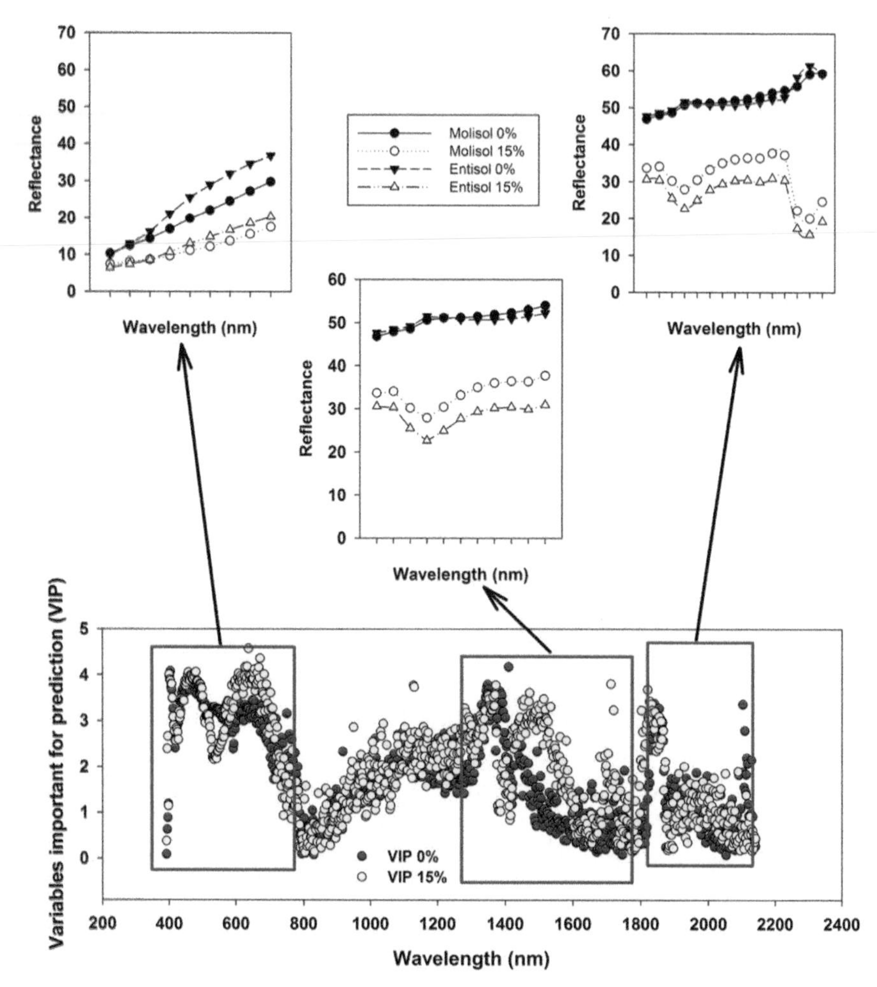

FIGURE 6.3 Relationship between the VIP and the visual aspect of the wavelengths in the region selected as the candidates for identifying different soil water contents. As an example, only two curves are shown corresponding to a molisol and an entisol with a soil water content of 0 and 15%, respectively.

6.3.1 VIP INFORMATION AS A SIMPLE INDEX TO ASSESS SOIL MOISTURE STATUS

To use the information obtained in the PLS regression, the authors built a simple subtracted formula based on the stage where the wavelengths identify the minimum reflectance, that is, the highest amount of water content used in this study. They set an arbitrary boundary limit at WC25, and a value of average reflectance was established for each one of the VIP selected regions, that is, the value for region 400–800 nm was different than the value for 1300–1700 nm or for 1800–2200 nm (Table 6.2).

TABLE 6.2 Average Values Calculated from the Spectral Reflectance Measured in the Database that Correspond to 25 g (100 g^{-1}) of Soil Water Content.

Wavelength region (nm)	Average value of reflectance
400–750	11.86
750–780	15.47
1260–1400	22.89
1450–1750	20.51
1760–2020	21.94
2020–2120	14.51

Thus, this unique value was used as a baseline to calculate the differences among the reflectance, corresponding to the different water content. The range of the differences to create the classes for 0–25% of WC was set in accordance with the measured data and it is displayed in Table 6.3. Because trying to isolate values of soil water content between 15 and 0% was considered impractical, and all the differences below 15% of water content were named as class 0%. This criterion assumes that when the soil contains less than 15% of water content, it is necessary to irrigate. Beyond this limit, the soil could be too dry to sustain crop growth and could lead to a crop failure.

TABLE 6.3 Simple Index Based on the Differences Calculated Through the Data Set on the Basis of the Reflectance Measured at 25% Soil Water Content.

Classes for soil water content	Rank of differences
25%	0
20–25%	10–20
15–20%	21–30
0%	>30

Figure 6.4 represents the use of the index on a farm of 247 acres. Each pixel in the map corresponds to 0.056 acre, approximately. As an example, the land units were identified in the map; the soils are Entisols with more

than 78% of sand content, where the land unit labeled as A corresponds to a dune. The hill slopes comprise long-sloped areas and at the end of these areas are located a complex soil map units that combine eventual flooding zones with wetlands and lagoons.

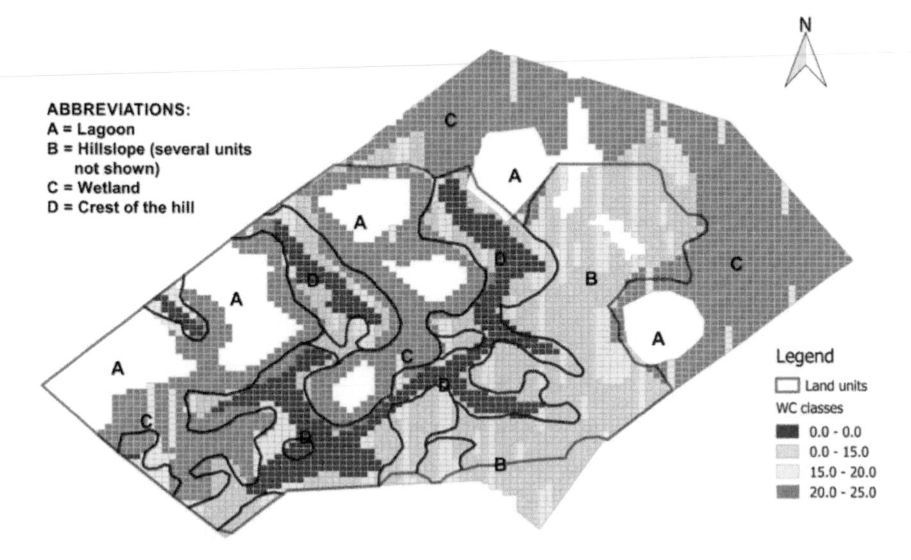

FIGURE 6.4 Example of the application of the index created from the differences in spectral information.

Despite several minor misclassifications, the simple index can isolate three principal zones with an acceptable precision. The low regions close to the lagoons are wet soils, which were identified as Class 25%. The crests of the hills (dunes) were well determined as class with minimum soil water content on the farm. It is also evident that to divide the intermediate zones (from 20 to 15% WC) are somehow not easy to detect and classify. However, because those values of soil moisture cannot determine risks for crops, for this simple approach, the separation could be considered acceptable. In addition, it is easy to realize that the classes could be substantially improved by using site-specific data instead of the general information obtained from our NAPT data set. This study provides an important tool for daily operations, because one can create an individual signature for each field under irrigation by using unmanned aerial vehicles or other proximal sensor devices carrying multispectral cameras.

6.4 SUMMARY

Quantifying soil water content (WC) with vis-NIR reflectometry relies on existing overtones of water absorption bands at 1450, 1950, and 2950 nm which result from resonance in the molecular vibration of the illuminated water molecule. Strong NIR absorption bands of water are found around 1400–1440 nm and between 1900 and 1950 nm, and have often been applied to quantitative analysis of water content. Wavelength bands related to water have also been utilized in NIR reflectance with remote sensing applications to determine WC and water status of plants. Several other multiple uses were found for this non-invasive technique, like measuring soil texture, bulk density, or soil OC content. However, a better understanding of the different vis-NIR spectra signature of WC and possible applicable solutions through transformations, filtering, and calibration strategies are still needed.

Spatial variability of texture and moisture within agricultural field is also suggested to affect the prediction accuracy of other soil properties during on-the-go measurement. The increasing advance in the use of unmanned aerial vehicle (UAV) for multiple uses and the possibility to carry multispectral cameras is expanding as it does the need to develop proper ways for data mining and analysis. The goal of this study was to describe and assess a methodology for extracting information from spectral data to determine soil moisture content useful for irrigation for different soils with varying textures and OC contents. The information extracted with PLS was used to create a simple easy-to-use index that can identify soil moisture status with relative precision for everyday farming operations.\

KEYWORDS

- **absorption band**
- **partial least-square regression**
- **QGIS application**
- **reflectance**
- **soil bulk density**
- **soil chromophore**
- **spectral application**
- **spectral curve**

REFERENCES

1. Bellon-Maurel, V.; McBratney, A. Near-infrared (NIR) and MId-infrared (MIR) Spectroscopic Techniques for Assessing the Amount of Carbon Stock in Soils—Critical Review and Research Perspectives. *Soil Biol. Biochem.* **2011,** *43*, 1398–1410.
2. Ben-Dor, E. Quantitative Remote Sensing of Soil Properties. In *Advances in Agronomy*; Elsevier/Academic Press: Amsterdam, The Netherlands; 2002; Vol. 75, pp 173–243.
3. Ben-Dor, E.; Taylor, R. G.; Hill, J.; Dematte, J. A. M.; Whiting, M. L.; Chabrillat, S.; Sommer, S. Imaging Spectrometry for Soil Applications. In *Advances in Agronomy*; Elsevier/Academic Press: Amsterdam, The Netherlands; 2008; Vol. 97, pp 321–398.
4. Condit, H. R. The Spectral Reflectance of American Soils. *Photogr. Eng.* **1970,** *36*, 955–966.
5. Miller, W. P.; Miller, D. M. A Micropipette Method for Soil Mechanical Analysis. *Commun. Soil Sci. Plant Anal.* **1987,** *18*, 1–15.
6. Mouazen, A. M.; Karoui, J.; Baerdemaeker, D.; Ramon, H. Characterization of Soil Water Content Using Measured Visible and Near Infrared Spectra. *Soil Sci. Soc. Am. J.* **2006,** *70*, 1295–1302.
7. Rannar, S.; Geladi, P.; Lindgren, F.; Wold, S. A PLS Kernel Algorithm for Data Sets with Many Variables and Few Objects, II: Cross-validation, Missing Data and Examples. *J. Chemomet.* **2006,** *9*, 459–470.
8. Reeves, J. B.; Follett, R. F.; McCarty, G. W.; Kimble, J. M. Can Near or Mid-infrared Diffuse Reflectance Spectroscopy be Used to Determine Soil Carbon Pools? *Commun. Soil Sci. Plant Anal.* **2006,** *37*, 2307–2325.
9. Rienzi, E. A.; Mijatovic, B.; Mueller, T. G.; Matocha, C. J.; Sichora, F. J.; Castrignano, A. Prediction of Soil Organic Carbon Under Varying Moisture Levels Using Reflectance Spectroscopy. *Soil Sci. Soc. Am. J.* **2013,** *78*, 958–967.
10. Rossel, R. A. V.; Behrens, T. Using Data Mining to Model and Interpret Soil Diffuse Reflectance Spectra. *Geoderma* **2010,** *158*, 46–54.
11. Rossel, R. A. V.; Webster, R. Predicting Soil Properties from the Australian Soil Visible-near Infrared Spectroscopic Database. *Eur. J. Soil Sci.* **2012,** *63*, 848–860.
12. Walkley, A.; Black, C. A. An Estimation of the Degtjareff Method for Determining Soil Organic Matter and a Proposed Modification of the Chromic Acid Titration Method. *Soil Sci.* **1934,** *37*, 29–37.
13. Wold, S. PLS for Multivariate Linear Modeling. In *Chemometric Methods in Molecular Design;* Wiley-VCH Verlag GmbH & Co, Deutsch, 1995; Vol. 2, 334 p.

CHAPTER 7

LONG-TERM EFFECTS OF CONVERSION FROM NATURAL FOREST TO TEA CULTIVATION ON SOIL PROPERTIES: HUMID NORTHERN BLACK SEA REGION†

TURAN YÜKSEK[1,*], CEYHUN GÖL[2], FILIZ YÜKSEK[3], and ESIN ERDOGAN YÜKSEL[4]

[1]*Department of Landscape Architecture, Faculty of Fine Arts, Design, and Architecture, Recep Tayyip Erdogan University, 53100 Rize, Turkey*

[2]*Department of Forest Engineering, Çankırı Karatekin University, 18200 Çankırı, Turkey*

[3]*Pazar Forest Management Directorate, 53300 Pazar, Rize, Turkey*

[4]*Department of Forest Engineering, Artvin Çoruh University, 08000 Artvin Turkey*

**Corresponding author. E-mail: turan53@yahoo.com; turan.yuksek@erdogau.edu.tr*

CONTENTS

†Modified version and printed with permission from Turan Yuksek, Gol, C.; Yuksek, F.; Yuksel, E. E., "The effects of land-use changes on soil properties: The conversion of alder coppice to tea plantations in the Humid Northern Black sea Region." Open Access Article in *Afr. J. Agric. Res.* 2009, *4*(7), 665–674, www. academicjournals.org/AJAR

ABSTRACT

The present study showed that land use has a significant influence on the chemical and physical properties of soils and hydrological processes in the research area. One major impact of cultivation is the reduction of total porosity and saturated hydraulic conductivity due to reduced abundance and activity of soil organisms. The increased surface runoff, reduced FC (due to soil loss), and evaporation caused an increase in water yield in the catchment area used for agriculture. The sustainability of existing plantations may be improved by greater organization of agricultural activities and the adoption of agro-ecological land-use zoning.

7.1 INTRODUCTION

In most parts of the world, agriculture is the primary cause of land-use change. Much of the pressure on conversion of forests to agricultural use comes from increasing population growth and developmental demands. During the 1990s, 14.6 million ha of forest per year was converted into agricultural or urban usage. However, 5.2 million ha per year was gained in plantations, reforestation, and natural forest expansion, for a net loss of 9.6 million ha of forest per year.[17] Land-use conversion affects both the amount and spatial pattern of forest habitat, which in turn can affect the ecological function and future development of remaining forest lands.

The impacts of land conversion are broadly categorized as environmental, economic, and social and can be both positive and negative. For example, habitat fragmentation and transportation corridors can create migration barriers or inhospitable habitats for wildlife and interfere with other ecological processes.[13] Small ownership parcels also complicate management and cooperation at landscape and watershed scales.[41] Within a landscape, hydrology reflects the balance between independent factors of geology and climate and dependent factors including topography, soils, and vegetation.[24] These dependent factors are interrelated, and a sudden change in one of these can cause adjustment to the others. On seemingly uniform slopes, hydrology is determined by interactions with vegetation and soil properties, affecting processes such as infiltration response and patterns of water penetration.[40] It is important to analyze soil characteristics within landform elements rather than isolated pedons.[38] Because soil physical and hydrological properties are related to vegetation type, vegetation conversion can alter these properties. The physical structure of the vegetation canopy and roots affects rainfall disposition by controlling how water is channeled into and through the soil.[32] Vegetation type has been shown to alter soil hydrological characteristics, including infiltration capacity, hydraulic conductivity, and water retention.[22]

The eastern part of the Black sea region is characterized by a rolling topography and is highly dissected by small streams. Therefore, only 2% of the total land area of the east Black sea region is suitable for agriculture. Much of the productive lands located in the coastal area have been converted into urban settlements, industry, and an airport.[43] Tea (*Camelia sinensis* L.) is an important cash crop in the northern part of the Black sea region, and because of its economic value, many farmers have replaced their traditional annual and food crops with tea. Moreover, since 1952, many farmers have converted their private coppice forest land into tea cultivation. Tea acreage continues to increase each year in the northern part of the Black sea region. Although tea plantations remain productive for long periods (25+ years), yet tea yields tend to decline in later years. This drop in productivity is traditionally attributed to natural aging of plants.

The total area under tea cultivation in Turkey is approximately 77,000 ha, of which approximately 65% (50,000 ha) is located in Rize. The land area in Rize under tea cultivation increased from 2800 ha in 1951 to 49,969 ha at present, representing 12.74% of the total land area in Rize.[3] Of 9255 ha of tea plantation, it is in land which is suitable for land-use classes (according to Turkish Land Use Classification System: between I and IV), while the rest of tea cultivation area is attributed to unsuitable land.[2] Degradation in soil quality is often associated with the type of intensive land-use involved in tea production. Any degradation of the soil can be expected to adversely affect the stability of system. Evaluation of soil quality changes during long-term tea production, which could therefore help to improve the sustainability of tea cultivation in the study area. Yüksek et al.[43] studied the effects of converting alder (*Alnus glutinosa* L. Gaertner subsp. barbata) forests into tea plantations and possible changes in soil characteristics and erosion rate. They found that the risk of soil erosion actually was doubled after the conversion. Similar results have been reported in the Pazar area, where deforestation and establishment of tea plantations has led to more frequent and larger flood events, causing greater damage to life and property.[44]

Tea (*C. sinensis* L.) is a globally important crop and is unusual because it both requires an acid soil and it acidifies soil. Tea stands tend to be extremely heavily fertilized in order to improve yield and quality, resulting in significant potential for diffuse pollution.[23] Assessing land-use-induced changes in soil systems is therefore essential for addressing the problem of agro-ecosystem transformation and sustained land productivity.[39]

The aim of this study was to determine the long-term effects of conversion from forest coppice (*A. glutinosa* L. Gaertner subsp. barbata) into tea cultivation (*C. sinensis* L.), in terms of changes in soil physical, hydrological, and some chemical properties. The trends observed were used to evaluate the impacts of tea cultivation on soil attributes and as indicators of the sustainability of the tea cultivation management. A small agricultural catchment located in Pazar watershed in Rize, Turkey, was chosen as the study area.

7.2 MATERIALS AND METHODS

7.2.1 SITE DESCRIPTION

The study was conducted in Pazar watershed in Rize, located in the northeastern part of the Black sea region of Turkey. The study area is located at 40°52′45″N and 40°45′27″E, at 40–150 m above mean sea level. The mean annual temperature is 14°C with an average annual precipitation of 2019 mm.[4] The soils of the area are classified as yellow red podsol according to the International Soil Classification System (ISCS).[2,6] The rock mass is extensively volcanically disrupted and the main material is andesite.[42]

The forest type within the study area is mainly composed of *A. glutinosa* L. Gaertner subsp. barbata.[43] Approximately 5 ha of alder stand was clear-cut. Terraces were then constructed of 40–60 m in length and 0.6–0.9 m in width, and soils in the terraces were tilled up to 0.6 m. Fertilizer (500 kg/ha NPK: 2:1:1) was applied to tea plantation areas during March every year. Tea was harvested three times during the vegetation season (in May, July, and September), and plantations were weeded before harvesting. The area adjacent to the tea plantation is mainly composed of alder stands (*A. glutinosa* L. Gaertner subsp. barbata) with small numbers of *Rubus platyphyllos*, Urtica sp., *Frangula alnus* Miller beneath the tree canopy and annual forage plants.[42] The alder forest has been clear-cut every 20 years since 1925, leaving old root sprouts to regrow naturally. The annual forage plants were harvested. Leaves were then collected for animal bedding. At the end of the coppicing cycle, there are 2–4 sprouts per stool, which belong to one or two diametric classes (8–15 and 15–35 cm). This type of traditional management has been conducted for at least 80 years in the study area. The mean average seedbed was 160 unit per ha and mean density (Ø > 5 cm dbh) was 1085 trees per ha in the study area.

7.2.2 EXPERIMENTAL DESIGN AND SOIL SAMPLING

The experimental design at each site was a randomized complete block with four replications in each study area. The experimental plots taken were from neighboring forest and tea cultivation regions with an area of 400 m^2 (20 m × 20 m). Soil samples were collected from forest and tea-cultivated sites (soils were collected from near the planting holes of tea plantations on terrace). Four disturbed and four undisturbed soil samples were taken randomly at a soil depth of 0–10, 10–30, and 30–50 cm in each plot in study area. Plots in the two sample areas had the same physiographic conditions such as landscape position and slope (%). The undisturbed soil samples were taken by using a steel core sampler of 100 cm^3 volume (5 cm in diameter and 5 cm in height). A total of 96 soil samples (2 land-use types × 4 replicates × 4 soil pits × 3 soil depths) were collected in April and May, 2007.

7.2.3 LABORATORY ANALYSIS

The particle size distribution was determined by the Bouyoucos hydrometer method.[7] The field capacity (FC) and permanent wilting point (PWP) were measured using pressure membrane and pressure plate extractors. The plant available water (PAW) content was calculated from the difference between the FC and the PWP.[28] The dry bulk density (D_b) was determined by the core method.[21] The particle density (D_p) was determined by the pycnometer method.[18] The total porosity (S_t) was calculated from the following equation[18]:

$$S_t(\%) = [100x\left(1 - \frac{D_b}{D_p}\right)] \quad (7.1)$$

where S_t is the total pore space, D_b the bulk density, and D_p the soil particle density.[18]

A wet sieving method was used to determine the water-stable aggregates (WSA).[27] The saturated hydraulic conductivity (K_{sat}) was measured by the falling-head method according to Klute and Dirksen.[29] The soil penetration resistance (SPR)[8] was measured for 0–40 cm soil depth. Measurements were recorded at depth intervals of 5 cm, using a manual (hand-pushed) 13 mm diameter cone (30°) penetrometer and 20 measurements were made in each plot. Soil pH was determined in a soil water mixture (1:2.5 by volume).[26] Electrical conductivity (EC) of the saturation was measured by the method developed by Rhoades.[36] The concentration of soil organic matter (SOM) and soil organic carbon (SOC) were determined by the method described by Nelson et al.[34] Total nitrogen (TN) was determined by the Kjeldahl method.[10] Carbon-to-nitrogen ratio is a ratio of the mass of carbon to the mass of nitrogen in a substance. It can, among other things, be used in analyzing sediments and compost. A useful application for C/N ratios is as a proxy for paleoclimate research, having different uses whether the sediment cores are terrestrial-based or marine-based. Carbon-to-nitrogen ratios are an indicator for nitrogen limitation of plants and other organisms and can identify whether molecules found in the sediment under study come from land-based or algal plants. Examples of devices that can be used to measure this ratio are the CHN analyzer and the continuous-flow isotope ratio mass spectrometer (CF-IRMS). However, for more practical applications, desired C/N ratios can be achieved by blending common used substrates of known C/N content, which are readily available and easy to use. The <http://www.klickitatcounty.org/solidwaste> has developed the *Compost Mix Calculator* from C:N ratio. In this study, C:N ratio was calculated from the following equation:

$$C:N\,ratio = \left(\frac{SOC}{TN}\right) \tag{7.2}$$

where C is carbon, N is total nitrogen, SOC is soil organic carbon, and TN is total nitrogen.

7.2.4 STATISTICAL ANALYSIS

Statistical analysis was performed using SPSS software (version 11.0 for Windows). Soil properties were grouped and summarized according to the land uses and soil depths. Statistical differences were tested using two-way analysis of variance (ANOVA) following the general linear model (GLM) procedure within SPSS. Duncan's significance test was used for mean separation when the ANOVA showed a statistically significant difference ($p < 0.05$). Mean values found for all properties are shown in relevant tables. The data were analyzed for correlation using SPSS.

7.3 RESULTS

The GLM showed that land use and soil depth were significant factors in determining sand and clay content, PAW, D_b, D_p, S_t, K_{sat}, pH, SOM, SOC, and TN ratios ($p < 0.005$). The EC and C:N values differed according to the land-use type, whereas the silt, FC, PWP, and WSA values differed depending on the sampling depth. The combination of these two factors also had a significant interactive effect on clay content, S_t, K_{sat}, EC, SOM, SOC, and C:N ratios ($p < 0.05$) (Tables 7.1 and 7.2).

TABLE 7.1 Two-way ANOVA for Physical Properties of Soil for Two Land Uses and for Three Soil Depths.

	df	Sand	Clay	Silt	FC	PWP	PAW	D_p	D_b	S_t	WSA	K_{sat}
		Physical properties of the soil										
Land use (LU)	1	0.000	0.000	0.082	0.166	0.822	0.027	0.000	0.000	0.000	0.718	0.000
Depth (D)	2	0.004	0.020	0.013	0.000	0.000	0.000	0.000	0.000	0.000	0.000	0.000
LUxD	2	0.071	0.013	0.468	0.578	0.522	0.590	0.476	0.176	0.009	0.241	0.000

LU: land use, D: depth, FC: field capacity, PWP: permanent wilting point, PAW: plant available water, D_p: particle density, D_b: bulk density, S_t: total porosity, WSA: water-stable aggregate, and K_{sat}: saturated hydraulic conductivity.

TABLE 7.2 Two-way ANOVA for Chemical Properties of Soil for Two Land Uses and for Three Soil Depths.

	Df	pH	EC	Sal	SOM	SOC	TN	C:N ratio
		Chemical properties of the soil						
Land use (LU)	1	0.000	0.002	0.070	0.000	0.000	0.000	0.034
Depth (D)	2	0.011	0.187	0.038	0.000	0.000	0.000	0.321
LUxD	2	0.543	0.000	0.038	0.011	0.001	0.625	0.000

EC: electrical conductivity, Sal: salinity, SOM: soil organic matter, SOC: soil organic carbon, and TN: Total N.

7.3.1 SOIL TEXTURE AND WATER CHARACTERISTICS

The soils within the tea plantation site are of sand-clay loam (SCL) texture, and the alder coppice soils are of loamy sand (LS) texture. The highest sand content was present in the AC soil at a depth of 30–50 cm, and the lowest sand content was found in the TC soil at a depth of 0–10 cm. Average silt amount was the highest in the TC soil (in all three depth levels). The clay content in forest soil changed irregularly depending on the depth. The highest clay content was present in the second depth level and the lowest clay content was present in the first depth level. The change in the clay content between the first and the second levels were statistically significant (Table 7.3). The highest FC was seen in the top soil at the AC site, and the lowest FC and PAW content were observed at a depth of 30–50 cm in the TC soil (Table 7.3).

7.3.2 BULK DENSITY (D_B), PARTICLE DENSITY (D_P), SPR, TOTAL POROSITY (S_T), WSA, AND SATURATED HYDRAULIC CONDUCTIVITY (K_{SAT})

The D_b and D_p values in the AC and TC soils showed statistically significant increase with greater depth. The highest D_b (1.18 g cm^{-3}) was observed at the 30–50 cm soil depth for TC site, and the lowest D_b (0.84 g cm^{-3}) was observed at 0–10 cm soil depth for the forest top soil (Table 7.3).

TABLE 7.3 Land Use and Soil Depth Effects on Soil Physical Properties (Mean ± SE).

Soil property	Land uses	Depth (cm)			
		0–10	10–30	30–50	Overall
Sand (%)	Forest	70.35 ± (2.20)	67.19 ± (1.45)	71.03 ± (1.57)	69.52 ± (1.03)a
	Tea plantation	58.25 ± (1.34)	60.10 ± (1.08)	65.96 ± (1.40)	61.44 ± (0.84)b
	Overall	64.30 ± (1.61)a*	63.64 ± (1.06)a	68.49 ± (1.11)b	
Clay (%)	Forest	11.93 ± (1.30)	17.01 ± (1.15)	13.63 ± (1.07)	14.19 ± (0.72)b
	Tea plantation	22.17 ± (1.17)	21.26 ± (0.99)	18.56 ± (0.75)	20.66 ± (0.59)a
	Overall	17.05 ± (1.19)ab	19.13 ± (0.82)a	16.09 ± (0.75)b	
Silt (%)	Forest	17.70 ± (2.02)	15.89 ± (0.70)	15.33 ± (0.84)	16.31± (0.76)
	Tea plantation	19.57 ± (0.79)	18.62 ± (0.60)	15.42 ± (0.91)	17.87± (0.50)
	Overall	18.63 ± (1.08)a	17.26 ± (0.50)ab	15.38 ± (0.61)b	
FC (vol.%)	Forest	33.19 ± (0.75)	31.79 ± (0.75)	29.01 ± (0.65)	31.33 ± (0.46)
	Tea plantation	31.70 ± (0.82)	31.79 ± (0.75)	28.04 ± (0.56)	30.51 ± (0.46)
	Overall	32.45 ± (0.56)a	31.79 ± (0.52)a	28.52 ± (0.43)b	
PWP (vol.%)	Forest	16.40 ± (0.37)	17.54 ± (0.38)	16.36 ± (0.42)	16.76 ± (0.23)
	Tea plantation	16.21 ± (0.38)	18.14 ± (0.41)	16.17 ± (0.40)	16.84 ± (0.25)
	Overall	16.30 ± (0.26)b	17.84 ± (0.28)a	16.26 ± (0.29)b	
PAW (vol.%)	Forest	16.79 ± (0.49)	14.05 ± (0.40)	12.60 ± (0.37)	14.48 ± (0.33)
	Tea plantation	15.49 ± (0.65)	13.65 ± (0.37)	11.87 ± (0.25)	13.67 ± (0.32)
	Overall	16.14 ± (0.41)a	13.85 ± (0.27)b	12.23 ± (0.22)c	
D_p (g cm^{-3})	Forest	2.45 ± (0.02)	2.54 ± (0.03)	2.62 ± (0.02)	2.54 ± (0.01)b
	Tea plantation	2.57± (0.03)	2.60 ± (0.03)	2.75 ± (0.03)	2.64 ± (0.02)a
	Overall	2.51 ± (0.02)b	2.57 ± (0.02)b	2.68 ± (0.02)a	
D_b (g cm^{-3})	Forest	0.840 ± (0.03)	0.95 ± (0.03)	0.98 ± (0.02)	0.93 ± (0.01)b
	Tea plantation	1.02 ± (0.01)	1.06 ± (0.02)	1.18 ± (0.03)	1.09 ± (0.01)a
	Overall	0.93 ± (0.02)c0	1.01 ± (0.02)b	1.08 ± (0.02)a	

TABLE 7.3 *(Continued)*

Soil property	Land uses	Depth (cm)			
		0–10	10–30	30–50	Overall
S_t (%)	Forest	59.45 ± (1.18)	52.40 ± (0.75)	50.41± (0.91)	54.09 ± (0.74)a
	Tea plantation	51.91 ± (0.87)	44.78 ± (0.96)	37.69 ± (0.95)	44.79 ± (0.92)b
	Overall	55.68 ± (0.94)a	48.59 ± (0.85)b	44.05 ± (1.20)c	
WSA (%)	Forest	70.65 ± (1.27)	66.15 ± (0.79)	65.50 ± (0.99)	66.93 ± (0.69)
	Tea plantation	69.40 ± (1.32)	65.35 ± (0.79)	64.45 ± (0.71)	67.23 ± (0.59)
	Overall	70.02 ± (0.91)a	66.25 ± (0.55)b	64.97± (0.62)b	
K_{sat} (mm h^{-1})	Forest	40.64 ± (0.77)	22.53 ± (0.67)	10.92 ± (0.28)	24.70 ± (1.63)a
	Tea plantation	16.33 ± (0.44)	4.85 ± (0.28)	4.31 ± (0.19)	8.49 ± (0.74)b
	Overall	28.49 ± (1.99)a	13.69 ± (1.46)b	7.61 ± (0.55)c	

LU: land use, D: depth, FC: field capacity, PWP: permanent wilting point, PAW: plant available water, D_p: particle density, D_b: bulk density, S_t: total porosity, WSA: water-stable aggregate, K_{sat}: saturated hydraulic conductivity, and SE: standard error.

*Values followed by different letters implies that values are significant at $P = 0.05$.

The particle density in the tea soils showed a statistically significant variation based on depth. The average total porosity was 37.69–51.91% in the tea soils and 50.41–59.45% in the forest soils (Table 7.3). Both the AC and TC soils showed a statistically significant variation in soil pore volume, according to sampling depth ($p < 0.007$). The average WSA content ranged from 64.45 to 69.40% in the tea soils and from 65.50 to 70.65% in the forest soils. The WSA values of the forest and tea soils showed a statistically significant decrease with depth (Table 7.3). Figure 7.1 indicates that SPR levels in the tea and forest soils increased with depth. The relationships were nonlinear and are defined by following equations:

Tea cultivation:

$$SPR = 0.8166\,Loge(X) + 0.8081, R^2 = 0.985 \tag{7.3}$$

AC cultivation:

$$SPR = 0.0096(X^2) + 0.636(X) + 0.3085, R^2 = 0.0.831 \tag{7.4}$$

The SPR value for tea soils showed statistically significant increase with depth. The average K_{sat} values were 10.92–40.64 mm/h in the forest soils and 4.31–16.33 mm/h in the tea soils. If the depth level factor was omitted, theww difference between the K_{sat} values of the AC and TC soils were found to be statistically significant (Table 7.3).

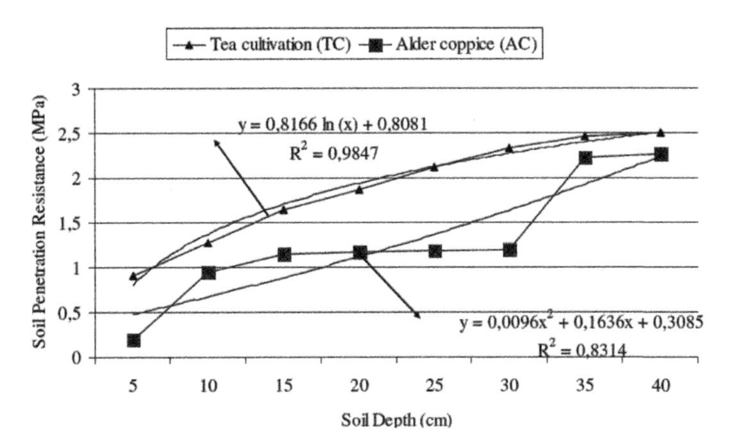

FIGURE 7.1 SPR (*Y*, MPa) versus soil depth (*X*, cm) for alder coppice forest and tea cultivation in the study area.

The mean SOC levels varied significantly with soil depth (Fig. 7.2). At a depth of 0–10 cm, the dominant source of carbon was humus, while at a depth of 30–50 cm, the dominant source of carbon is inorganic calcium and magnesium carbonates. For each incremental soil sampling depth, SOC levels for TP soils declined by 25–35%, compared with a decline of 32% in AC soils. These results are similar to those for TN.

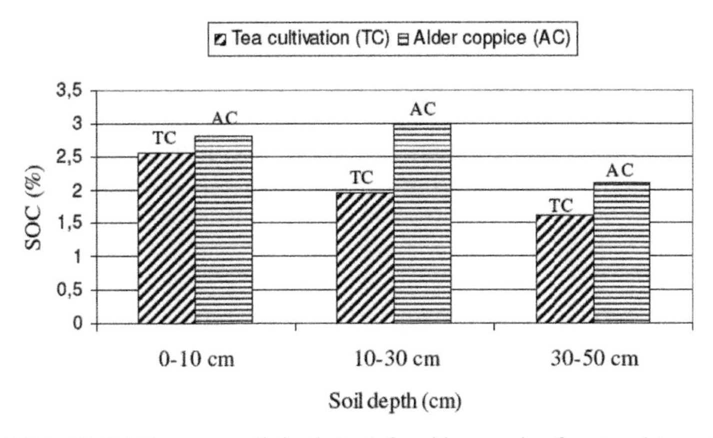

FIGURE 7.2 SOC (%) versus soil depth (cm) for alder coppice forest and tea cultivation in the study area.

7.3.3 pH AND EC

The soil pH in both the forest and tea soils was found to be higher soils at deeper depths. EC dropped with depth in the tea soils, whereas it was first increased and then decreased in the forest soils. The differences among pH and EC values of the forest and tea soils were statistically significant according to depth (Table 7.4).

TABLE 7.4 Effects of Land Use and Soil Depth on Soil Properties (Mean ± SE).

Soil property	Land uses	Soil depth (cm)			Overall (mean ± SE)
		0–10	10–30	30–50	
pH (1/2.5 H_2O)	Forest	4.35 ± (0.09)	4.41 ± (0.06)	4.57 ± (0.06)	4.44 ± (0.04)a
	Tea plantation	3.73 ± (0.03)	3.86 ± (0.06)	3.88 ± (0.06)	3.82 ± (0.03)b
	Overall (mean)	4.04 ± (0.06)b*	4.13 ± (0.06)ab	4.23 ± (0.06)a	
EC ($dS\ m^{-1}$)	Forest	0.26 ± (0.01)	0.38 ± (0.01)	0.34 ± (0.01)	0.33 ± (0.01)a
	Tea plantation	0.36 ± (0.01)	0.27 ± (0.00)	0.26 ± (0.02)	0.29 ± (0.01)b
	Overall (mean)	0.31 ± (0.06)	0.32 ± (0.01)	0.30 ± (0.01)	
Salinity (%)	Forest	0.010± (0.00)	0.010 ± (0.00)	0.010 ± (0.00)	0.010 ± (0.00)
	Tea plantation	0.012 ± (0.001)	0.010 ± (0.00)	0.010 ± (0.00)	0.011 ± (0.00)
	Overall (mean)	0.011 ± (0.000)	0.010 ± (0.00)	0.010 ± (0.00)	
SOM (%)	Forest	5.14 ± (0.22)	4.28 ± (0.18)	2.95 ± (0.14)	4.12 ± (0.15)a
	Tea plantation	4.06 ± (0.14)	2.71 ± (0.18)	2.39 ± (0.10)	3.05 ± (0.12)b
	Overall (mean)	4.60 ± (0.15)a	3.50 ± (0.17)b	2.67 ± (0.09)c	
SOC (%)	Forest	2.81 ± (0.13)	2.97 ± (0.09)	2.09 ± (0.09)	2.62 ± (0.08)a
	Tea plantation	2.57 ± (0.11)	1.94 ± (0.10)	1.62 ± (0.06)	2.04 ± (0.07)b
	Overall (mean)	2.69 ± (0.08)a	2.45 ± (0.10)b	1.86 ± (0.06)c	
Total N (TN) (%)	Forest	0.25 ± (0.01)	0.22 ± (0.01)	0.19 ± (0.01)	0.22 ± (0.01)a
	Tea plantation	0.21 ± (0.00)	0.20 ± (0.01)	0.15 ± (0.01)	0.19 ± (0.00)b
	Overall (mean)	0.23 ± (0.00)a	0.21 ± (0.01)b	0.17± (0.01)c	
C:N ratio	Forest	11.45 ± (0.31)	13.68 ± (0.58)	11.20 ± (0.35)	12.11± (0.28)a
	Tea plantation	12.42 ± (0.69)	9.91± (0.64)	11.11± (0.59)	11.14 ± (0.38)b
	Overall (mean)	11.93 ± (0.38)	11.80 ± (0.52)	11.15± (0.33)	

EC: electrical conductivity, SOM: soil organic matter, SOC: soil organic carbon, TN: total N, and SE: standard error.

*Values followed by different letters implies that values are significant at $P = 0.05$.

7.4 DISCUSSION

A significant change in the sand and clay ratios is not expected with the change in the land usage type under normal conditions. However, when the TC site was converted into cultivation, terraces were formed at an approximate width of 60–90 cm. Stones larger than 10 cm were removed from the terraces, and tea seedlings were set after discombobulating the soil. Beans and corps were seeded in the newly set tea terraces, and the soil was hoed. It can be concluded that these operations mixed the original topsoil within the soil profile, thereby altering texture and structure of the soil.

Many previous researchers have concluded that discombobulating the soil, particularly the top part of the soil profile, changes the texture of soil.[35] Brye et al.[11] reported that land leveling significantly altered soil particle-size fractions. In the tea plantations converted from forest, the soil protection ability against erosion is lower compared to alder plantation and as a result, some of the topsoil had been removed by surface flow. This is another factor which leads to changes in the soil texture. In the top soil of the TC site, the volume weights and the penetration values of the soil were significantly higher, while the total porosity and the net structure of the pores were destroyed, due to heavy field traffic and hoeing. As a result, water penetration into the soil was slower, and the FC and PAW values of the soil were decreased. The FC and PAW in the tea and forest soils were decreased with increased soil depth, partly due to the decrease in SOM at greater depths. As the SOM decreases, the water retention and available water capacities of the soil decrease.[25] The construction of terracing using human power and the application of high levels of organic fertilizers in the initial years of the tea plantation help to increase the PAW capacity. The use of organic amendments and the application of additional organic matter in the terraces increase the water-resistant aggregate stability of the soil. In these terraces, the volume weight of the soil decreases compared to terraces in which no organic amendments is used; and the water-holding capacity, PAW, and hydraulic conductivity increase.[35]

The herbaceous plants growing beneath the alders are also harvested and used as animal food or animal bedding. These practices reduce the organic inputs and cause a loss of soil nutrients within the forest ecosystem and have negative effects on the soil properties. Le Bissonnais[30] mentions that the aggregate stability of soil increases in line with the SOM. Yüksek[42] mentioned that the water capacity of the soils decreases and the water flow increases as a result of using the organic wastes collected from the forest for different purposes.

In the upper levels of the tea soils, dense trampling and soil processing decrease the total pore volume especially in the top soils and cause the volume weight to increase. Under normal conditions, as a result of heavy field traffic and cultivation operations, the porosity in the tea top soils is expected to be low and the volume weight is expected to be high. The organic fertilizers applied to the tea terraces improve the structure of top soil to some extent. This avoids reduction in the pore volume and causes the volume weight to increase further. Most of the previous researchers have concluded that compost and animal fertilizers applied to the top soil decrease the volume weight[1] and increase the WSA and total porosity.[5]

Organic amendments decrease soil bulk density due to the "*dilution effect*" of the added organic matter with the denser mineral fraction, and by influencing soil aggregation, which can lead to greater porosity.[19] The application of organic fertilizers to the tea cultivation site is a factor in the increase of WSA. Campbell et al.[12] mentioned that the WSA content of the agricultural soils may increase as a result of adding organic fertilizer addition. Another possible reason is that the relatively high clay content and low organic matter content of the tea plantations may have increased the wettability of soil aggregates causing the aggregates to suffer more slaking on sudden wetting. It is well known that there is a positive correlation between SOM and aggregate stability.[14]

Soil hydraulic properties, soil hydraulic conductivity function, and water retention characteristics are affected by soil texture, bulk density, soil structure, and organic carbon content. Many of these factors are strongly influenced by land use and management even though the soil classification may be the same. As a result of the soil processes applied at the TC site, with the deformation of the soil texture and structure, the macropores and the pore network were destroyed.

Jamming and the increase of D_b caused the K_{sat} values to decrease. The root structure and the low root density compared with the forest soils also contributed to the low K_{sat} values. The temporal change in land use and management, or natural disturbances and cycles such as diurnal and seasonal changes can affect soil hydraulic properties. Soil compaction caused by human trampling, wheel traffic, or animal grazing can destroy large pores and therefore reduce saturated or near-saturated hydraulic conductivity.[16] The depth-dependent C:N values in the tea soils were first decreased and then increased, whereas in the forest soils they first increased and then decreased. The depth-dependent clay contents in the forest soils first decreased and then increased, whereas in the tea soils they decreased linearly. The change in land-use type and the cultivation practices applied in the tea terraces reduced the organic material content, especially in the tea top soils. However, the regular application of organic fertilizer can maintain the SOM content of the tea top soils. This prevents the decrease in organic material content in the tea top soils to some extent.

Land-use change[15] and long-term cultivation may lead to changes in SOM quantity and quality.[9] For each incremental soil sampling depth, SOC levels for TP soils declined by 25–35%, compared with a decline of 32% in AC soils. These results are similar to those for TN. This decline was due to declining humus levels with increasing soil depth. Soil carbon loss first occurs predominantly by mineralization after conversion of virgin land to cultivation, followed in subsequent years by soil erosion as the dominate soil carbon loss process.[20] Since the organic wastes collected from the forest and stable fertilizers are applied to the tea terraces, the TN content in the tea top soils was high. Also, with the help of the organic wastes applied to the tea terraces, the removal of N and P with surface flow was partially prevented. This was effective in maintaining a relatively high TN value. Indeed, by applying mulch to the soil, the amount of N carried with surface flow is reduced.[31]

Navarrete and Tsutsuki[33] reported that land conversion decreased the soil carbon and nitrogen content. It may be concluded that the humidity of the research region (annual average rainfall approximately 2000 mm) and the soil management process applied to the terraces caused some leaching of N. Different types of fertilizers were applied to the tea cultivation soils. In the first 15 years after conversion from alder forest, compost + stable fertilizer were applied; in the next 10 years, stable fertilizer was applied; in the final 30 years, chemical fertilizers (ammonium, sulfate, ammonium nitrate, and NPK) were applied. The initial application of compost during the first years prevented acidification of the tea soils. However, the large amounts of chemical fertilizers applied in subsequent years and the effects of leaching increased the acidity level of the tea soils.

In general, a strong positive relationship between the clay content of the soils and their water-holding capacity has been reported.[37] In this study, the clay content in the forest soils first increased with depth and then decreased. In other words, there was a positive relationship between the clay content and EC. In the tea soils, the clay content and the FC decreased with depth. It is likely that this decrease caused a drop in the EC values. Also, it may be noted that the application of compost to the tea soils increased EC. The movement of electrons through bulk soil is complex. Electrons may travel through soil water in macropores, along the surfaces of soil minerals (i.e., exchangeable ions), and through alternating layers of particles and solution.[37]

Vegetation type and quality has a large influence on the variations of soil hydraulic property. Although the alder coppice waterside thicklets cannot protect the soil as well as the natural old-growth forest cover, they perform better than the tea cultivations.[43] Also, their root system (thin and thick roots), their greater underground biomass, and their ability to penetrate deeper into the soil than the tea roots could increase the macroporosity of the soil and facilitated improved pore meshes (pore network). This might cause the alder soils to have better infiltration, saturated hydraulic conductivity, and water-holding capacity compared to the tea soils.

7.5 SUMMARY

In the last century, the conversion of natural ecosystems into agricultural production is one of the primary factors in environmental degradation. As in most parts of the world, forest soils in the northeast of Turkey are being seriously degraded and destroyed due to extensive agricultural activities. This study investigated the effects of changes in land-use on some soil properties in Rize, Turkey. Two adjacent sites were studied: One had been converted 60 years previously from alder coppice to tea cultivation (TC); and the other remained as alder coppice (AC). The experimental design at each site was a randomized complete block with four replications in the study area. Four disturbed and four undisturbed soil samples were taken randomly at soil depths of 0–10, 10–30, and 30–50 cm in each plot. When the alder coppice was converted into tea cultivation, the bulk density (D_b) was increased from 0.84 to 1.02 g/cm^3, SPR increased from 0.94 to 1.27 MPa, SOM decreased from 5.14 to 4.06%, and saturated hydraulic conductivity (K_{sat}) decreased from 40.64 to 16.33 mm/h at 0–10 cm depth of soil. According to soil depth steps, the mean PAW, S_t, K_{sat}, SOM, and TN content were decreased linearly in alder coppice (AC) and tea cultivation (TC). The results indicated that the change in land use and introduction of cultivation had a significant effect on soil properties.

KEYWORDS

- **alder coppice**
- **penetration resistance**
- **permanent wilting point**

- **saturated hydraulic conductivity**
- **stable aggregates**
- **tea cultivation**
- **water-stable aggregates**

REFERENCES

1. Aggelides, S. M.; Londra, P. A. Effects of Compost Produced From Town Wastes and Sewage Sludge on the Physical Properties of a Loamy and Clay Soil. *Bioresour. Technol.* **2000**, *71*, 253–259.
2. Anonymous. *Available Land Properties of City of Rize* (in Turkish); Publications of General Offices of Village's Foundation, City Report No. 53; Ankara, Turkey, 1993, pp 1–108.
3. Anonymous. *The Past and Present Day of Tea* (in Turkish). http://www.biriz.biz/cay/turkcay.htm#icf (accessed Jan 16, 2006).
4. Anonymous. *Some Climatic Data of Pazar in the Year between 1970–2007* (in Turkish); Turkish State Meteorological Service, Ankara, Turkey, 2007, pp 1–11.
5. Aoyama, M.; Angers, D. A.; N'Dayegamiye, A. Particulate and Mineral-Associated Organic Matter in Water-Stable Aggregates as Affected by Mineral Fertilizer and Manure Applications. *Can. Soc. Soil Sci.* **1999**, *79*, 295–302.
6. Aydınalp, C.; Fitz Patrick, E. A. Classification of Great Soil Groups in the East Black Sea Basin According to International Soil Classification Systems. *J. Cent. Eur. Agric.* **2004**, *5*(2), 119–126.
7. Bouyoucos, G. Hidrometer Method Improved for Making Particle Size Analysis of Soils. *Agro. J.* **1962**, *54*, 464–465.
8. Bradford, J. M. Penetrability. In *Methods of How Farmers Assess Soil Health and Quality*; Klute, A., Ed.; Soil Science Society of America: Madison, WI, 1986; Vol. 50, pp 229–236 (*J. Soil Water Cons. Anal., Part I).*
9. Brady, N. C.; Weil, R. R. *The Nature and Properties of Soils*, 13th edition; Prentice Hall: Upper Saddle River, NJ 07458, 2002; p 960.
10. Bremner, J. M. Total Nitrogen. In *Methods of Soil Analysis, Part 2: Chemical Microbiology Properties;* Black, C. A., Evans, D. D., White, J. L., Ensminger, L. E., Clark, P. E., Eds.; Soil Science Society of America: Madison, WI, 1965; pp 1149–1178.
11. Brye, K. R.; Slaton, N. A.; Savin, M. C.; Norman, R. J.; Miller, D. M. Short-term Effects of Land Leveling on Soil Physical Properties and Microbial Biomass. *Soil Sci. Soc. Am. J.* **2003**, *67*, 1405–1417.
12. Campbell, C. A.; Selles, F.; Lafond, G. P.; Biederbeck, V. O.; Zenter, R. P. Tillage-Fertilizer Changes: Effect on Some Soil Quality Attributes Under Long-Term Crop Rotations in A Thin Black Chernozem. *Can. J. Soil Sci.* **2001**, *81*, 157–165.
13. Chazan, D.; Cotter, A. A. *Evaluating the Impacts of Proposed Land Conversion: A Tool for Local Decision-Making, Center for Sustainable Systems;* Report No. CSS01-02. University of Michigan, 2001.

14. Chenu, C.; Le Bissonnais, Y.; Arrouays, D. Organic Matter Influence on Clay Wettability and Soil Aggregate Stability. *Soil Sci. Soc. Am. J.* **2000**, *64*, 1479–1486.
15. Desjardins, T.; Barros, E.; Sarrazin, M.; Girardin, C.; Mariotti, A. Effects of Forest Conversion to Pasture on Soil Carbon Content And Dynamics in Brazilian Amazonia. *Agric. Ecosyst. Environ.* **2004**, *103*, 365–373.
16. Drewry, J. J.; Paton, R. J. Soil Physical Quality Under Cattle Grazing of A Winter-Fed Brassica Crop. *Aust. J. Soil Res.* **2005**, *43*(4), 525–531.
17. FAO (United Nations Food and Agriculture Organization). *Global Forest Resources Assessment 2000*, FAO Forestry Paper 140, FAO, Rome, Italy, 2001, p 482.
18. Flint, A.; Flint, L. E. Particle Density. In *Laboratory Methods, Methods of Soil Analysis, Part 4: Physical Methods*, Dick, W. A., Ed.; Soil Science Society of America: Madison, WI, 2002; pp 229–240.
19. Garnier, P.; Ezine, N.; De Gryze, S.; Richard, G. Hydraulic Properties of Soil–Straw Mixtures. *Vadose Zone J.* **2004**, *3*, 714–721.
20. Gregorich, E. G.; Carter, M. R.; Anger, D. A.; Monreal, C. M.; Ellert, B. H. Towards A Minimum Data Set to Assess Soil Organic Matter Quality in Agricultural Soils. *Can. J. Soil Sci.* **1994**, *74*, 367–385.
21. Grossman, R. B.; Reinsch, T. G. The Solid Phase, Bulk Density And Linear Extensibility. In *Laboratory Methods, Methods of Soil Analysis, Part 4: Physical Methods;* Dick, W. A., Ed.; Soil Science Society of America: Madison, WI, 2002; pp 201–228.
22. Gutierrez, J.; Sosebee, R. E.; Spaeth, K. E. Spatial Variation of Runoff and Erosion Under Grass and Shrub Cover on A Semiarid Rangeland. In *Watershed Management-Planning for the 21st Century*, Ward, T. J., Ed.; American Society of Civil Engineers: San Antonio, TX, 1995; pp 11–20.
23. Han, W.; Kemmitt, S. J.; Brookes, P. C. Soil Microbial Biomass and Activity in Chinese Tea Gardens of Varying Stand Age and Productivity. *Soil Biol. Biochem.* **2007**, *39*, 1468–1478.
24. Horton, R. E. Drainage-Basin Characteristics. *Eos Trans. Am. Geophys. Union* **1932**, *13*, 350–371.
25. Hudson, B. D. Soil Organic Matter and Available Water Capacity. *J. Soil Water Conserv.* **1994**, *49*(2), 189–193.
26. Karaöz, O. Analyze Methods of Some Chemical Soil Properties (pH, Carbonates, Salinity, Organic Matter, Total Nitrogen, Available Phosphorus). *Rev Faculty For. Univ. Istanbul* **1989**, *39*(B3), 64–82 [in Turkish].
27. Kemper, W. D.; Rosenau, R. C. Aggregate Stability and Size Distribution. In *Methods of Soil Analysis, Part I: Physical and Mineralogical Methods*. Klute, A., Ed.; Soil Science Society of America: Madison, WI, 1996; pp 425–442.
28. Klute, A. Water Retention: Laboratory Methods. In *Methods of Soil Anal. Part 1: Physical and Mineralogical Methods*; Klute, A., Ed.; Soil Science Society of America: Madison, WI, 1986; pp 635–662.
29. Klute, A.; Dirksen, C. Hydraulic Conductivity and Diffusivity: Laboratory Methods. In *Methods of Soil Analysis, Part I: Physical and Mineralogical Methods*; Klute, A., Ed.; Soil Science Society of America: Madison, WI, 1986; pp 687–734.
30. Le Bissonnais, Y. Aggregate Stability and Measurement of Soil Crustability And Erodibility: I. Theory and Methodology. *Eur. J. Soil Sci.* **1996**, *47*, 425–437.
31. Linde, D. T.; Watschke, T. L.; Jarrett, A. R. Surface Runoff Comparison Between Creeping Bentgrass and Perennial Ryegrass Turf. *J. Turfgrass Manag.* **1998**, *2*, 11–33.

32. Martinez-Meza, E.; Whitford, W. G. Stemflow, Throughfall and Channelization of Stemflow by Roots in Three Chihuahuan Desert Shrubs. *J. Arid Environ.* **1996**, *32*, 271–287.
33. Navarrete, I. A.; Tsutsuki, K. Land-use Impact on Soil Carbon, Nitrogen, Neutral Sugar Composition and Related Chemical Properties in A Degraded Ultisol in Leyte, Philippines. *Soil Sci. Plant Nutr.* **2008**, *54*(3), 321–331.
34. Nelson, D. W.; Sommers, L. E. Total Carbon, Organic Carbon, and Organic Matter. In *Methods of Soil Analysis. Part 2,* Page, A. L., et al., Eds.; Soil Science Society of America: Madison, WI, 1996; pp 961–1010.
35. Querejeta, J. I.; Roldán, A.; Albaladejo, J.; Castillo, V. Soil Physical Properties and Moisture Content Affected by Site Preparation in the Afforestation of A Semiarid Rangeland. *Soil Sci. Soc. Am. J.* **2000**, *64*, 2087–2096.
36. Rhoades, J. D. Salinity: Electrical Conductivity and Total Dissolved Solids. In *Methods of Soil Analysis, Part 3: Chemical Methods*, Page, A. L., et al., Eds.; Soil Science Society of America: Madison, WI, 1996; pp 417–435.
37. Rhoades, J. D.; Manteghi, N. A.; Shouse, P. J.; Alves, W. J. Estimating Soil Salinity from Saturated Soil-Paste Electrical Conductivity. *Soil Sci. Soc. Am. J.* **1989**, *53*, 428–433.
38. Slater, B. K.; McSweeney, K.; Ventura, S. J.; Irvin, B. J.; McBratney, A. B. A Spatial Framework for Integrating Soil-landscape and Pedogenic Models. In *Quantitative Modeling of Soil Forming Processes*; Bryant, R. B., Arnold, R. W., Eds.; Soil Science Society of America: Madison, WI, 1994; pp 69–185.
39. Tchienkoua, M. W.; Zech, W. Organic Carbon and Plant Nutrient Dynamics Under Three Land Uses in the Highlands of West Cameroon. *Agric. Ecosyst. Environ.* **2004**, *104*, 673–679.
40. Wagenet, R. J.; Hutson, J. L.; Bouma, J. Modeling Water and Chemical Fluxes as Driving Forces of Pedogenesis. In *Quantitative Modeling of Soil Forming Processes;* Bryant, R. B.; Arnold, R. W., Eds.; Soil Science Society of America: Madison, WI, 1994; pp 17–35.
41. Yüksek, T. Investigations on Soil Erodibility and Some Properties of the Soils Under Different Land Use Types in Pazar Creek Watershed, Near Rize-Turkey. Ph.D. Thesis, Karadeniz Technical University, Sci. Inst. Trabzon, Turkey, 2001, p 204 (in Turkish).
42. Yüksek, T.; Kalay, H. Z. In *Comparison of Some Soil Properties between Tea Cultivation and Adjacent Alder (Alnus glutinosa L. Gaertner subsp. barbata (C.A. Meyer)) Stands in Kesikköprü Village,* Pazar, Turkey. Final Proc., 2nd National Blacksea Forestry Congress, Artvin, Turkey, 2002, pp 780–789 (in Turkish).
43. Yüksek, T.; Kalay, H. Z.; Yüksek, F. Land Use Problems in Pazar Creek Watershed. *J. Sci. Instrum. Süleyman Demirel Univ.* **2004,** *8*(3), 121–127 (in Turkish).

PART III

Technological Interventions for Soil and Water Conservation

CHAPTER 8

YIELD AND WATER PRODUCTIVITY ANALYSIS OF TOMATO CROP UNDER WATER STRESS CONDITIONS IN ETHIOPIA

MICHAEL ESHETU BISA[1,*], FITSUME YEMENU DESTA[2], and ELENA BRESCI[3]

[1]*Department of Water Resource and Irrigation Engineering, Arba Minch University, POB 21, Arba Minch, Ethiopia*

[2]*School of Environmental and Rural Sciences, University of New England, POB 2350, 137 Mann St. Armidale, NSW, Australia*

[3]*Department of Agricultural, Food and Forestry Systems, University of Florence, Via San Bonaventura 13, 50145 Firenze FI, Italy*

[*]*Corresponding author. E-mail: michaeshe@gmail.com*

CONTENTS

ABSTRACT

By irrigating only the first two growth stages of tomato crop, the best water productivity was found without significant yield reduction for a specific growing period (from January 23 to May 22, 2013) under similar local climatic conditions. With this irrigation application, irrigated lands can be increased to 111.4% with the same amount of water. In this study, the rainfall contribution was 41.1% of CWR at mid growing stage and 26% in the later stage. Under such rainfall conditions, irrigating in the mid and/or late growing stages did not show any significant differences in yield.

8.1 INTRODUCTION

Ethiopia's food production is highly dominated by small-scale farmers, who largely depend on rain-fed and traditional agricultural practices.[3,4,5,14] According to Ndaruzaniye,[14] climate variability renders Ethiopian farmers highly vulnerable as was observed during past persistent periods of drought during 1972–1973, 1984, and 2002–2003. Vegetable production by smallholder's farmers and floriculture farm by private sectors expanded rapidly around the central parts of Ethiopia, particularly in research area Ada'a district. Tomato is one of the most adopted cash crops by local farmers to generate income. Although irrigation has long been practiced at different farm levels of management practice, yet there are no efficient or well-managed irrigation water practices. There is very little or no information regarding appropriate management of irrigation water and crop management practices for the rapidly expanding small-scale irrigation farms in the area.

Deficit (or regulated deficit) irrigation is a method of maximizing water-use efficiency (WUE) for higher yield per unit of irrigation water application. In this method, the crop is exposed to a certain level of water stress either during a particular period or throughout the whole growing season. It is expected that crop-yield reduction will be insignificant compared with the benefits gained through diverting the saved water to irrigate other crops.[11] Deficit irrigation can also be defined as the application of less water than is required for potential ET and maximum yield, resulting in the conservation of limited irrigation water.[13]

Under conditions of scarce water supply and drought, deficit irrigation can lead to greater economic gains by maximizing yield per unit of water. Deficit irrigation[6] has the fundamental goal to increase WUE. Fereres and Soriano[9] recently reviewed deficit irrigation research and concluded that the level of irrigation supply should in most cases be 60–100% of full evapotranspiration (ET) requirements to improve water productivity. Studies have shown that deficit irrigation significantly increased grain yield, ET, and WUE as compared with rain-fed winter wheat.[15] However, this approach requires precise knowledge of crop response to water as drought tolerance varies considerably by growth stage, species, cultivars, and also farmers competence.

According to Eugenio et al.,[7] deficit irrigation in tomato crop showed stomata conductance values lower than full irrigation and saved a substantial amount of water maintaining reasonable marketable yield. During canopy development and much of the flowering period, irrigation needs to be sufficient to ensure fast canopy growth and yet not so much as to cause excessive leaf growth and the associated dropping of flowers and young fruits.

This research was conducted to identify tomato growth stages that were sensitive to water stress, to determine the critical time for irrigation application in the case of limited water resources and to determine the productivity of water for tomato crop under water stress conditions.

8.2 METHODOLOGY

The experiment was conducted in the Debrezeit Agriculture Research Center (DZARC), Ethiopian Institute of Agriculture Research (EIRA). DZARC is located at 45 km far from Addis Ababa, Ethiopia. The geographical location of the site is 08°45¢49.5²N latitude, 038°53¢51.6²E longitude and at an elevation of 1900 m asl in central Ethiopia. Andosol and vertisol (black cotton) groups of soil type dominate the area, with silty loam to heavy clay soil texture. The experiment was accomplished in vertisol (black cotton) with high clay content, swelling when it gets moisture and cracks at drying. The soil has average bulk density 1.12 g/cm^3 and field capacity and permanent welting point in volume basis 43.8 and 30%, respectively. The experiment site is located immediate to the climatic station.

8.2.1 EXPERIMENT LAYOUT

Field experiment was conducted to identify the stages that are most sensitive to soil moisture stress. The tomato (*Solanum lycopersicum* var. Chali) has adopted very well in medium-altitude area. The test crop was planted on January 25, 2013. The treatments were:

- Check: Irrigated at the optimal irrigation schedule (irrigating when the total available moisture was 40% of field capacity based on FAOWAT 8)[2].
- Depriving irrigation water application during any of one growth stage.
- Depriving irrigation water application during any of the two growth stages.
- Depriving irrigation water application during any of the three growth stages.
- Depriving irrigation water application during all growth stages.

The entire experiment was carried out on 48 plots of 2.4 × 3 m each. Recommended plant and row (0.4 × 1 m) spacing has been used for growing the crop. Each experimental treatment had a fertilized plot with recommended fertilizer application (200 kg days after planting (DAP) and 200 kg urea fertilizer). The experiment was then laid out in a randomized complete block design (RCBD) with three replications (Table 8.1). All cultural practices were conducted in all treatments according to the recommendations for the area.

TABLE 8.1 Treatment Combinations.

Treatments	Growth stages			
	Initial	Development	Midseason	Maturity
T1	1	1	1	1
T2	0	1	1	1
T3	1	0	1	1
T4	1	1	0	1
T5	1	1	1	0
T6	0	0	1	1
T7	0	1	0	1
T8	0	1	1	0
T9	1	0	0	1
T10	1	0	1	0
T11	1	1	0	0
T12	0	0	0	1
T13	0	0	1	0
T14	0	1	0	0
T15	1	0	0	0
T16	0	0	0	0

Note: 1 means irrigated and 0 means not irrigated during the crop growth stages.

Irrigation water was applied per treatment to refill the crop root zone depth close to field capacity. The amount of irrigation water in each irrigation application was measured using 3-in. Parshall flume manufactured by Melkasa agricultural research center (Fig. 8.1). Soil moisture content before and after irrigation was monitored gravimetrically at 15-cm depth interval up to the maximum root depth.

FIGURE 8.1 Research study area with irrigation water measuring parshall flume.

8.2.2 IRRIGATION SCHEDULING

To organize the irrigation scheduling in the research, the potential ET was estimated in the area. There are a number of empirical and semiempirical formulas to calculate potential ET from meteorological data. FAO Penman–Monteith method is now the standard method for computation of reference evapotranspiration (ETo[2]) by Allen et al.[1] By using the CROPWAT 8 software developed by FAO in 2006, the monthly ETo of the area was computed from minimum and maximum temperature, relative humidity, wind speed, and sunshine hour data.

The monthly long-term ETo data obtained from the CROPWAT program was then fitted to different standard frequency distribution models using a computer-based routine (VTFIT) package developed by Richard Cooke Virginia Polytechnic Institute in 1993. The distribution that best fitted the data was used to determine ETo occurrence at 80% probability level.

The crop ET[1] was then calculated by multiplying ETo by K_c (a coefficient expressing the difference in ET between the cropped and reference grass surface). K_c is the crop coefficient under standard conditions (well wetted free from disease crop land). The coefficient integrates differences in the soil evaporation and crop transpiration rate between the crop and the grass reference surface.[1] The K_c value of every crop largely depends on the crop growth stage and regional climate. The K_c values of tomato crop at different stages were obtained from FAO CROPWAT software.

The irrigation scheduling in each treatment plot was carried out on the basis of crop water requirement (CWR) of tomato at the different stages (initial, development, mid, and late stages) of the growing period and the soil water-holding capacity. The irrigation criteria were to irrigate when the soil moisture was depleted to critical level (40% of total available water, TAW, for tomato) which is recommended in FAO CROPWAT 8. The irrigation was applied to replenish crop water use. The net irrigation requirement was 60% of TAW multiplied by the root depth at specific growth stage. The gross irrigation was calculated by dividing net irrigation by application efficiency.

8.2.3 *WATER PRODUCTIVITY*

Water productivity measures how a system converts water into goods and services.[12] It was calculated by dividing the unit product (in this chapter, the yield of tomato) by the amount water used to produce a unit product.

8.2.4 *YIELD RESPONSE FACTOR*

FAO addressed the relationship between crop yield and water use in the late 1970s proposing a simple equation,[16–18] where relative yield reduction is related to the corresponding relative reduction in ET according to Eugenio et al.[7] Specifically, the yield response as a function of ET is expressed as follows:

$$\left(1-\frac{Y_a}{Y_x}\right)=K_y\left(1-\frac{ET_a}{ET_x}\right), \tag{8.1}$$

where Y_x is the maximum yield, Y_a the actual yield, ET_x the maximum evapotranspiration, ET_a the actual evapotranspiration, and K_y the yield response factor representing the effect of a reduction in ET on yield loss.

8.3 RESULTS AND DISCUSSION

Throughout the growing period, expected rainfall from long-term meteorological data was 208.5 mm. Actual rainfall was less than the expected rainfall (Table 8.2). The rainfall was concentrated in the mid and later part of the season.

TABLE 8.2 The Actual and Expected Rainfall During Growing Period of Tomato.

Month	Expected RF* (mm/month)	Actual RF* (mm/month)
January	7.9	0
February	32.2	0
March	47.7	42
April	58.2	78.3
May	62.5	43.64
Total	208.5	163.94

*Rainfall depth (mm).

8.3.1 YIELD RESPONSE FOR WATER STRESS AT DIFFERENT GROWING STAGES

Treatment 1 (T1) included full irrigation in all stages, based on the crop water demand to relate with the stress at different growth stages. There were no significant differences in yield among irrigation during all growth stages but only the first two stages because of considerable rainfall contribution at mid and later stages. By irrigating only the first two stages, the total harvested yield was 423.82 kut/ha (1.00 kut = 100 kg) higher than 90% of full irrigation by using 42% of the total irrigation water, which could save 3342.8 m^3/ha. With these rainfall conditions, either irrigation or no irrigation in late and mid stages did not show any significant differences in yield. Table 8.3 shows the yield response for a combination of different stages under water stress conditions. This helps farmers to choose the best combination based on the water availability, the irrigated area, and capitals.

8.3.2 YIELD RESPONSE FACTOR

In T1 with irrigation at all stages, water was applied at the maximum evapotranspiration ET_x to give the maximum yield Y_x, so that the growth water used and total yield harvested from this treatment were taken as reference (ET_x and Y_x, respectively) values. The rainfall evidence in mid and late growing seasons specially benefited the three-fourths of treatments. The sum of irrigation application and rainfall during the growing stage give the

TABLE 8.3 Tomato Yield Response to Water Stress at Different Growing Stages.

TRT	Growth stages				Gross irrigation (m^3/ha)	Mean yield (kut/ha)	Rank	
	Initial	Devel.	Mid.	Maturity				
T1	1	1	1	1	6342.9	469.07	A	1
T2	0	1	1	1	5585.7	346.46	ABCDE	6
T3	1	0	1	1	4485.7	265.35	DEFGH	10
T4	1	1	0	1	4114.3	396.95	ABC	3
T5	1	1	1	0	4114.3	392.57	ABCD	4
T6	0	0	1	1	3914.3	193.96	FGH	13
T7	0	1	0	1	4471.4	367.78	ABCDE	5
T8	0	1	1	0	4471.4	321.6	BCDEF	7
T9	1	0	0	1	2257.1	289.44	CDEFG	8
T10	1	0	1	0	3371.4	258.96	EFGH	11
T11	1	1	0	0	3000.0	423.82	AB	2
T12	0	0	0	1	1685.7	151.46	H	16
T13	0	0	1	0	2800.0	201.11	FGH	12
T14	0	1	0	0	3357.1	264.57	CDEFGH	9
T15	1	0	0	0	1142.9	188.13	GH	14
T16	0	0	0	0	571.4	175.97	GH	15

Note: The value with the same letter indicates there is no significant difference between treatments. Kut refers to 100 of kg; and TRT refers treatment.

actual water use in each treatment (ET_a). The rainfall started after the quarter of mid stage that means every treatment gets 103.2 mm in the mid stage and 60.7 mm in the later stage. Based on this, irrigating the first two stages gave highest proportion of water saving compared to the proportion to yield reduction. Generally from Figure 8.2 stressing both or one of the first two growing stages, one can observe the maximum yield reduction compared to the proportion of water saved due to water deficit.

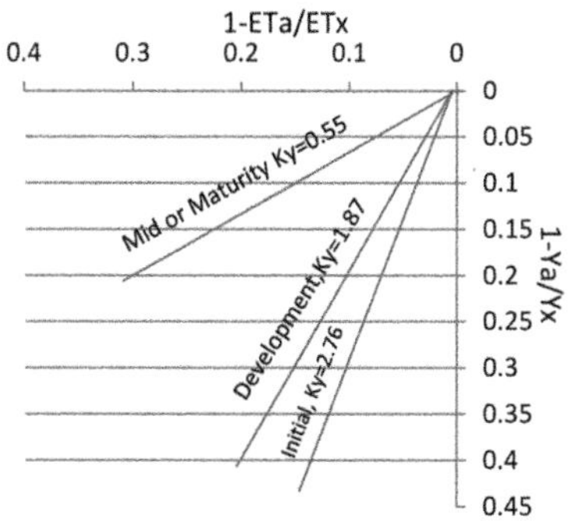

a. Depriving irrigation water application during any of one growth stage

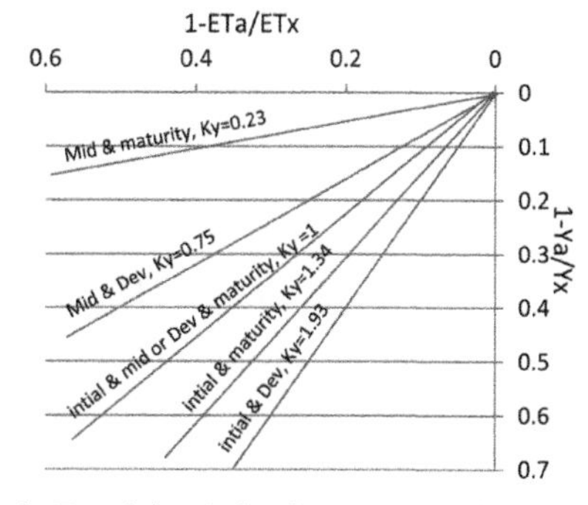

b. Depriving irrigation water application during any of the two growth stages

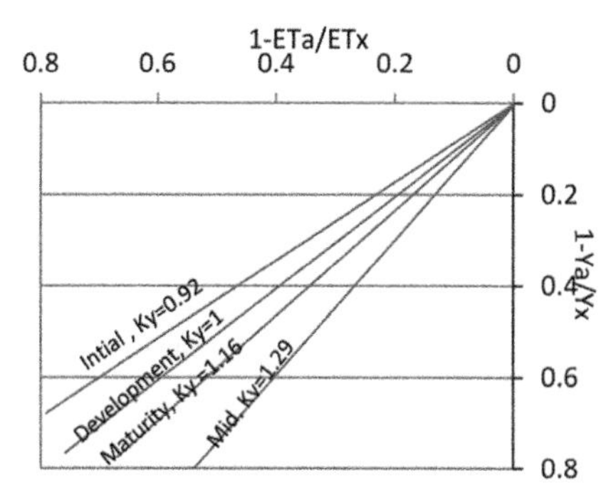

c. Irrigating water application only in one of the plan growth stage

FIGURE 8.2 The yield response factors of different stage soil moisture stress conditions.

In Figure 8.2(a), the effects of soil moisture stress on one of the growing stages are compared. Water stress at initial growth stage had highest yield reduction compared with stressing at other stages. The corresponding yield response factor value of 2.87 indicates loss of 2.87 unit of yield to safe 1 unit of water. Therefore, it is not economical to cause stress at initial or development stage. If we compare water stress at any of the two growth stages in Figure 8.2(b), one can observe maximum yield reduction against the consequent water saving, while stressing at initial and development growth stage. Water stressing at mid and maturity growth stages is best water-saving tool with minimum yield reduction compared to depriving at any of the two growth stages. When the available water resource is too low or irrigation design to irrigate only one of the plant growth stages (Fig. 8.2(c)), water applying for only any of one growth stages had increased the yield response in the following sequence: initial, development, maturity, and mid growth stages. Thus, irrigating only during initial growth stage is most effective to save water for tomato production compared with irrigating in any of one of the growth stages.

8.3.3 WATER PRODUCTIVITY

The treatment T16 (irrigating only at transplanting time for crop establishment with rain water contribution) gave the highest water productivity (Table 8.4). However, it shows only the yield from crop water use of 1 m^3. To decide the benefits of one treatment, one must consider profitability of the final produce in the treatment. Irrigating the only first two stages provides 14.13 kg yield per 1 m cube water. This is approximately two times greater than irrigating at all stages (T1) without significant yield reduction.

TABLE 8.4 Water Productivity of Tomato Crop at Different Soil Moisture Conditions.

TRT I.D.	TRT	Average yield (kut/ha)	Gross irrigation (m^3/ha)	Water productivity (kg yield/m^3 water)
T1	1111	469.07	6342.9	7.40
T2	0111	346.46	5585.7	6.20
T3	1011	265.35	4485.7	5.92
T4	1101	396.95	4114.3	9.65
T5	1110	392.57	4114.3	9.54
T6	0011	193.96	3914.3	4.96
T7	0101	367.78	4471.4	8.23
T8	0110	321.6	4471.4	7.19
T9	1001	289.44	2257.1	12.82

TABLE 8.4 *(Continued)*

TRT I.D.	TRT	Average yield (kut/ha)	Gross irrigation (m^3/ha)	Water productivity (kg yield/m^3 water)
T10	1010	258.96	3371.4	7.68
T11	1100	423.82	3000.0	14.13
T12	0001	151.46	1685.7	8.98
T13	0010	201.11	2800.0	7.18
T14	0100	264.57	3357.1	7.88
T15	1000	188.13	1142.9	16.46
T16	0000	175.97	571.4	30.79

Note: Gross irrigation amount is the total amount of water applied to the field including common irrigation during transplanting.

8.4 SUMMARY

The experiment discusses effects of water stress on tomato crop at different growth stages to identify its sensitivity to water stress at different growth stages, the critical time for irrigation application in the case of limited water resources, and the level of water productivity under water-stressed conditions.[8–10] The 16 treatments included a combination of soil moisture stress at different growth stages, in RCBDs with three replications.

The long-term meteorological data of the area was collected from national meteorology agency and potential ET was analyzed by Cropwat 8.0 model, for estimating CWRs. The soil physical properties and soil moisture content, before and after irrigation, were analyzed in the laboratory. The irrigation application in all treatments was done based on the soil water-holding capacity and water requirement of tomato. Based on water requirement, the irrigation scheduling of all treatments was planned.

During the growing season (From January 23 to May 22, 2013), 103.2 and 60.7 mm of rainfall was recorded in the mid and in later growth stages, respectively. The total rainfall in growing season was smaller than the expected rainfall (208.5 mm, obtained from New_LocClim). The final results show that irrigating only in the first two growing stages saves more than 52% of water without significant yield reduction. Irrigating at mid or late growth stages did not gave significant differences in tomato crop yield.

KEYWORDS

- **CROPWAT**
- **CROPWAT8.0**
- **irrigation scheduling**
- **maximum evapotranspiration**
- **water productivity**
- **yield responses factor**

REFERENCES

1. Allen, R. G.; Pereira, L. S.; Raes, D.; Smith, M. *Crop Evapotranspiration: Guidelines for Computing Crop Water Requirements*; FAO Irrigation and Drainage Paper 56. Rome, Italy 1998, 230 p.
2. Andreas, P.; Karen, F. Crop Water Requirements and Irrigation Scheduling. In: *Irrigation Manual Module 4*, Water Resources Development and Management Office, FAO Sub-Regional Office for East and Southern Africa, Harare, 2002, 202 p. ftp://ftp.fao.org/docrep/fao/010/ai593e/ai593e00.pdf (accessed December 15, 2016).
3. Awulachew, S. B.; Merrey, D. J.; Kamara, A. B.; Van Koopen, B.; De Vries, F.; Penning, K.; Boelle, E. *Experiences and Opportunities for Promoting Small-scale/Micro Irrigation and Rainwater Harvesting for Food Security in Ethiopia*, IWMI Working Paper 98, 2005, 110 p. http://www.iwmi.cgiar.org/Publications/Working_Papers/working/WOR98.pdf. (Accessed on December 15, 2016)
4. Burney, L.; Carrigerm S.; Davis, J.; Larsen, H.; Rogers, P.; Seckler, D. *Taking an Integrated Approach to Improving Water Efficiency*, Global Water Partnership Technical Brief No. 4, 2006, 12 p.
5. Dessalegn, R. *Water Resources Development in Ethiopia: Issues of Sustainability and Participation*, Forum for Social Studies, Addis Ababa, Ethiopia, 1999, 80 p. http://www.ethiopians.com/Main_FSS_Paper1.htm (accessed December 15, 2016).
6. English, M. J. Deficit Irrigation: Observations in the Columbia Basin. *J. Irrig. Drain. Eng. (ASCE)* **1990,** *116*(3), 413–426.
7. Eugenio, N.; Marcella, M. G.; Giuseppe, G.; Antonio, D. Yield Response to Deficit Irrigation and Partial Root-zone Drying in Processing Tomato (*Lycopersicon esculentum Mill.*) *J. Agric. Sci. Technol.* **2012,** *A2*, 209–219.
8. FAO. *The State of the World's Land and Water Resources for Food and Agriculture Managing Systems at Risk,* The Food and Agriculture Organization of the United Nations and Earthscan, 2011, 130 p. ISBN: 978-92-5-106614-0.
9. Fereres, E.; Soriano, M. A. Deficit Irrigation for Reducing Agricultural Water Use. *J. Exp. Bot.* **2015,** *66*(8), 2239–2252.

10. Hillel, D. *Small-Scale Irrigation for Arid Zones: Principles and Options*. FAO, 1997, 115 p. ISSN 1020-0819: http://www.fao.org/docrep/W3094E/w3094e02.htm (accessed December 15, 2016).
11. Kirda, C. *Deficit Irrigation Scheduling Based on Plant Growth Stages Showing Water Stress Tolerance*, Deficit Irrigation Practices, FAO, Rome, 2002, 150 p. ISSN 1020-1203.
12. McGlade, J. *Measuring Water Use in a Green Economy: A Report of the Working Group on Water Efficiency to the International Resource Panel*, United Nations Environment Programme International Resource Panel, 2012, 90 p. ISBN: 978-92-807-3220-7.
13. Musick, J. T.; Jones, O. R.; Stewart, B. A.; Dusek, D. A. Water–yield Relationship for Irrigated and Dryland Wheat in the US Southern Plain. *J. Agron.* **1994,** *86*, 980–986.
14. Ndaruzaniye, V. *Water Security in Ethiopia: Risks and Vulnerabilities Assessment*, Global Water Institute for Africa Climate Change, Environment and Security, 2011, 140 p.
15. Oweis, T.; Zhang, H.; Pala, M. Water Use Efficiency of Rain Fed and Irrigated Bread Wheat in a Mediterranean Environment. *J. Agron.* **2000,** *92*, 231–238.
16. Steduto, P.; Hsiao, T. C.; Fereres, E.; Raes, D. *Crop Yield Response to Water*, FAO Irrigation and Drainage Paper 66, Rome, Italy, 2012, 133 p.
17. USDASCC. *Soil–Plant–Water Relationships,* United States Department of Agriculture Soil Conservation Service USDASCC. National Engineering Handbook 210.VI. NEH 15.1. 2nd Edition, 1991, 153 p.
18. WMO. *International Meteorological Vocabulary*, 2nd edition. *World Meteorological organization (WMO),* Geneva WMO No. 182, 1992, 141 p.

CHAPTER 9

USE OF COIR GEOTEXTILES FOR SOIL AND WATER CONSERVATION: CASE STUDIES FROM INDIA

SUBHA VISHNUDAS[1,*], K. R. ANIL[1,2], HUBERT H. G. SAVENIJE[3], and PIETER VAN DER ZAAG[4]

[1]*Faculty of Civil Engineering, Cochin University of Science and Technology, Cochin 22, India*

[2]*National Coir Research Management Institute, Trivandrum, Kerala, India*

[3]*Water Resources Section, UNESCO-IHE, Delft University of Technology, The Netherlands*

[4]*IWSG Department, Water Resources Section, UNESCO-IHE, Delft University of Technology, The Netherlands*

[*]*Corresponding author. E-mail: v.subha@cusat.ac.in*

CONTENTS

ABSTRACT

This chapter gives a brief description of coir and its properties, coir industry, in general, and the current scenario of coir in the state of Kerala, South India. Field investigations carried out by the authors and other researchers in the state of Kerala have been discussed in detail to prove the potential of coir geotextile for soil and water conservation for crop production.

9.1 INTRODUCTION

India is the largest producer of coir fiber (66% of the world production) of which two-thirds is contributed by Kerala. Only 10% of the coconut husk is used for fiber extraction out of total annual global production of coconuts. Only about 30% of this enters the world market. In Kerala, the main source of income is agriculture for 70% of the rural population. Agricultural productivity has been severely affected due to lack of water for irrigation during the summer season and soil erosion and flooding during the monsoon.[25,26,67] This demands for a sustainable solution to conserve soil and preserve water. At the same time, about half a million people are working in the coir industry of which about 80% are women in Kerala who depend on it for their daily bread.

These two issues were brought together in this chapter that narrates, through selected case studies, how coir geotextiles can be utilized for soil and water conservation in India and abroad, and particularly in Kerala, South India.

Many researchers have conducted detailed experimental research throughout Kerala using different types of coir geotextiles, for different topographical areas for effective soil and water conservation and also for crop production on sloping land. This would help to reduce the number of risers while contouring/terracing in the high-land region of Kerala and would help to increase the cultivable area and reduce labor costs. Apart from soil and water conservation,[14] it improves crop productivity by providing mulching and by reducing the weed problem. Field experiments show that coir significantly enhances soil moisture in the root zone and improves crop production. A detailed literature review has been carried out in this chapter, which could serve as a comprehensive reference for the effective use of coir geotextiles for soil and water conservation in Kerala and elsewhere in the world with similar geographical and climatic conditions.

In seeking appropriate technology, simple and low-cost technologies are more acceptable for farmers than expensive and labor-intensive conservation techniques. Farmers require technologies, which they can easily understand and implement on their farms without the need for public subsidies.[70] Hence authors conducted a participatory research using coir geotextiles in the mid-land region of Kerala and proved that it is possible to develop an alternative technological option using locally available materials, incorporating indigenous technology and local labor, providing income-generation activities and cost-effective sustainable solutions for water management and crop production in Kerala.

9.2 COIR—GLOBAL SCENARIO

Coir fiber is obtained from the outer husk of the coconut fruit. Coconut is grown all over the world except Europe and Australia. All parts of this palm tree are useful to man. Nuts of the coconut tree are used extensively to make coconut oil from the palm than the fibers and thus contribute to the market economy. In addition, tender coconut is considered to be a health drink and is given to patients with diarrhea. The husk fiber is commonly used as a material for making rope and coconut matting. In India, coir has remained as one of the traditional items, which have been exported for decades.

9.2.1 SOME STATISTICS OF COCONUT AND COIR

About 91 countries of the world grow coconut, and spread over 15 Asian and Pacific Coconut Community (APCC) countries, 8 Asian countries, 13 Pacific countries, 24 African countries, and 13 American countries.[6] From 1961 to 2009, Indian coconut production showed a growth of 204.93% which is much higher than that of the world growth rate of 158.20%. India with an annual production of 12,685 million nuts occupies third place in the world production after Indonesia and the Philippines. The share of India in the world coconut production is about 16.28% of the total production and 17.07% of the area harvested. The four Asian countries (India, Sri Lanka, Indonesia, and the Philippines) have recorded 76.26% of the area harvested and 79.39% of the world production of coconut. Therefore, these four countries have the potential to produce the raw material required for the coir industry and even they could develop monopoly of the coir industry in the world.[22] India and Sri Lanka together contribute to the world production of coconut by 19.65%, but their share in the world production of coir was 63.11% during 2009. India alone contributed to 45.84% to the coir production in 2009 and continues to be the major producer of coir in the world.[22]

9.2.2 PRODUCTION OF COCONUT IN INDIA

It is concentrated in South Indian states of Kerala, Tamil Nadu, Karnataka, and Andhra Pradesh. Major share of it is cultivated in the low-land regions of Kerala. These four South Indian states together grab 90% of the area under coconut and 91% of the coconut production in India and they are in fact the hub of coconut production in India. Kerala ranks first in the national production of coconuts with its share in the area under cultivation of 41.58 and 36.89% in production.[17]

The coconut palm is called the *Kalpavriksha* of Kerala State, South India. It is believed that the name Kerala is derived from the Sanskrit word *Kera* for coconut.[9] The state accounts for more than 75% of the total coconut production. Hence, Alleppey in Kerala became a hub for the coir industry in India, producing beautiful coir mats in various designs, matting, geotextiles/ geo-fabric for soil bioengineering, and garden articles. The byproduct of coir industry, coir pith, is increasingly being used as a soil conditioner and manure.[17]

9.3 COIR INDUSTRY

Coir fiber is the outer covering of fibrous material of a matured coconut and is termed coconut husk that is the reject of coconut fruit. Coir is natural, biodegradable, and environment-friendly. It is tough and durable, versatile and resilient, resistant to flame and fungi, provides insulation, and helps sound modulation and so it captures both domestic and foreign markets.

Coir industry is an agro-based traditional industry, which originated in Kerala state and proliferated to the other coconut-producing states in India like Tamil Nadu, Karnataka, Andhra Pradesh, West Bengal, Maharashtra, Assam, and Tripura. It is an export-oriented industry and has greater potential to enhance exports by value addition through technological interventions and diversified products like coir geotextiles. The acceptability of coir products has increased rapidly due to its "environment-friendly" image. The total production of *coir fiber* in the country during the year 2012–2013 was 536,185 MT and 536,800 MT during 2013–2014.[18] Out of the total annual global production of coconuts, only 10% of the coconut husk is used for fiber extraction. Out of this, only about 30% enters the world trade. The exports in the form of fiber and yarn from producing countries are used for value addition in the importing countries. Sri Lanka is the highest exporter of coir fiber followed by Thailand and India.[67,69]

Coir fiber making in Kerala dates back to the 11th and 12th centuries as it is mentioned in the chronicles of Arab writers and European traders.[9] It was used for making ropes, carpets, and mattings. The first factory to manufacture coir products was started by a European entrepreneur in 1859 at Alleppey district in Kerala. Coir industry is being one of the major traditional industries in the state and second only to agriculture as a source of employment in Kerala. It provides employment to around half a million people, of which 84% are women and earns foreign exchange to the tune of 3 billion Indian rupees per annum (1 US$ = 60.00 Rs.). It is a highly labor-intensive industry but the productivity levels in the industry are low. The average earning of the workers is insufficient to maintain even a subsistence standard of living. With one million hectare under coconut cultivation, this accounts for 45% of the net cropped area in the State. At present, industry consists of about 10,000 tiny and small units. Public sector undertaking and cooperatives play a dominant role in the state's coir industry. The production and processing methods in the coir industry continue to be traditional.[18]

Kerala state has more than two-thirds of the total units. The white coir fiber produced in Kerala is of superior quality to brown coir fiber produced in other States, mainly Tamil Nadu. The cost of white fiber is double to the cost of brown fiber. The total output of coir and coir products (other than rubberized coir) in India is estimated to be around 15 billion Indian rupees including exports of 3.5 billion Indian rupees.[27] The coconut palm comprises of a white meat which has a total percent by weight of 28% surrounded by a protective shell and husk which has a total percent by weight of 12 and 35, respectively. The husk from the coconut palm comprises of 30% weight of fiber and 70% weight of pith material. The fibers are extracted from the husk by several methods such as retting, which is a traditional way, decortications, using bacteria and fungi, mechanical and chemical process, for the production of building and packaging materials, ropes and yarns, brushes and padding of mattresses, and so on.[48]

9.4 COIR FIBER

Coir is a natural fiber and is an organic polymer. It consists mainly of lignin, tannin, cellulose, pectin, and other water-soluble substances.[29] It contains more lignin than all other natural fibers such as jute, flax, linen, cotton, and so on. It is the lignin component that is responsible to a large measure for the strength of the natural fibers and hence it is reputed to be the strongest of all known natural fibers. Again, high lignin content also enables coir to degrade much more slowly than in other natural fibers.[12] Hence it is said to be a lignocelluloses polymeric fiber with 45% lignin and 43% cellulose.[9,29] As it is a natural fiber, its composition varies slightly from sample to sample depending on the source which mainly depending on the maturity degree of the fruit from which the husk is got, the soil condition and climate of the region of cultivation of the palm, the treatment methods, the curing, the climatic conditions, and so on. Chemical composition of coir fiber is shown in Table 9.1.[9]

TABLE 9.1 Chemical Composition of Coir Fiber.

Constituents	Percentage of coir fiber
Ash	2.22
Cellulose	43.44
Hemi cellulose	0.25
Lignin	45.84
Pectin and related compounds	3.00
Water soluble	5.25

Coir or coconut fiber belongs to the group of hard structural fibers. Coir is much more advantageous in different applications for erosion control, reinforcement and stabilization of soil, mulching, and is preferred to any other natural fibers.[44] The coir fiber is elastic enough to twist without breaking and it holds a curl as though permanently waved.[9,59] They are less sensitive to UV radiation due to leaching out of photo-sensitive materials from its surface during the retting processes. Coir fiber is tough and has relative stability against environmental factors like temperature and moisture fluctuations, climatic factors like sunshine, wind, and wave.[29,67,69] In addition to these properties, it is a highly crystalline structure with a spiral angle of the microfibers ranging from 30 to 45°. This leads to a greater extensibility than in most other natural fibers. The coir fiber is relatively waterproof than other natural fibers and is the only natural fiber resistant to damage by salt water.[27] Table 9.2 shows the comparative properties of coir with few natural fibers.[9]

The mechanical properties are rather impaired if nuts attain full maturity. The best use is to leave the nut to ripe for about 10 months. If the nuts are harvested earlier, the husk will be too spongy, weak, and soft, whereas if it is fully ripened, the husks will be tough and the fiber will be coarse. The coir fibers are normally 50–350 mm long and diameter is about 0.1–0.6 mm. The tensile strength of coir increases with increasing diameter of the fiber. For fibers with diameters in the range 0.2–0.45 mm, the mean tensile strength varies from 170 MN/m^2 for retted to 150 MN/ m^2 for unretted fibers. It has a low tenacity value (a unit used to measure the strength of a fiber or yarn), but the elongation is much higher compared to other natural fibers. The percentage elongation at break is about 33%. Coir retains much of its tensile strength when wet.[10,66,68]

TABLE 9.2 Comparative Properties of Coir with Natural Fibers.

Fiber	Coir	Sisal	Jute
Cellulose/lignin content (%)	43/45	67/12	61/12
Density (g/cm^3)	1.40	1.45	1.3
Elastic modulus (GN/m^2)	4–6	9–16	—
Elongation at break (%)	15–40	3–7	1–1.2
Spiral angle (°)	30–45	10–22	8.1
Tenacity (MN/m^2)	131–175	568–640	440–533
Texture	Smooth, tough, cylindrical, and twisted fiber	Long, rough, and twisted fiber	Soft and resilient fiber

Coir is hygroscopic. It can intake considerable amount of water. The water absorption rate varies from 12 to 25% under 65 and 95% humidity, respectively. When coir was fully soaked in water, it absorbs 40% of moisture. This hygroscopic property helps to retain soil moisture in field applications. When the coir gets degraded in the soil over time, it adds fertility to the soil. The degradation of coir in soil depends on the medium of embedment, the climatic conditions, and is found to retain 80% of its tensile strength after 6 months of embedment in clay. Mainly, coir fiber shows better resilient response against synthetic fibers by higher coefficient of friction. The percentage of water absorption increases with an increase in the percentage of coir. Tensile strength of coir-reinforced soil (oven-dry samples) increases with an increase in the percentage of coir.[10,29] Thus, the main advantages of coir fiber highlighted are: it is a renewable resource and CO_2-neutral material; the fiber is abundant, non-toxic in nature, biodegradable, low density, and very cheap. It has a high degree of retaining water and rich in micronutrients. The fibers, instead of going to be waste, are explored for new uses, which in turn provide gainful employment to improve the standard living condition of individuals.[10,48]

9.5 COIR FIBER EXTRACTION

Depending on the process of extracting the fibers from the husk, coir is classified into two types: *white coir* and *brown coir*. White coir is produced from husks of mature green coconuts by subjecting the husk to a retting process of 6–8 months in water, followed by beating of the rutted husks with mallets against a log by manual labor for thrashing out the pith. On the other hand, brown coir is produced from dry/semi-dry coconut husks by mechanical process.[10]

Brown fibers are obtained when the husks of coconut are soaked in pits or in nets in a slow-moving body of water to swell and soften the fibers. The long bristle fibers are separated from the shorter mattress fibers underneath the skin of the nut, a process known as wet-milling. The mattress fibers are then sifted to remove dirt and other rubbish and then dried in the sun and packed into bales. Some mattress fiber is allowed to retain more moisture so that it retains its elasticity for "twisted" fiber production. The coir fiber is elastic enough to twist without breaking and it holds a curl as though permanently waved. Twisting is done by simply making a rope of the hank of fiber and twisting it using a machine or by hand. The longer bristle fiber is washed in clean water and then dried before being tied into bundles or hunks. It may then be cleaned and "hackled" by steel combs to straighten the fibers and remove any shorter fiber pieces. Coir bristle fiber can also be bleached and dyed to obtain hanks of different colors.[17]

White fiber: The coconut husks are submerged in a river or in any water body for up to 10 months. During this time, microorganisms break down the plant tissues surrounding the fibers to loosen them and this process is known as retting. Segments of the husk are then beaten by hand to separate out the long fibers which are subsequently dried and cleaned. Cleaned fiber is ready for spinning into yarn using a simple one-handed system or a spinning wheel.[13,17]

Brown fiber is relatively inferior in terms of quality. It is mainly used for ropes, rubberized coir, and in upholstery. The extracted fibers are then spun into yarn of different counts and grammage. The yarn is classified in terms of type of fiber, color (natural), twisting, and spinning. The yarn is then converted into mats in handlooms, semi-automatic looms, or power looms. The scorage of yarn differs among different types of geotextiles. The scorage of the yarn is the number of strands that can be laid close to each other without overlapping in a length of 0.9 m (1 yard). Figures 9.1–9.4 show the different processes involved in the yarn making of coir.

FIGURE 9.1 Dehusking from coconut shell.

FIGURE 9.2 Retting of coconut husk.

FIGURE 9.3 Fiber extraction by beating husks after retting.

FIGURE 9.4 Twining coir fibers for yarn making.

Apart from this coir, cut fiber is also recognized as a high air-fill porosity material that mixes any soil or peat. It can be used as a ground cover for good watering that prevents direct intensive sunlight. The length of cut fibers varies from 1 to 20 mm with a diameter of 1.6 μm and a density of 1.4 g/cm^3. Coir fibers make up only about one-third of the coconut pulp. The other two-thirds is called the pith or dust, which is biodegradable but takes almost 20 years to decompose. It has a density very much lower than that of coir fiber. The specific gravity of air-dried pith is only about 0.1 compared to 1–1.5 for fiber. The pith is more water-absorbing. It absorbs about 600–800% by weight of water, whereas for fiber it ranges from 10 to 40% only. Once considered as waste material, pith is now being used as mulch, soil treatment, and a hydroponic growth medium.[9]

9.6 GEOTEXTILES IN GENERAL

Soil is a natural resource of great importance. It is living, dynamic, material, non-renewable on the human time scale and plays many key roles in terrestrial ecosystems.[49] Soil is the basis of plant growth, providing a variety of mineral nutrients and water for plants, and plays a crucial role in food production of human life.[56] Due to urbanization, climate change, and change in land-use pattern, much of the natural resources are in the verge of degradation. Soil moisture is the vital factor controlling suitability of soils for crop production. Soil properties, such as moisture relations, fertility, and nutrient availability, are the crucial factor for crop production. The interaction of soil properties on crops is influenced by clay content, temperature, moisture content, and oxygen availability within the soil.[50,57] There are many strategies, such as *geotextiles*, mulches, and conversation tillage; soil amendments used to maintain good soil properties and optimize use of water resources for crop production.

Geosynthetics, wherein geotextiles is a part, are used in a wide variety of applications. There are about nine categories, namely: geotextiles, geogrids, geonets, geomembranes, geoforms, geocells, drainage/infiltration cells, geocomposites, and geosynthetic clay liners.[50,62] By definition, geotextiles are a "permeable geosynthetic comprised solely of textiles".[8] Depending on the functions they serve, they are being used in different applications. Of these, geotextiles form one of the largest groups of geosynthetic material.

Geotextiles is a woven knitted structure made of natural or synthetic textile fiber, which has found extensive use in soil stabilization, erosion control, and other applications. Literature shows that geotextiles have been used since 1950. There are two principal types of geotextiles: woven and non-woven.[9,62] They transmit fluids across or in-plane or both but can retain soil particles.

Woven geotextile: It is characterized by two sets of interlacing threads at right angles to each other. One set of thread is known as warp run along the length of the fabric and the other known as weft running perpendicular to the warp. The woven fabrics are made from spun yarn. Depending on the pattern of interlacement of warp and weft threads, different types of woven fabrics such as plain, twill, and so on are produced. The plain weave fabric gives maximum interlacement between the warp and weft threads, thereby imparting maximum dimensional stability, rigidity, and strength to the fabrics.[9,62]

Non-woven geotextile: It is a textile structure produced by bonding or interlocking staple fibers, monofilament or multifilament accomplished by mechanical, chemical, thermal, or solvent means or combinations thereof. Each non-woven manufacturing system generally includes four basic steps, namely: fiber preparation, web formation, web bonding, and post-treatment They are desirable for surface drainage and erosion control.[9,62]

Almost all synthetic geotextiles available are made from polypropylene or polyester formed into fabrics. The geotextiles are mainly used in six broad functions, namely: drainage, filtration, separation, reinforcement, protection, and barrier. All of these functions have been used one way or the other in the application of soil and water conservation using geotextiles.

9.6.1 DRAINAGE

Geotextile is used to remove excess fluid either to accelerate consolidation or relieve pore pressure build up which can cause instability. It allows the liquid to flow with minimum soil loss within the plane of the geotextiles. The drainage function of geotextile may be for temporary purpose or permanent as a drain. The in-plane permeability must be several times higher than the soil and the area of flow adequate.

9.6.2 FILTRATION

When the excess fluid is removed, the soil particles should not be carried with it as it will cause instability and the voids which may lead to piping. The soil should be retained and only the fluid should be drained away.

9.6.3 SEPARATION

Geotextiles function to prevent mutual mixing between two layers of soil having different particle sizes or different properties. The pressure on retained material on one side is greater than the fluid pressure which must be supported by frame work as in silt fence.

9.6.4 REINFORCEMENT

Due to their high soil fabric friction coefficient and high tensile strength, heavy grades of geotextiles are used to reinforce earth structures, allowing the use of local fill material. Geotextiles provide additional strength to soils to enable steep slopes and soil structures to be constructed, and allow construction over weak and variable soils.

9.6.5 PROTECTION

A geotextile is used to protect the soil from erosion due to water flow, the road surface from attrition of traffic, or as a surface protector. Also a geotextile can be placed as an interface to lower the stress on weaker layers and reduce deformation.

9.6.6 BARRIER

It isolates one material from another. The most frequent use of this function is in landfills where impermeable linings prevent contamination of surrounding soils.

9.7 COIR GEOTEXTILES

Modern living compels man to avoid the indiscriminate use of plastics and such ecologically damaging materials. Industries that manufacture these products generate countless tons of chemical wastes that pollute the air, land, and water. Large-scale abuse of fertile forestland has led to heavy loss of precious topsoil. The undigested plastics prevent recharge of groundwater and chokes drainage. The consequent environmental degradation is threatening the very existing of all forms of life in this planet. However, there is an increase in awareness almost the world over the urgency to maintain the much needed biological balance. It is imperative that any developmental effort should go hand in hand with ecological equilibrium so as to sustain it. And a return to natural living has been accepted as a safe way to solve many of these environmental problems.[50]

The export of coir geotextiles from India was to the extent of 2140.69 tons valued at US$ 2.34 million in the year 2002–2003, while in 2006–2007, the quantity of coir geotextiles consigned was equivalent to 3044.5 tons with a value of US$ 3.35 million. The transparent growth was 35.53% with a compounded annual growth rate of 7.10%.[17]

Coir geotextiles, a generic member of the geosynthetic family, are made from coconut fibers. Like their polymeric counterparts, coir geotextiles are developed for specific application as per functional requirements. Again, it is abundantly available, and a renewable natural resource with an extremely low decomposition rate and a high strength compared to other natural fibers. Coir is woven into thick textiles which are applied like blankets on the ground in erosion-prone areas. Coir geotextiles are durable, absorb water, resist sunlight, facilitate seed germination, and are 100% biodegradable. These blankets have high strength retention and a slow rate of degradation, and hence it retains for several years in field applications. Thus, in geosynthetic field, coir geotextiles have become an ecologically acceptable material. The characteristics of coir fibers which are different from other natural fibers are that the cell length to diameter ratio (*l*/*d*) is nearly 15, as compared with flax (1700) and also have high modulus and low failure strain. Therefore, to increase stiffness, proper preparation and surface modification is required to produce a cost-effective matrix.[12] Coir geotextiles are mainly of two types: woven and non-woven.

Woven coir geotextiles are mats made of plain, chemically leached, dyed with coir fiber or coir yarn. There are different types like fiber mats, loop mats, rope mats, and so on. They are woven by power and handloom with treadle plain or treadle basket weaves. Loop fabrics are commonly specified as coir geotextiles, is a product made with loop construction manufactured in rolls for use as geofabrics for soil erosion control and stabilization. Such a mat has one tight chaining and one slack chain formed in the weaving process. Mesh mats are made by laying coir yarn in a crisscross manner between a number of nails fixed on a frame and knotting the intersecting points with coir yarn. They are available in different mesh openings ranging from 3 to 25 mm.[9]

Non-woven coir geotextiles are made from loose fibers, which are interlocked by needle punching or rubberizing. Manufacturing of non-woven coir geotextiles involves the process of mechanical bonding by needle punching. The extracted fiber in bale form is opened and passed through bale openers, garneted, and formation of webs by cross-lapping process/air laying process. The cross-lapped webs are then needle punched in the needle punching looms with or without the introduction of scrim fabric. The scrim fabric has to be strong and at the same time light in weight. Different combinations and weights could be produced as per specific end-use requirements. The properties of non-woven fabrics depend on various bonding methods like thermal bonding and needle punching besides fiber arrangement and fiber properties.[35] When fabric is subjected to tensile load, the deformation of the fabric takes place. The main structural parameter, that is, the curl of a fabric, is the ratio of the fiber segment to the shortest distance. The relationship of the curl and its direction affect the fabric strength.

In order to standardize the tensile behavior and biodegradability characteristics of coir, studies were initiated for the first time in Indian Institute of Technology, Delhi. The work included an evaluation of the physical and engineering characteristics and the biodegradability behavior of coir/jute geotextiles in different soil environments.[10,51]

9.8 BUREAU OF INDIAN STANDARD SPECIFICATIONS: COIR GEOTEXTILES

Coir nettings/geotextiles are produced under 10 specifications [H2M1–H2M10] by the Bureau of Indian Standards (BIS), which vary in weight from 400 to 1400 g/m and mesh sizes from 1 to 2.5 cm, respectively. There are four Indian Standards published by Bureau of Indian Standards (BIS) besides the constructional details for mesh matting used as coir geotextiles.

9.8.1 IS 15869: 2008 TEXTILES—OPEN WEAVE COIR GEOTEXTILES (BHOOVASTRA) SPECIFICATIONS

This standard prescribes constructional details and other requirements of open weave coir *bhoovastra* (CBV) of three different grades used in the prevention of erosion of soil and reinforcement of paved and unpaved roads. According to IS 15869-2008, an open weave coir geotextile is woven fabric of two treadles weave in construction made from coir yarn in which the warp and weft strands are positioned at a distance to get a mesh (net) effect of 1″, 3/4″ and 1/2″ square. The open weave coir geotextile shall have the following grades based on the mass: (a) Grade I—400 g/m^2; (b) Grade II—700 g/m^2; (c) Grade III—900 g/m^2. Figure 9.5(a)–(c) shows the three different grades of coir geotextiles.

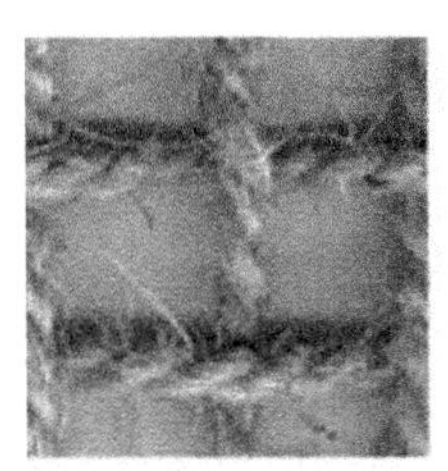

a. Grade I-400g/m^2

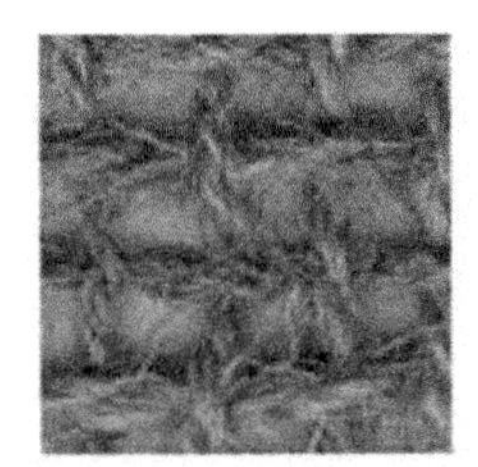

b. Grade II-700gm/m^2

c. Grade III-900gm/m^2

FIGURE 9.5 (a–c) Three grades of coir geotextiles as per BIS.

9.8.2 IS 15868 (PARTS 1–6): 2008 NATURAL FIBER GEOTEXTILES (JUTE GEO TEXTILES AND COIR GEOTEXTILES (BHOOVASTRA) METHOD OF TEST

Standard (Part 1) specifies a method *for the determination of the mass per unit area* of all natural fiber geotextiles for identification purposes and for use in technical data sheets. And also the method is applicable to all natural fiber geotextiles: coir geotextiles, jute geotextiles, and erosion control blankets (ECBs).

Standard (Part 2) prescribes method *for the determination of the thickness of geotextiles* at specified pressures and defines at which pressure the nominal thickness is determined. The method is applicable to all types of natural fiber geotextiles.

Standard (Part 3) prescribes method *for the determination of the percentage of swell in water* of geotextiles. The method is applicable to all types of natural fiber geotextiles.

Standard (Part 4) prescribes method *for the determination of the water absorption capacity* of geotextiles.

Standard (Part 5) details a procedure *for the determination of the smoldering resistance* of degradable rolled erosion control products.

Standard (Part 6) specifies method *to determine the mesh size by projecting the geotextile* through an overhead projector (OHP). This method is suitable for meshes having large opening sizes.

9.8.3 IS 15871 (2009): USE OF COIR GEOTEXTILES (COIR BHOOVASTRA) IN UNPAVED ROADS GUIDELINES

This standard prescribes the guidelines of coir-woven geotextiles (*bhoovastra*) suitable for application in unpaved roads including the selection of coir-woven geotextiles (*bhoovastra*) and installation methods.

9.8.4 IS 15872 (2009): APPLICATION OF COIR GEOTEXTILES (COIR-WOVEN BHOOVASTRA) FOR RAIN WATER EROSION CONTROL IN ROADS, RAILWAY EMBANKMENTS, AND HILL SLOPES GUIDELINES

This standard prescribes the code for the guidelines of woven coir geotextiles (*bhoovastra*) suitable for application in slopes of road and railway embankments and also in hill slopes including the selection of woven coir *bhoovastra* and installation methods.

9.9 PROPERTIES OF COIR GEOTEXTILES

The tensile strength of natural geotextiles[65] can be taken as that corresponding to 200 mm wide × 100 mm length specimen at a deformation rate of 10 mm/min, for all practical purposes.[51] The woven coir geotextiles have more tensile strength in the machine direction than in the cross-machine direction. The tensile strength is influenced by number of yarns. The non-woven coir geotextiles have more tensile strength and tensile elongation in the machine direction than in the cross-machine direction. The behavior of non-woven coir geotextiles is influenced by the presence of the type of stitching, yarn used, coir/jute netting, and the coir web weight.[10,50,51] The variance in tensile strength and tensile elongation of the non-woven coir geotextiles is generally higher in comparison to woven coir geotextiles. The tensile strength ranges from 59.50 to 64 kN/m for woven coir geotextiles compared to 2.8–4.5 kN/m for non-woven specimens. The failure strain ranges from 55 to 60% and secant modulus varies from 0.98 to 1.15 kN/m in woven coir geotextiles. In non-woven coir geotextiles, the failure strain ranges from 8 to 12% and secant modulus varies from 0.75 to 1.25 kN/m. For woven coir geotextiles under wet conditions, the strength was reduced by 36% and the failure strain was increased by 21% when compared to dry specimen. For non-woven geotextiles under wet conditions, the tensile strength decreases by about 61% and the failure strain increases by 30% than that of the dry specimen.

Regarding compressibility, non-woven geotextiles have more compressibility than woven coir geotextiles.[10] Thus, woven geotextiles exhibit higher strength compared to non-woven products, and tensile elongation at failure is higher for the non-woven geotextiles. The ratio of width to length (aspect ratio) does not have any influence on the tensile strength. With regard to the durability study, coir has a life span of 2–3 years.[51] Coir has the highest tensile strength of any natural fiber and retains much of its tensile strength when wet. It is also long lasting, with infield service life of 4–10 years.[21] Tests conducted by Schurholz[53] on jute, sisal, coir, and cotton over a prolonged period of time in highly fertile soil maintained at high humidity (90%) and moderate temperature revealed that coir retained 20% of its strength even after 1 year, whereas cotton degraded in 6 weeks and jute degraded in 8 weeks. According to Schurholz,[53] coir can better withstand traction effect due to flooding than any other natural fabric. Alternate wetting and drying of coir yarns did not accelerate the degradation of coir samples as it was observed that coir retained 30% of its original strength even after 1 year.[12,68] Water absorption varies from 12 to 25% under 65 and 95% humidity. When coir geotextile is fully soaked in water, it absorbs 40% of moisture.[10] This hydroscopic property helps to retain soil moisture in irrigation application increasing water efficiencies. When the coir gets degraded in the soil over time, it adds fertility to the soil.[66,68]

9.10 APPLICATIONS OF COIR GEOTEXTILE

9.10.1 SOIL BIOENGINEERING

Soil erosion is a naturally occurring process that affects all landforms. In agriculture, soil erosion refers to wearing away of topsoil in the field by the natural physical forces of water and wind or through forces associated with farming activities such as tillage.[32] This type of erosion threatens to sustain global population with food and fiber and is closely linked to economic vitality, environmental quality, and human health concerns. Roughly 75 billion tons of fertile topsoil is lost worldwide from agricultural systems every year. Erosion results in degradation of soil productivity in a number of ways: it reduces the efficiency of plant nutrient use, damages seedlings, decreases plant's rooting depth, reduces the soil's water-holding capacity, decreases its permeability, increases runoff, and reduces its infiltration rate. The sediment deposited by erosive water as it slows can bury seedlings and cause the formation of surface crusts that impede seedling emergence, which may decrease crop yield. The combined effects of soil degradation and poor plant growth often result in even greater erosion later on.[5]

Erosion-induced loss in soil productivity is one of the major threats to global food and economic security, especially to resource-poor farmers in the tropics. It not only diminishes the quality of soil resources, but also makes difficult in gaining a livelihood from the land. Through reduction in soil productivity, it affects outputs such as crop yields derived from the renewable natural resource systems of the biosphere. By affecting the net primary production of vegetation, erosion is the driver for a number of critical environmental, economic, and social issues in developing and developed countries.[43]

To protect erosion of seeds and seedlings from unprotected sites by surface runoff and winds is costly and also difficult to attain since all previous attempts to establish vegetation on the slopes have to be repeated. Hence a protective covering on soil is essential which resists soil erosion, retains runoff, and facilitates establishment of vegetation on the surface. By protecting the surface, these covering materials dissipate the energy of raindrop impact, increase infiltration by reducing surface sealing, and reduce the velocity of overland flow. In addition, they help to reduce intense solar radiation, suppress extreme fluctuations of soil temperature, reduce water loss through evaporation, and increase soil moisture, which can assist in creating ideal conditions for plant growth.[60,74]

Since the early 1980s, natural fiber materials have been used in the field of soil bioengineering to solve erosion problems. The German Bundersanstalt fur Material Prufung (GBFMP) and other important testing institutes in Germany tested such materials for 6 years before allowing them to enter the market place. Since mid-1987, woven coir geotextiles have been used extensively in Germany. They have expanded from there to most of Western Europe and North America.[54] Coir geotextiles are found to provide protection against soil erosion to the various types of slopes that has been demonstrated and documented by the Coir Board.[4] The ability of coir fibers to absorb water and to degrade with time is one of its prime properties, which give it an edge over synthetic geotextiles for erosion control purposes. The drappability factor of coir geotextiles (due to their flexibility) allows them to conform closely to the terrain. Coir matting has an open area of 40–70%. Hence, it allows growth of grass and provides a large number of miniature porous check dams per square meter of soil. It slows down and catches runoff so that sediment settles and water either passes through the matting or percolates into the underlying soil. As geotextiles degrade, they provide mulch and conserve moisture for plant growth. On impact with an unprotected soil surface, raindrops loosen the soil particles, causing an incremental movement of the suspended particles down slope. Soils are susceptible to erosion by flowing water even at very low flow rates. If the energy of falling rain can be absorbed or dissipated by vegetation or some other soil cover or surface obstruction, the energy transfer to the soil particles will be reduced and hence soil erosion. When geotextiles are used, they absorb the impact and kinetic energy of raindrops and reduce surface runoff. Also seeds and vegetation are protected from being washed away.[3]

9.10.2 GULLY PLUGGING

Gully erosion reduces the area of land available for farming. Its destructive manifestations include scouring of the land, deposition of sediment on growing crops, worsening of soil quality in the accumulation zone, general drying of the affected regions by interrupting the groundwater table, silting of river channels, and pollution of surface waters. Though most of the erosion damage caused to farmland is ascribed to sheet erosion, the losses due to gully erosion are also considerable and what is worse, they are irreversible. The majority of the gullies developed back in the past when extensive farming prevailed.[20] A case study in Kerala which is being described in Section 9.13 shows how coir geotextiles has been effectively used for gully plugging.

9.10.3 MULCHING

Soil evaporation is an important component of the water balance in natural and cultivated systems. It is estimated that 50–70% of annual precipitation returns to the atmosphere without any benefit to biomass production.[31] The reduction of soil evaporation is essential to increase the water-use efficiency of agricultural crops. The use of mulching materials is an efficient way to reduce the exchange of water vapor between the soil surface and the atmosphere. Consequently, the evaporation of water from a mulched soil decreases relative to a bare soil, and more water is available for beneficial crop transpiration.[30,52] The type, amount, or thickness of the mulching materials, and the atmospheric evaporative demand determine the rate of soil drying.[64] Mulching with impervious materials such as plastic films minimizes the evaporation of water from the soil surface, but prevents the entry of rainfall into the root zone of crops. In contrast, mulching with porous materials allows the entry of rainfall, but soil evaporation increases over that of impervious materials. Therefore, the benefits of the different types of mulching materials for water conservation are weather-dependent and rely on the balance between the water entering the soil from rainfall and irrigation, and the water leaving the soil by evaporation and transpiration.[75]

In an agricultural country like India, heavy rainfall enables growth of weeds which may later compete with crops for sunlight nutrients and moisture. One among the most important factors that interferes with the production potential of soil is the cropping system that leaves the soil surface exposed periodically. This encourages weed growth, thereby reducing crop yield. Though the introduction of herbicides has revolutionized the agriculture, the residual effects of these herbicides on the succeeding crop or the possible interactions of their residues with other pesticides or cultural practices may pose problems. Technologies should therefore be evolved in an eco-friendly and sustainable way, not merely aiming at short-term gains. Experiments undertaken at Soil Conservation Research Station (SCRS), Konni, Kerala, demonstrated that the use of different types of coir geotextiles as mulch can be effectively used as mulching material.[1]

9.11 APPLICATIONS OF COIR GEOTEXTILES: A REVIEW

From the previous sections, it is clear that coir is an abundant, renewable natural resource with an extremely low decomposition rate and a high strength compared to other natural fibers. When coir is woven into thick textiles, it can be applied like blankets on the soil surface in erosion-prone areas.[22]

Soil degradation by erosion is one of the world's most serious environmental problems, causing extensive loss of cultivated and potentially productive soil and crop yields.[23,34,45] It has been estimated that some 6000 million tons of soil per year have been washed off the croplands of India.[23] It is postulated that biological geotextiles can act as a complementary measure for temporary prevention of deflation and to temporarily increase moisture storage, creating better conditions for re-vegetation.[40] Few successful case studies in the field application of coir geotextiles around the world are mentioned below.

Sotir[58] illustrated case studies of river bank stabilization using coir geotextiles in the USA. In Longfellow Creek Bypass channel, coir geotexiles with selected plants were used to stabilize trapezoidal channel slopes. Results show that the use of coir geotextiles in and along streams and river bank protection and for the establishment of healthy riparian zones for aquatic enhancement appears to be a viable alternative. Cammack[16] reported the use of coir geotextiles in the Noora basin in Australia, for causeway protection to prevent wave-lap erosion in saline water condition. He also reported the successful case study of Gooburrum main canal bank protection using coir geotextiles. White[73] reported various control techniques adopted by the Illinois Department of Conservation for the control of stream bank erosion of the Crow Creek. Coir geotextiles were found to be the most effective and environmentally sound biotechnical application to effectively enhance our environment.[38] His case studies revealed that coir geotextiles and coco logs were widely used for river training works in Korea. Schurholz[55] has illustrated various field trials in Germany using coir geotextiles. It includes stabilization of a creek bed and its bank using woven geotextiles, river bank stabilization, and re-vegetation of shore lines by sedimentation. Oosthuizen[47] has conducted a comparative field study of the use of natural fiber nettings sisal and coir in controlling erosion in South Africa. They reported that coir netting withstood the extreme wet condition that prevailed in the area of study, and strength was decreased only by 56% after 6 months of trial.

9.11.1 OTHER APPLICATIONS OF COIR

Coir dust (CD) is the spongy, peat-like residue from the processing of coconut husks (mesocarp) for coir fiber. It consists of short fibers (<2 cm) of around 2–13% of the total and cork-like particles. Until recently, it was the only part of the coconut tree that had no real value. CD is found in most countries where coconuts are grown. CD strongly absorbs liquids and gases due to the presence of mesocarp tissue, which gives it high surface area–unit volume. CD is also hydrophilic. Asiah[7] conducted studies on physical and chemical properties of coconut CD and oil palm empty fruit bunch (EFB) and the growth of hybrid heat tolerant cauliflower plant using CD and oil palm EFB. In his studies with coconut CD and oil palm EFB, it was reported that the physical properties showed both the EFB and CD provided optimum plant growth conditions at the start of the growing the period, the readily available water value for CD was 34%, whereas that for EFB was 19%. The results also indicated higher nutrient contents in CD than in EFB. Despite a high initial electrical conductivity value for CD than EFB, the hybrid cauliflower plant dry weights and total leaf area for CD-grown plants were doubled grown in EFB. CD was more suitable growing medium for growing hybrid cauliflower compared to EFB. With Indian coir, which had an electrical conductivity of 7 dS/m, tomato seedlings grew larger and faster than in coir or peat with an acceptable salt content.[24]

9.12 COIR GEOTEXTILES FOR EROSION CONTROL: CASE STUDIES FROM KERALA

The state of Kerala is a narrow strip of land that lies in the southwest of India sloping from Western Ghats in the east to Arabian Sea in the west. It is located in the southern part of India, having a land area of 38,863 km^2. The state has a total coastal length of 560 km and has an average width of 67.5 km. It has a tropical climate with a unique topographical setting. It is divided into three physiographic zones parallel to the coastal line: high lands, midlands, and lowlands.

The highland slopes of the Western Ghats are characterized by steep slopes. They rise to an average height of 900 m with some peaks reaching over 1800 m. The rainfall on the area drains toward the lowlands with little resistance. Tropical forests occupy this area and there has been considerable reduction in forest area during the last few decades. The midland is characterized by low hills and valleys, forming the unique watersheds of Kerala with streams flowing through the valleys.[71] The lowlands consist of coastal belts, which receive all the water from the upper reach, and are subject to flooding during the monsoon, followed by drought in summer.

The bulk of the rainfall in Kerala is received in the two monsoon seasons from June to November. The following 6 months are relatively dry with little summer rain. This skewed distribution over the year leads to water scarcity during the summer months. The average rainfall on the state is 3000 mm/year, of which 60% is obtained during the southwest monsoon and 25% during the northeast monsoon. The remainder results from summer showers.[57] In Kerala, land and water management is the most neglected part of water resource development. The entire state is seriously prone to water shortage especially for agriculture during the summer season. As per Census

2011, 52.3% of the population lives in rural areas of which more than 70% depend on agriculture for their livelihood. Hence scarcity of water is mostly affecting the rural poor.

9.12.1 APPLICATION OF A COIR GEOTEXTILE IN LOWLAND REGION OF KERALA

Vembanad Lake, which is the largest lake in Kerala, has been separated into several fragments by bunds and dykes resulting in the formation of numerous sub-lakes. These sub-lakes have been converted into paddy fields. The sub-region, which is thus reclaimed from the Vembanad Lake and its surrounding marshy land, is called Kuttanad.[63]

Kuttanad is a highly complex, dynamic, and unique rice-growing agro-climatic tract of Kerala lying 0.5–2.5 m below MSL, bears strong resemblance to the Netherlands in terms of geographical features. It extends between 9°8′N and 9°52′N latitude and 76°19′E and 76°44′E longitude. Most of the areas in Kuttanad are water logged almost throughout the year and subjected to flood during the monsoon period. The organic matter transported from the high ranges makes Kuttanad a unique ecosystem in the world due to its location near equator, equitable temperature regime, high rainfall, and high solar radiation throughout the year similar to the Philippines in the tropic. Kuttanad is said to be the traditional rice bowl of Kerala. It has an area of about 53,777 ha of low-lying reclaimed lands, from shallow stretches of the Vembanad Lake, and is exclusively under paddy cultivation.[36] This is perhaps the only region in the world where farming is done at 1.5–2 m below sea level. Inland waterways, which flow above land level, are an amazing feature of this region. Kuttanad soil resembles peat soil. It is acidic [pH 4.5–6] and saline with medium in organic matter content. It is poor in available nutrients but rich in calcium. River-borne alluvial soil is found at 1–2 m below the sea level, which is generally clay loam in texture, high acidity, fair amount of organic matter, but poor in available nutrients. Average temperature in the area ranges between 23.5 and 31.20°C with a relative humidity of 89%.[41] For over 2000 years, the mud walls have been made to protect the stream banks in India for preventing them from causing soil erosion, which have to be constructed every year after the flood.

In Kuttanad, coir geotextile was used to serve as an effective filter and reinforcing material for the clay dykes for early consolidation of clay.[39] Every year, farmers need to construct and strengthen bunds which require labor and money. The Thanneermukkom barrier was constructed across the

Vembanad Lake in order to prevent salinity intrusion during dry season and also to retain the freshwater from rivers flowing into the lake. This bund is being opened and closed every year 1 month before and after the harvesting to maintain flooding in the paddy fields. Each time before paddy cultivation starts, the water has to be drained out and the shutters of Thaneermukkom barriage are closed to prevent the entry of seawater. Temporary clay dykes constructed around the paddy fields separate it from water bodies. Preparation of paddy fields for cultivation involves construction of outer dykes that must be strong enough to protect the paddy fields from the surrounding rivers and water bodies. This necessitates construction of stronger clay dykes, which can last for comparatively longer period of time.

Coir geotextile is used as an external reinforcing material for the clay dykes. Coir logs laid at the bottom along the length of the dyke provide proper anchorage to the geotextile netting (Coir *logs are massive and flexible coir products with a dense packing of coir fibers*). Good-quality bamboo poles cut in lengths of 1.5 m are used for lateral support of the clay–coir dyke. Two parallel stretches of clay dyke, each of length 10 m, were identified for comparative study: one with and the other without coir net side protection. Penetration resistance, time-settlement response, and vertical and horizontal displacement were monitored in both dykes for a period of 1 year since construction. The study revealed that coir geotextile is very useful in the wetland environment on account of its basic functions of reinforcement, filtration, and drainage. Also, its advantages such as biodegradability, environmental compatibility, and economy are other added factors for improving its utility in various applications. After 1 year of monitoring, it was observed that with the draining out of flood water, the coir-protected dyke offers more penetration resistance compared to the non-protected dyke. Thus, compared to conventional clay dyke, coir geotextile-protected dyke is least affected by the effect of flooding. This technology can be applicable in hydrologically similar low-lying regions where temporary clay dykes are used to drain out excess water and to prevent the ingress of salt water.

Similar studies were conducted by Sharma and Jose[53] in Kuttanad, to utilize the coir geotextiles, as reinforcement to the mud wall for a high-velocity stream that was causing floods in the entire area of paddy cultivation despite putting up mud walls every year causing damage to rice crops and to allow vegetation to become established for providing sustainable protection against soil erosion. The project was commenced in April 1999. Two treadle basket weave fabric [warp3 × weft3], Vaikom × Vaikom (*coir collected from Vaikom region to manufacture geotextiles*) made from coir yarn was chosen for the construction of mud wall on a high-velocity stream bank near the Mancombu Rice Research Station in Kuttanad, along with bamboo poles for providing support. Due to slow biodegradation of coir geotextile, the vegetation could sustain and provide extra support to the mud wall. The strength of the soil has been found to increase in course of time as the organic skeleton has remained in place in compressed form that acts as a filter cake providing sustainable protection to the streambank. Local vegetation grown over the embankment has been providing extra protection against erosion of mud wall. This treatment opens up new avenues for the application of coir geotextiles that could be applied in low-lying areas all over the world. The tremendous strength and biodegradability of coir makes it suitable to various new areas of application in the soil bioengineering. Figure 9.6(a)–(d) shows the application of coir geotextiles in mud wall protection in Kuttanad, lowland area of Kerala.[52]

(a) The broken mud wall

(b) Application of coir geotextile using bamboo poles

(c) Close up view of applied coir geotextile

(d) Reinforced mud wall with grown vegetation after 8 years of construction

FIGURE 9.6 (a–d) Application of coir geotextile in lowland area in Kerala.

9.12.2 APPLICATION OF A COIR GEOTEXTILE IN MID-LAND REGION OF KERALA

The study was conducted in Amachal watershed in the Trivandrum District, in the Western Ghat region of Kerala, India, to test the effectiveness of using coir geotextiles for bank stabilization.[67,68,70] The watershed lies in the midland region between 8°28′57″ and 8°29′44N, and 77°6′26″ and 77°7′16E. The watershed is characterized by moderately sloping to steep hills intervened by very gently sloping valleys. The area experiences a humid tropical climate with two distinct monsoons (northeast and southwest) and an average mean annual temperature of 26.50°C. The relative humidity varied from 62 to 100%.[28] The southwest monsoon commences by the first week of June and continues up to September, and the northeast monsoon sets in by the middle of October and extends up to December. Annual rainfall amounts to 1500 mm/year. Peak rainfall in the experimental period was observed in the month of October (429 mm/month) followed by June (243 mm/month). Rainfall events are generally of high intensity and short duration especially in the southwest monsoon. This rainfall typically is in the form of an evening shower with a clear sky during the day. This watershed experiences severe water scarcity during the dry period from January to May. Agriculture is the main source of income of the inhabitants in this watershed.

A village pond in the watershed was selected for the field experiment. This pond is the source of water for irrigation in this watershed. The side banks of this pond get eroded even during summer showers. During monsoon, the side banks of these ponds erode and the ponds get silted up. The same silt from the pond is subsequently used to restore the side banks but it is often eroded before vegetation can establish. Hence continuous maintenance is required for deepening and desilting of ponds to maintain its water-holding capacity. Neither the local government nor the community may have enough funds for these labor-intensive works. Ultimately, the ponds get filled up and deteriorate and the area becomes subject to water shortage during the summer season and even during dry spells. The aim of this experiment was to study the effectiveness of coir geotextiles for embankment protection and to provide an alternative, cost-effective option to reduce soil erosion, increase vegetation growth, and increase soil moisture availability.

The type of soil is silty sand. The capacity of the pond is 48 m × 123 m × 2.1 m. The pond has a natural depression on one side; the water level in the pond fluctuates from season to season. The slope of the embankment is 70°. The height of the exposed slope of the embankment is about 3 m. The length of the embankment varies from 3.10 to 3.50 m. Erosion is caused by both rainfall and runoff. The limitation for providing a gentle slope to the embankment is that three sides of the pond are surrounded by existing village roads and the other side is a pedestrian road. Beyond the road on two sides, there are existing irrigation canals.

The coir was laid just before the onset of the monsoon. Trenches of 30 × 30 cm were dug at the top of the slope to anchor the geotextile. Rolls of the matting were first anchored in the top trench and then unrolled along the slope, and anchoring was done using bamboo pins cut to a length of 25–30 cm, spaced in a grid of 1 m spacing. Each roll was given minimum overlap of 15 cm. At the bottom, matting was rolled in two layers and anchored to hold the soil eroded if any and also to reduce the intensity of runoff.

Soil moisture retention, establishment of vegetation, biodegradation of coir, and nutrient losses were observed for a season in control plot (CP), plot with coir geotextiles (CG), and plot with coir geotextiles planted with *Axonopus compressus* (CGG). This variety of grass has been widely used in this area as fodder to feed cattle. Length of grass, weed intensity, uniformity, and density of grass has been considered for vegetation growth. A group of 60 people were selected randomly from the user community living within the vicinity of the pond for monitoring and evaluation. The user community themselves developed indicators for the qualitative evaluation. This included length of grass, color of grass, uniformity of grass, density of grass, and soil erosion.

In 9 months, vegetation was well established and the slope was stabilized in the area covered with geotextiles. Average length of the grass of the same species as that in CGG was measured from all the plots to compare the length of the grass. Soil moisture in CGG was 21% higher than in the control plot during the dry period. In CG, soil moisture was less than in CGG. In CP, the density and uniformity of vegetation was much less along with the occurrence of soil erosion and runoff. Growth of vegetation in CGG showed greater values than in CG. The control plot showed the lowest value.

In CGG, vegetation was established well before it started at CG and CP. The coir retained 19% of the strength of a fresh sample after 9 months. After 7 months, it was observed that tensile strength of geotextiles was reduced by about 70%. By that time, a sustainable erosion control measure by the establishment of vegetation was observed in the CGG and CG plots, whereas erosion persisted in the control plots. Hence the increase in the rate of degradation during the period did not affect the effectiveness of coir geotextiles as an erosion control measure. Soil samples from the surface (top soil) were periodically collected from the field and tested in the laboratory for nitrogen, phosphorous, potassium, and organic carbon. In all the plots, it was seen that loss in NPK and organic carbon was higher in CP than in the plots treated with coir geotextiles. Figure 9.7(a)–(f) shows the various methods for the laying of coir geotextiles on the sides of pond for embankment protection.

FIGURE 9.7 (a–f) Application of coir geotextiles in midland region of Kerala.

The specialty of this study was that the research was conducted with people's participation. Through joint experimentation with the people, participation was enhanced substantially contributing to a project's success. By participatory research, beneficiaries received training and experience in the design, implementation, and evaluation of experiments. In this way, their capacity for innovation was substantially increased.[15,33] The study revealed that the technology should enhance yields for farmers to accept soil conservation technologies. The increase in yield convinced the farmers of the value of soil conservation. If the yields have increased or costs have decreased, artificial incentives are not required. On the other hand, if yields have not been increased, no artificial incentive will make the adoption of the technology sustainable. The results showed that coir geotextiles can be effectively used for the embankment protection of the irrigation pond and also, through experimentation with people, people can visualize directly the impact of the introduced technology. In addition, it helped to develop innovation capacity both for individuals and communities.

9.12.3 APPLICATION OF A COIR GEOTEXTILE IN HIGH-LAND REGION OF KERALA

Balan[11] conducted studies in Pullangode plantation area, at Nilambur in Malapuram district, Kerala. The site chosen was in a rubber plantation area of 225 m above mean sea level on the foot of Western Ghats. Erosion of the fertile top soil is the vexing problem in the plantation area. It leads to the loss of mature plants and also causes ecological imbalance. Due to continuous erosion of hill slopes, planters had to abandon some of these areas. Rainfall is abundant with rain for nearly 10 months of the year and some dry spell in the rest of 2 months. The top fertile soil is silty sand. It varies in 10–15 cm thick with an organic content of 10.3% and pH 5.8. Outcrops of rock in the form of large boulders occur all along the slope (Fig. 9.8).

An area of 20 × 31 m along the slope was chosen for the study. Two types of coir mattings were used namely H_2M_8 (Type A—small aperture *as per the BIS specifications*) and H_2M_6 (Type B—larger aperture, *as per the BIS specifications*). The site slopes are 66° in the top portion and 49° in the bottom portion. Coir geotextiles of Type A were used at the top half portion and that of Type B were used in the bottom half portion. The area was seeded with very limited quantity of *penisettum* grass. Observation was carried out periodically after the onset of monsoon. Both local varieties of plants along with penisettum grass were fully developed and covered nearly 30% of the treated areas after 1 month. After 2 months, vegetation has grown up and covered nearly 55% of the area. Due to the visual establishment of vegetation, plantation owners decided to grow cashew plantation in the treated area. Cashew siblings were planted in the area at 3 m center to center through the geotextiles. After 7 months, vegetation had grown up by about

1–2 m high and covered 90% of the area. The average moisture content of the top soil was 17% after the monsoon. The tensile strength of the coir geotextiles was reduced by 56% only after 7 months of installation. This showed that coir geotextiles were capable of preventing surficial erosion of particles along the surface of slopes and helped in sedimentation of soil on previously exposed rock surfaces apparently through the action of a series of check dams.[50]

FIGURE 9.8 Laying of geotextiles on hill slope.

Anil[4] conducted an experimental study using multi-slot devices and it was shown that at 40% slope, the soil loss of a control plot was 12% greater than the soil loss of a plot treated with coir geotextiles. Of the total volume of 4.89 m^3 of water received as rainfall per square meter, 0.02 m^3 was absorbed by coir geotextile, 0.15 m^3 was lost as runoff and the remaining 4.72 m^3 was assumed to infiltrate into the soil. The plot size was 25 m long along the slope and 5 m wide.

Balan[10] concluded that when coir was embedded in soil, it retained 43% of its strength at a pH value of 11 and 60% at a pH value of 3. Degradation was faster between the pH values of 6 and 8. The strength retained was 34 and 26%, respectively. But the moisture absorption capacity of the geotextile was increased as degradation advanced. After 1 year, the moisture absorption of the degraded geotextile was 2.5 times that of the fresh sample. This property is of particular advantage in enhancing soil moisture and vegetation growth. Figure 9.9 shows the experimental set up in 40% sloping land in Konni research station.

FIGURE 9.9 Slope land cultivation on 30% slope with CG, CGC, and CP using multi-slot device.

At National Coir Research Management Institute (NCRMI) in Trivandrum and Konni Soil Conservation Research Station (KSCRI), Anil also conducted studies with different slopes (20, 30, 40, and 50% slopes) with three treatments: control plot (CP), coir geotextile (CG), and coir geotextile with crop (CGC). Coir geotextiles with different mesh openings were used in all these slopes to test the potential of coir geotextiles in reducing runoff and soil loss. Figure 9.10 shows the experimental set up for 40% slope at Konni research station. This enabled reduction in the number of risers while contour/terracing such lands, thereby increasing the cultivable area and reducing labor costs. Results indicated that geotextiles considerably reduced soil and water loss due to slopes. In his unpublished data for 50% slope, the control plots showed erosion of about 14% more than slopes treated with geotextiles. Control plot recorded a maximum organic matter loss of 19,300 kg/ha, 2500 kg/ha in the geotextile with crop plot and 1400 kg/ha in the geotextile alone plot. *Brinjal* (Eggplant) crop was used for the experimental study. Average yield of crop was 1.73 kg/plant in the second

FIGURE 9.10 Land cultivation on 40% sloping area with CG, CGC, and CP using multi-slot device.

to fourth months after planting and 1.08 kg/plant in the fifth and seventh months after planting. A yield of 0.48 kg/plant was obtained in the ninth month after planting. Figures 9.11 and 9.12 show the slope land cultivation with coir geotextile with eggplant crop. Hence it was illustrated that perennial vegetable crops can be successfully cultivated on slope lands of 50% slope. In plots with geotextile and crop, a total volume of 28,567.9 m^3/ha was received as rainfall, 108.24 m^3/ha was absorbed by coir geotextiles, and 1152.4 m^3/ha was lost as runoff, remaining water had infiltered into the soil due to the establishment of vegetation. Multi-slot devices were used to measure runoff and soil loss. The rate of runoff was on par in the geotextile alone plot, whereas runoff measured was 2047 m^3/ha in the control plot. Plots with geotextile alone took 9 months to stabilize, whereas plots with geotextile and crop took 11 months.

Vishnudas[67,72] conducted study in the Kumbazha watershed in the high-land region of Kerala, in Pathanamthitta District (9°51′20″N, 76°13′54″E). It is in the Western Ghat region where 50% of the total geographical area is covered by forests. Seventy-five percent of the population has agriculture as the main source of income. The climate is humid tropical with two monsoons. The temperature varies from 23 to 39°C. The relative humidity varies from

FIGURE 9.11 Eggplant cultivation on 40% slope and coir geotextile.

FIGURE 9.12 Eggplant cultivation on 40% slope with CGC.

62 to 100%. It has an undulating topography, and hills have steep gradients. The main crops raised in this region are paddy, rubber, coconut, and tapioca. In some regions, cashew, pineapple, and vegetables are cultivated. Tapioca has been the staple food of this region over the last two decades for the small scale and poor farmers living upstream. Due to soil erosion, presently this tuber crop is not recommended in this region.

In Kerala, terraces are made initially with contour bunds constructed with dry rubble packing of 75 cm to 1.2 m high on slopes. These are constructed in such a way that the lower bund is level with the mid-slope between two bunds, so that a natural terrace forms after a few years of cultivation. By this time, risers become deteriorated due to erosion and top soil is washed away. The maintenance of these structures is normally done by constructing earthen embankments on top of risers as shown in Figure 9.13.[66] But these structures breech during heavy rainfall. Hence the conventional method does not help to enhance vegetation growth or productivity from the slope. A small initial movement in this unstable slope can trigger further soil water movement resulting in soil erosion and landslides. Thus, cultivation in the slope land becomes difficult for poor farmers. Since slopes of more than 20% require physical measures for slope stabilization, in this study risers of the terraces were eliminated and slopes were treated with coir geotextiles.

FIGURE 9.13 Conventional bench terraces with dry rubble packed bunds and earthen bunds.

The site was identified in one of the farmer's plot to ensure acceptance and practice of soil conservation by the farmers. Three plots were selected for conducting the experiment.[66] The well-demarcated plots were first leveled and debris was removed. Slopes of the risers were shaped to 40% slope and terraces were leveled with a gradient of 3%. The size of the plot is 5 m along the slope and 25 m wide. The terrace width was kept at 5 m (Fig. 9.14). The soil is forest loam.

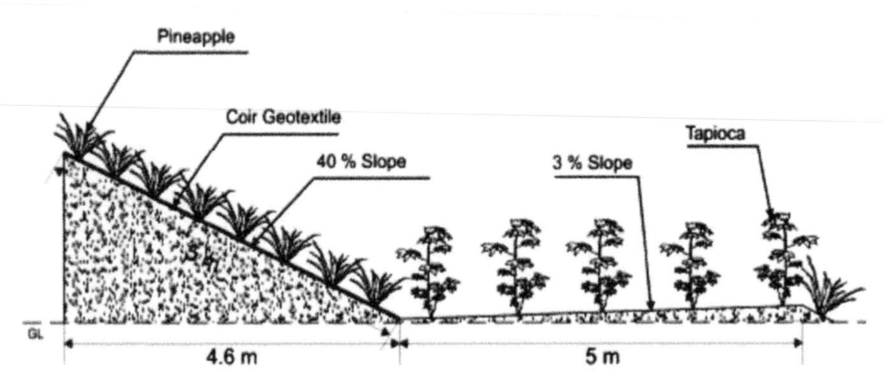

FIGURE 9.14 Cross-section: slope land cultivation, alternate conservation technique using coir geotextile.

Tapioca, being the staple food, was selected as a food crop for the terrace. The crop for the riser was selected based on the following factors: it should not cause any damage to the slope, the slope should not require maintenance for a minimum of 4 years, it should provide an income to the farmers, it should withstand drought, and be adequate for the highland region. Considering these factors, pineapple (*Ananas comosus*) was selected for slope land cultivation. This crop can withstand drought because of its ability to retain water in the leaves which is used during these periods. Tapioca was planted at a spacing of 1 m × 1 m. Care was taken while planting pineapple suckers not to disturb the weft and wrap of the geotextile as it may cause erosion. The soil moisture contents at 10, 20, 30, and 40 cm depth were measured in the treatment plot and control plot using thetaprobe.

The moisture content at 20, 30, and 40 cm were higher in CGC and CG than in CP. At 10 cm depth during the dry season, the moisture content in CGC was 32% more than that of CP. Hence even in the summer season, crops sustained well without irrigation. The higher percentage in moisture content in the treated plot is due to the mesh opening in the geotextile. It provides a large number of miniature porous check dams per square meter of the soil. It slows down and catches runoff so that sediment settles and water passes through the matting and infiltrates into the underlying soil. As the geotextiles degraded, they acted as mulch and conserved moisture for plant growth. The hygroscopic property of the geotextile also contributed to the increase in soil moisture. Hence from the experiment, it was observed that coir geotextile can retain moisture in the root zone and promote cultivation on the slope land.[72] Figure 9.15 shows the photograph of pineapple suckers in slope laid with coir geotextiles and tapioca in terraces.

FIGURE 9.15 Pineapple suckers planted on slope laid with coir geotextiles and tapioca on terraces.

9.13 COIR GEOTEXTILES FOR GULLY PLUGGING: CASE STUDY FROM KERALA

Contour bunds and gully plugs are important and are economically viable methods for soil and water conservation. With little economic assistance, even poor farmers can treat their field. A well-maintained gully plug creates a flat, fertile, and moist field, where high-value crops and trees can be grown. Examples of gully plugs are clearly visible in many developing countries such as India, Pakistan, Ethiopia, and so on. When gully plugs are done, agricultural production can be increased, and farmers can shift to high-value crops.

This study highlights the outcome and effectiveness of contour bunds and gully plugs using coir geotextiles soil and water conservation measures in Kerala, during 2001. Contour bund helps in reducing soil erosion and increasing water retention capacity of soil. As gullies expand, storm runoff increases, with declines in infiltration, groundwater, base flow, and evapotranspiration. In this process of channel entrenchment, groundwater table may be lowered, resulting in declining base flows and conversion of perennial streams to intermittent or ephemeral flow. Lands adjacent to entrenched gullies have reduced moisture available for plant growth, as the water table is lowered.[46]

Balan[11] conducted a study in Malanadu Development Society's Site at Kambammedu in Idukki District. Though the site is a rain shadow area, when rainfall occurs, the intensity of rain is very heavy causing severe erosion. The soil profile at the site consists of sandy silt for a depth of 1.25 m followed by disintegrated rock. Owing to the terrain and the type of surface soil, rainwater runs away very fast and percolation of water to the ground is very less and often leads to the formation of gullies in the area. In order to increase the percolation rate of water and thereby recharging the groundwater table, several measures were taken using coir geotextiles. Construction of contour bunds and gully plugging was done in this region (Figs. 9.16 and 9.17).

FIGURE 9.16 Coir geotextiles along contour bunds.

FIGURE 9.17 Coir geotextiles for gully plugging.

Coir mesh matting of H2M5 with BIS specification of coir geotextiles having an aerial density of 740 gsm with a mesh opening size of 11.7 × 13.2 mm made of conventionally spun vycome yarn was used along the benches and natural slopes to prevent surficial soil erosion and to activate the growth of locally available fodder grass. Same type of mesh matting was used for the prevention of soil erosion along a slope of 35°. In both cases, coir geotextiles of H2M5 have performed effectively. Vegetation has taken place within 3 weeks of laying the coir geotextiles and very thick vegetation has formed within 5 weeks in these areas. Thus, coir geotextiles made of conventionally spun vycome coir yarns are ideal for soil erosion control along benches with slopes of 35° in sandy silt soil. Again, H2M5 matting was used to protect the side of streams running in the same region.

After the application of coir geotextiles, the side erosion of the streams was controlled and it was observed that 85% protection has been obtained using coir geotextiles along the streams compared to the bare side of streams. Wide gullies of 3–4 m width at the top and 0.5–0.8 m wide at the bottom were formed owing to the specific terrain conditions. The height of gullies averaged to about 1.2–1.8 m with a length of 50–100 m. Gully plugging was done at an interval of 10 m along the gullies by using mesh matting of H2M3 having 875 gsm and 7.5 mm × 7.5 mm opening size. Beyond the first rainy season after the installation, the gullies on the upstream side had shown a siltation of 45 cm and that of the downstream side shown a siltation of 10 cm. Through this experimental study, it was proved that coir geotextiles can be effectively used for gully plugging.

9.14 COIR GEOTEXTILES FOR MULCHING: CASE STUDY FROM KERALA

Mulching has become an important practice in agriculture. Use of mulch depends on crop type, cropping system, and production environment, which enable to enhance crop growth by varying soil temperature and conserving soil moisture.[37] Plastics are most widespread mulching materials, and especially black polyethylene is used almost everywhere due to its cost and effectiveness. Removal and disposal of these plastics at the end of the growing season is costly and environmentally unsound.[19] Hence, several research studies have been conducted for the replacement of plastic by biodegradable materials.[61] Growth of weed in subtropical and tropical countries is very rapid especially in humid regions. In India, heavy rainfall enables growth of weeds which may later compete with crops for sunlight nutrients and moisture. This is mostly affected by orchard and horticultural crops.

Anil[1] conducted a field study in Kerala using coir geotextiles in mulching. Comparative study was carried out with mulching with rubberized coir, control plot, and mulching with transparent polythene. The soil is forest alluvium falling under the textural class forest loam. The experimental area was leveled, and the mulching materials were spread and fixed intact. The test crops used were okra and pineapple (Fig. 9.18). Soil samples were analyzed for moisture content before and after mulching. Observations were taken at fortnight intervals on growth characters and yield attributes. Observations on weed count were also recorded. Weed count was noted randomly from a quadrate of 25 cm^2. The height of plants was found to be significantly influenced by mulching throughout the growth period. All mulching materials were found beneficial in increasing the plant height. Rubberized coir and transparent polythene suppressed the weed growth up to a period of 8 months. Mulching with transparent polythene enhanced the soil temperature, whereas rubberized coir lowered soil temperature. All mulched treatments had influenced in increasing soil moisture.

For the okra crop, height of plant 12 WAS (*weeks after sowing*) was 62.5 and 65 cm in mulching with rubberized coir and transparent polythene, respectively. In the control plot, the height was 40 cm. At 10 and 12 WAS, the maximum number of leaves were 19 in plants treated with rubberized coir, which was on par with transparent polythene treatment of 18 leaves. In the control plot, it showed a minimum value of 10. Plants mulched with transparent polythene flowered earlier and this was on par with rubberized coir. The maximum fruit yield was 35.23 tons/ha from plots mulched with transparent polythene followed by 30.35 tons/ha in rubberized coir. Control plot had the least yield of 10.21 tons/ha. Weed intensity was higher in the control plot compared to almost negligible in all mulched plots.

FIGURE 9.18 Okra (left) and pineapple (right) under mulching.

For the pineapple crop, results revealed that mulching had no significant influence on length of fruits. Girth of fruits was significantly influenced by mulching. Rubberized coir was significantly superior to all other treatments and gave maximum girth of 37.9 cm against 33.5 cm for transparent polythene and 30.5 for the control plot. Rubberized coir recorded a maximum fruit weight of 1.4 kg, followed by 1.3 and 1.2 kg for transparent polythene and control plots, respectively. The intensity of weed growth was found to be least in rubberized (29 Nos./m^2), whereas it was 67.8 Nos./m^2 in transparent polythene and 391 Nos./m^2 in the control plot. Soil moisture retention was higher in rubberized coir (= 22.31%) compared to 18.65% in transparent polythene and 16.52% in the control plot. Transparent polythene recorded the highest temperature of 34°C throughout the crop period owing to the storing and warming up of heat energy in the soil and lowest temperature was 31°C in the rubberized coir. Thus, it is proved that rubberized coir can be effectively used for mulching in okra and pineapple crops with increased soil moisture and decreased weed intensity and soil temperature.

9.15 COIR GEOTEXTILES FOR RIVER BANK PROTECTION: CASE STUDIES FROM KERALA

During the monsoon season, the river banks are eroded and engulfed by the river itself. The river width will increase and simultaneously depth will decrease. Every year, the same situations arise in most of the rivers. The agricultural land has been decreasing through river bank erosion, which is a national disaster in all over the world. In Bangladesh, bamboo bandalling structures were constructed near the prone area of the Jamuna River for its erosion protection. The sediment available in the flowing water in the river is deposited downstream and between the constructed bamboo bandalling structures due to the comparatively lesser flow velocity behind and in between the bandals. This method was proved to be cost effective.[42]

During July 2002, Anil[2] conducted a field study near the Sugarcane Research Station, Thiruvalla, Kerala. Manimala river flows along its boundary and faces acute riverbank erosion during monsoon months. In this study, Cocologs (*scrap of coir fiber stuffed into a woven net of cylindrical in shape*) and coir geotextiles were used. Spurs were constructed using cocologs at an interval of 20 m. These spurs were projected into the river at an angle of 45°. The cocologs were placed horizontally in between the coconut logs piled to the riverbed (Fig. 9.19a–c). The number of cocologs varied from spurs to spurs depending on the condition and depth of riverbed; and the height of the spurs was maintained uniformly from the water level. The coir geotextiles of H2M6 was placed connecting all the spurs 5 m from the riverbank. Cocologs of 30 cm diameter and length ranging from 5 to 10 m were used. Cocologs were stacked one above the other, from riverbed upwards. After installation of spurs, in a span of 2 months, it was observed that a considerable amount of soil was deposited. A total sediment deposit of 19.64 kg/m^2 was attained during the season. Variation in water current increased with depth of water in river. This did not show any linear relationship. Maximum velocity of water current of 212 cm/s was observed corresponding to a depth of water in the river 621 cm. Silt suspensions in water were maximum when the depth of flowing water in the river was maximum. Silt suspension of 6 g/l was recorded when the depth of water was 621 cm. Again silt suspension in the flowing water did not show any linear relationship with sedimentation. Within a period of 14 months, natural vegetation was established on rear side of spurs predominant with reed and locally grown vegetation which consolidated the soil along the riverbank. After 30 months, the spurs constructed with cocologs and coconut logs started biodegradation by that time, and strong vegetation was established along riverbank capable of arresting the erosion from bank completely.

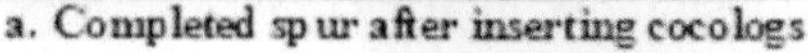

a. Completed spur after inserting cocologs

b. Vegetation establishment on the rear side of the spurs

c. Sediment deposition on the rear side of the spur

FIGURE 9.19 (a–c) River bank protection in Thiruvalla, Kerala.

9.16 SUMMARY

This chapter provides a detailed literature review on coir and its properties, and field applications of coir geotextiles in the area of soil and water conservation for crop production. The study reveals that coir is an ecologically friendly material, which can be used worldwide for soil and water conservation as an environmentally sound material and an economically viable solution for farmers especially in developing countries where coir is abundantly available.

The experiment in the highland region aimed at providing an alternative for bench terraces to stabilize the slopes for cultivation. From the results, it is evident that the slopes treated with geotextile have the highest moisture retention capacity followed by geotextiles alone and then the control plot. The application of geotextile on slope land increases moisture availability in the soil and enhances infiltration. Since the slopes were stabilized with the application of geotextiles, sediment deposits on the terraces due to erosion were minimized and hence cultivation is possible both on the slopes and the terraces. As the poor and marginal farmers occupy the highland region, this method provides an economically viable option for income generation and food security along with slope stabilization. This method can also be applied to wasteland cultivation in the highland region. The result of the participatory research carried out in the Amachal watershed proved that science and technology are socially integrated. It has been shown that actively involving people in experiments enhances their ability in selecting new technologies.

In the study conducted for mulching with rubberized coir products, the moisture content of the soil had substantially increased as well as soil temperature had reduced significantly. Unlike other conventional soil and water conservation structures, slope lands treated with coir geotextiles itself provides land for cultivation. Thus, the land used for conservation measures can also be effectively utilized. Therefore, coir geotextile provides a livelihood and an important source of food security for many farmers in Kerala. Hence these technologies can be adopted elsewhere in the world where similar geographical and climatic features prevail and where coir is economically viable or abundantly available.

ACKNOWLEDGMENT

The authors acknowledge the help and guidance given for writing this chapter by Dr. K. Balan, Retired Professor, Kerala University, who had initiated the study in Kerala using coir geotextiles in the early 1990s. The first author gratefully acknowledges NUFFIC, the Netherlands, for the financial assistance given for conducting the field study in Kerala.

KEYWORDS

- **biodegradation**
- **BIS code**
- **coir**
- **coir industry**
- **coir geotextiles**

- **riverbank erosion**
- **soil and water conservation**
- **soil erosion**
- **soil**
- **spurs**
- **stream bank erosion**

REFERENCES

1. Anil, K. R.; George, R.; Rani, R. In *Application of Coir Geotextiles as Mulch,* Proceedings of the 42nd Conference and Expo of the International Erosion Control Association, Orlando, FL, USA, 2011, pp 231–238.
2. Anil, K. R.; George, R.; Rani, R. In *Eco-friendly River Bank Protection Technologies Using Coir Geotextiles,* Proceedings of the 42nd Conference and Expo of the International Erosion Control Association, Orlando, FL, USA, 2011, pp 276–283.
3. Anil, K. R. *Use of Coir Geotextiles for Soil and Water Conservation at Varying Slopes*; Technical Report to Coir Board: India, Kerala Agriculture University, Kerala, India, 2004, 50 p.
4. Anil, K. R. *The Study on Use of Geotextile for Soil and Water Conservation Under Varying Slopes*; Technical Report to Coir Board: Kerala Agriculture University, Kerala, India, 2006, 62 p.
5. Anthony Toby O'geen, U. C.; Schwankl, L. J. *Understanding Soil Erosion in Irrigated Agriculture*; Publication 8196, Division of Agriculture and Natural Resources, University of California, 2006, 43 p, ISBN-13-978-1-60107-389-1.
6. APCC. *Coconut Statistical Yearbook*; Asia and Pacific Coconut Community (APCC), Jakarta, Indonesia, 2007; 87 p.
7. Asiah, A. M.; Khanif, R. M. Y.; Marziah, M.; Shaharuddin, M. Physical and Chemical Properties of Coconut Coir Dust and Oil Palm Empty Fruit Bunch and the Growth of Hybrid Heat Tolerant Cauliflower Plant. *Pertanika J. Trop. Agric. Sci.* **2004,** *27*(2), 121–133.
8. ASTM D4439-15a. *Standard Terminology for Geosynthetics*. ASTM International, West Conshohocken, PA, 2015; 101 p.
9. Ayyar Ramanatha, T. S.; Nair, R.; Nair, B. N. *Comprehensive Reference Book on Coir Geotextiles*; Center for Development of Coir Technology, (CDOCT): Trivandrum, India, 2002; 82 p.
10. Balan, K. Studies on Engineering Behavior and Uses of Geotextiles with Natural Fibers. Ph.D Thesis, Indian Institute of Technology, Delhi, India, 1995, 52 p.
11. Balan, K. Case Studies on the Civil Engineering Applications of Coir Geotextiles. In *Case Histories of Geosynthetics in Infrastructure*; Mathur, G. N., Rao, G. V., Chawla, A. S., Eds.; CBIP Publications: New Delhi, India, 2003; pp 172–176.

12. Balan, K.; Rao, G. V. In *Durability of Coir Yarn for Use in Geomeshes,* Proceedings of the International Seminar and Techno Meet on Environmental Geotechnology with Geosynthetics, ASEG/CBIP Publications, New Delhi, 1996, 112 p.
13. Belas, A. K. Uses of Coir Fiber, Its Products and Implementation of Geo-coir in Bangladesh. *Daffodil Int. Univ. J. Sci. Technol.* **2007,** *2*(2), 33–38.
14. Benediktas, J.; Genovaite, J.; Fullen, M. A. A Field Experiment on the Use of Biogeotextiles for the Conservation of Sand-Dunes of the Baltic Coast in Lithuania. *Hung. Geogr. Bull.* **2012,** *61*(1), 3–17.
15. Bunch, R.; Lopez, G. *Soil Recuperation in Central America: Sustaining Innovation after Intervention*; Gatekeeper Series No. 55. International Institute for Environment and Development: London, 1999; 98 p.
16. Cammack, A. In *A Role for Coir Fiber Geofabrics in Soil Stabilization and Erosion Control,* Proceedings of the Workshop on Coir Geogrids and Geofabrics in Civil Engineering Practice, Coimbatore, India, 1988, pp 28–31.
17. Coir Board. *Fifty Seventh Annual Report of Coir Board (2010–2011)*; Ministry of Micro, Small and Medium Enterprises, Government of India, 2011, 88 p.
18. *Coir Board*; Sixtieth Annual Report of Coir Board (2013–2014). Ministry of Micro, Small and Medium Enterprises, Government of India, 2013, 79 p.
19. David, W. Evaluation of Biodegradable Mulches for Production of Warm-season Vegetable Crops. *Can. J. Plant Sci.* **2010,** *90*(5), 737–743.
20. Dvořák, J. Gully Control. Chapter 7. In *Developments in Soil Science*; *Dvorak, J., Novak, L.,* Eds.; Elsevier: The Netherlands; 1994; pp 233–237.
21. English, B. Filters, Sorbents and Geotextiles. In *Paper and Composites from Agrobased Resources*; CRC Press/Lewis Publishers: Boca Raton/FL, 1997; pp 403–425.
22. FAO. *Agricultural Statistics*; Food and Agriculture Organization of the United Nations Statistics Division (FAO), 2011.
23. Fullen, M. A.; Catt, J. A. *Soil Management: Problems and Solutions*; Arnold Publishers: London, 2004; 83 p, ISBN: 0-340-80711-3.
24. Geoff, C. *Coir Dust A Proven Alternative to Peat;* Horticultural Services: Grose Vale, NSW, 2011; 67 p.
25. George, J. K.; Sarma, U. S. In *Retted (White) Coir Fiber Nettings: The Ideal Choice as Geotextiles for Soil Erosion Control,* Proceedings of the IECA's 28th Annual Conference, Tennessee, USA, February 25–28, 1997, pp 67–76.
26. Girisha, C.; Sanjeevamurthy, S.; Gunti, R. Sisal/Coconut Coir Natural Fibers—Epoxy Composites: Water Absorption and Mechanical Properties. *Int. J. Eng. Innov. Technol.* **2012,** *2*(3), 166–170.
27. *Government of India (GOI)*; Annual Report, 2013–2014. Ministry of Micro, Small and Medium Enterprises Publication, India, 2013, 113 p.
28. *Government of Kerala*; Agriculture Statistics, Government of Kerala, India, 2003, 118 p.
29. Hejazi, S. M.; Zadeh, M. S.; Abtahi, S. M.; Zadhoush, A. A Simple Review of Soil Reinforcement by Using Natural and Synthetic Fibers. *Constr. Build. Mater.* **2012,** *30*, 100–116.
30. Hou, X. Y.; Wang, F. X.; Han, J. J.; Kang, S. Z.; Fena, S. H. Duration of Plastic Mulch for Potato Growth Under Drip Irrigation in an Arid Region of Northwest China. *Agric. For. Meteorol.* **2010,** *150*, 115–121.
31. Jalota, S. K.; Prihar, S. S. Bare-Soil Evaporation in Relation to Tillage. In *Advances in Soil Science*; Steward, B. A., Ed.; Springer-Verlag: New York, 1999, pp 187–216.

32. Jim, R. *Soil Erosion—Causes and Effects*; Fact Sheet, Ministry of Agriculture, Food and Rural Affairs, Queen's Printer for Ontario, Ontario, 2013; 5 p.
33. Johnson, N.; Lilja, N.; Ashby, J. Measuring the Impact of User Participation in Agricultural and Natural Resource Management Research. *Agric. Syst.* **2003,** *78*, 287–306.
34. Kertész, Á. The Global Problem of Land Degradation and Desertification. *Hung. Geogr. Bull.* **2009,** *58*(1), 57–59.
35. Kothari, V. K.; Patel P. C. In *Theoretical Prediction of Tensile Behavior of Nonwoven Geotextiles*, Proceedings of the Geosynthetics Asia. CBIP Publications, Bangalore, 1997, 223 p.
36. *Kuttanad Water Balance Study—Draft Final Report*; Kingdom of the Netherlands and Ministry of Foreign Affairs, 1999, 111 p.
37. Lamont, W. J. Plastics: Modifying the Microclimate for the Production of Vegetable Crops. *Hort. Technol.* **2005,** *15*(3), 477–481.
38. Lee, M. H. In *Soil Bioengineering—The Korean Experience*, Proceedings of the India International Coir Fair and Seminar, Coir Board, Kerala, India, 2005, pp 52–53.
39. Lekha, K. R.; Kavitha, V. Coir Geotextile Reinforced Clay Dykes for Drainage of Low-Lying Areas. *Geotext. Geomembr.* **2006,** *24*, 38–51.
40. Madarász, B.; Bádonyi, K.; Csepinszky, B.; Mika, J.; Kertész, Á. Conservation Tillage for Rational Water Management and Soil Conservation. *Hung. Geogr. Bull.* **2011,** *6*(2), 117–133.
41. Manorama, K. C. Morphological, Physical and Chemical Characterization of the Soils of North Kuttanad. Ph.D Thesis, Department of Soil Science and Agricultural Chemistry, College of Horticulture, Vellanikkara, Kerala, 1997.
42. Md, L. R.; Basak, B. C.; Md, S. O. In *Low Cost Techniques to Recover Agricultural Land through River Bank Erosion Protection*, ICID 21st International Congress on Irrigation and Drainage, Tehran, Iran, 2011, 65 p.
43. Michael, S. Erosion and Crop Yield. In *Encyclopedia of Soil Science*; Marcel Dekker: New York, NY, 2003; 87 p.
44. Mittal, S.; Singh, R. R. Improvement of Local Subgrade Soil for Road Construction by the Use of Coconut Coir Fiber. *Int. J. Res. Eng. Technol.* **2014,** *3*(5), 212.
45. Morgan, R. P. C. *Soil Erosion and Conservation*, 3rd Edition; Blackwell Publishing: Oxford, 2006; 237 p.
46. National Engineering Handbook Part 654. *Technical Gullies and Their Control*, Technical Supplement, 14P, 210-VI-NEH, 2007.
47. Oosthuizen, D.; Kruger, D. In *The Use of Sisal Fiber as Natural Geotextile to Control Erosion*, Proceedings of the Fifth International Conference on Geotextiles, Geomembranes and Related Products, Singapore, 1994, 53 p.
48. Pillai, S. M. In *Eco-friendly Plastics/Remedial Measures for Environment Sustainability*, Fourth International R&D Conference for Water and Energy for 21st Century, Coir Board Government of India, Aurangabad, India, January 28–31, 2003.
49. Powlson, D. S.; Gregory, P. J.; Whalley, W. R.; Quinton, J. N.; Hopkins, D. W.; Whitmore, A. P.; Hirsch, P. R.; Goulding, K. W. T. Soil Management in Relation to Sustainable Agriculture and Ecosystem Services. *Food Policy* **2011,** *36*, S72–S87.
50. Rao, G. V.; Balan, K. Coir Geotextiles—A Perspective. In *Coir Geotextiles—Emerging Trends*; Rao, G. V., Balan, K., Eds.; Kerala State Coir Corporation: Kerala, India, 2000; 32 p.
51. Rao, G. V.; Dutta, R. K. Characterization of Tensile Strength Behavior of Coir Products. *Electron. J. Geotech. Eng.* **2005,** *10*, 80–82.

52. Sarkar, S.; Pramanik, M.; Goswami, S. B. Soil Temperature, Water Use and Yield of Yellow Sarson (*Brassica napus*, var. *glauca*) in Relation to Tillage Intensity and Mulch Management Under Rainfed Lowland Ecosystem in Eastern India. *Soil Tillage Res.* **2007,** *93*, 94–101.
53. Sarma, U. S.; Jose A. C. In *Application of a Coir Geotextile Reinforced Mud Wall in An Area Below Sea Level at Kuttanad, Kerala*, 39th Annual Conference and Expo of the International Erosion Control Association—Environmental Connection: Drive Your Future, Orlando, FL, USA, 2008, pp 355–375.
54. Schurholz, H. In *Use of Woven Coir Geotextiles in Europe*, Proceedings of the International Conference on Coir Geotextiles Conference, USA, 1991, 251 p.
55. Schurholz, H. In *Use of Woven Coir Geotextiles in Europe*, Proceedings of the United Kingdom Coir Geotextile Seminar, West Midlands, 1992, 221 p.
56. Shengtao, X.; Lei, Z.; Neil, B. M.; Junzhen, M.; Qin Chen, J. Effect of Synthetic and Natural Water Absorbing Soil Amendment Soil Physical Properties Under Potato Production in a Semi-arid Region. *Soil Tillage Res.* **2015,** *148*, 31–39.
57. Sooryamoorthy, R.; Antony, P. *Managing Water and Water Users: Experiences from Kerala*; University Press of America: Lanham, MD, 2003; 112 p.
58. Sotir, R. B.; Simms, A. P. In *North American Coir Geotextiles Experiences in Combination with Soil Bioengineering*, Proceedings of the North American Coir Seminar by Coir Board, India, 1991, pp 131–133.
59. Stuti, M.; Sharmab, A. K.; Jain, P. K.; Kumar, R. Review on Stabilization of Soil Using Coir Fiber. *Int. J. Eng. Res.* **2015,** *4*(6), 296–299.
60. Sutherland, R. A.; Menard, T.; Perry, J. L. The Influence of Rolled Erosion Control Systems on Soil Moisture Content and Biomass Production: Part II—A Greenhouse Experiment. *Land Degrad. Dev.* **1998,** *9*, 217–223.
61. Tapani, H.; Paulina, P.; Korpela, A.; Jukka, A. Feasibility of Paper Mulches in Crop Production—A Review. *Agric. Food Sci.* **2014,** *23*, 60–79.
62. The Bombay Textile Research Association. *Hand Book of Geotextiles (Special Publication 8.2.34)*; Bombay Textile Research Association (BTRA): Bombay, 2012; 187 p, ISBN 978-81-7674-132-3.
63. Thomas, P. M. *Problems and Prospects of Paddy Cultivation in Kuttanad Region;* Project Report, Kerala Research Programme on Local Level Development (KRPLLD), Kerala, India, 2002, 156 p.
64. Tolk, J. A.; Howell, T. A.; Evett, S. R. Effect of Mulch, Irrigation, and Soil Type on Water Use and Yield of Maize. *Soil Tillage Res.* **1999,** *50*(2), 137–147.
65. Venkatappa, R. G.; Balan, K.; Dutta, R. K. Characterization of Natural Geotextiles. *Int. J. Geotech. Eng.* **2009,** *3*(2), 261–270.
66. Vishnudas, S.; Savenije, H. H. G.; Van der Zaag, P.; Anil, K. R.; Balan, K. Experiment Study Using Coir Geotextile in Watershed Management. *Hydrol. Earth Syst. Sci.* **2005,** *2*, 2327–2348.
67. Vishnudas, S. Sustainable Watershed Management Illusion or Reality: A Case of Kerala State in India. PhD Thesis. Eburon Academic Publishers, The Netherlands, 2006, 212 p, ISBN-13-978-90-5972-154-8.
68. Vishnudas, S.; Savenije, H. H. G.; Van der Zaag, P.; Anil, K. R.; Balan, K. The Protective and Attractive Covering of a Vegetated Embankment using Coir Geotextiles. *Hydrol. Earth Syst. Sci.* **2006,** *10*, 565–574.

69. Vishnudas, S.; Savenije, H. H. G.; Van der Zaag, P.; Balan, K.; Anil, K. R. In *Technology Option Using Coir Geotextile for Sustainable Land and Water Management*, Proceedings of the ICSTEP Conference, Vol. 1, India, 2006, pp 887–890.
70. Vishnudas, S.; Savenije, H. H. G.; Van der Zaag, P.; Anil, K. R.; Balan, K. Participatory Research Using Coir Geotextiles in Watershed Management—A Case Study in South India. *Phys. Chem. Earth* **2008,** *33*(1–2), 41–47.
71. Vishnudas, S.; Savenije, H. H. G.; Van der Zaag, P.; Kumar, A.; Anil, K. R. Sustainability Analysis of Two Participatory Watershed Projects in Kerala. *Phys. Chem. Earth* **2008,** *33*(1–2), 1–12.
72. Vishnudas, S.; Savenije, H. H. G.; Van der Zaag, P.; Anil, K. R. Coir Geotextile for Slope Stabilization and Cultivation—A Case Study in a Highland Region of Kerala, South India. *Phys. Chem. Earth* **2012,** *47*, 135–138.
73. White, W. In *Flood Plain Dynamics and Available Erosion Control Techniques*, Proceedings of the North American Coir Seminar, Coir Board, India, 1991.
74. Ziegler, A. D.; Sutherland, A.; Tran, L. T. Influence of Rolled Erosion Control Systems on Temporal Rain Splash Response—A Laboratory Rainfall Simulation Experiment. *Land Degrad. Dev.* **1997,** *8*, 139–157.
75. Zribi, W.; Aragüés, R.; Medina E.; Faci, J. M. Efficiency of Inorganic and Organic Mulching Materials for Soil Evaporation Control. *Soil Tillage Res.* **2015,** *148*, 40–45.

PART IV

Crop Management for Non-Conventional Use

CHAPTER 10

PRACTICES IN NITROGEN FERTILIZATION OF WHEAT (*TRITICUM AESTIVUM* L.): EFFECTS ON THE DISTRIBUTION OF PROTEIN SUB-FRACTIONS, AMINO ACIDS, AND STARCH CHARACTERISTICS

DIVYA JAIN[1], BAVITA ASTHIR[1,*], and DEEPAK KUMAR VERMA[2,*]

[1]*Department of Biochemistry, College of Basic Sciences and Humanities, Punjab Agriculture University, Ludhiana 141004, Punjab, India*

[2]*Department of Agricultural and Food Engineering, Indian Institute of Technology, Kharagpur 721302, West Bengal, India*

[*]*Corresponding author. E-mail: b.asthir@rediffmail.com; basthir@pau.edu; deepak.verma@agfe.iitkgp.ernet.in; rajadkv@rediffmail.com*

CONTENTS

ABSTRACT

Wheat grain proteins play a crucial role in forming the strong, cohesive dough that retains gas and produces light baked products. Glutens are important in baking quality because of their impact on water absorption capacity of the dough. Their elasticity and extensibility can affect wheat flour quality extensively. Albumin and globulin probably have critical role in flour quality. Starch serves as a source of carbon during yeast fermentation in bread making, in setting of the bread loaf, and in retrogradation during storage. Reports indicate that different N rates can cause changes in the total amount of different grain proteins. None of the past studies have reported the effects of different doses of nitrogen application on grain quality parameters such as starch (amylose and amylopectin) and proteins (albumin, globulin, gluten, and prolamin) sub-components of wheat.

10.1 INTRODUCTION

Wheat (*Triticum aestivum* L.) is an important cereal crop next to rice- and maize-cultivated world. It contributes to various end products used for improving human diets. It is mainly cultivated to obtain proteins, starch content, and yield, which need maximum amounts of nitrogenous fertilizers to receive its higher growth and productivity.

Wheat crop is important among the cereals, known for its excellent source of nutrition in terms of carbohydrates and proteins. Globally, wheat is being cultivated over an area of 227 million ha with a production of 610 million tons. In India, it has a total production of 78.4 million tons from 28.5 million ha, which is 12% of the world food production.[112] While in Punjab, it is cultivated over an area of 3.5 million ha with a production of 16,472 million tons. The most effective environmental factor for wheat grain quality is nitrogen fertilization.

Inadequate supply of available N frequently results in slow growth, depressed protein levels, poor yield of low-quality produce, and inefficient water use. Thus, nitrogen is the most important constituent of plant proteins and is required throughout the crop growth period from vegetative stage to subsequent harvesting. For N, ultimately proteins are regarded as a major sink which is available in the form of seeds.[90] The biological yield in the form of grain yield can be enhanced if N uptake and its utilization efficiency is increased. Reports in the literature indicate that nitrogen shortage is one of the main constraints limiting the productivity of major crops such as wheat.[11] N fertilization is not only important for nutritional improvement of wheat by increasing protein content, but also for gluten strength and starch composition.[154] It was also reported that amino transferases such as glutamate oxalo-acetate transaminase (GOT) and glutamate pyruvate transaminase (GPT) are responsible for biosynthesis of nitrogen-carrying amino acids and are responsive to N supply. Very few studies have been conducted to assess the effect of different doses of N on grain quality parameters of wheat.

This chapter reviews literature on "Nitrogen fertilization in wheat (*T. aestivum* L.): Effects on the distribution of protein sub-fractions, amino acids; and starch characteristics".

10.2 IMPORTANCE OF NITROGEN FERTILIZATION

Nitrogen fertilization has been a powerful tool in increasing grain yield especially for cereals such as wheat.[64–67,69,129] Although nitrogen is taken up slowly by the plants from natural sources, yet the majority of nitrogen used by plants comes from inorganic fertilizer. Most inorganic fertilizers contain nitrogen that is available in the basic forms, such as urea, ammonium (NH_4^+), and nitrate (NO_3^-). Fertilizers can be formulated to contain varying proportions of each of these forms by choosing different ingredients.[15] Since nitrogen is the main component in many biological compounds associated with crop yield capacity,[32] its deficiency constitutes one of the major limiting factors for cereals. However, overuse of N fertilizers is economically costly to the low-value crops in terms of both the price of N fertilizer and the potential loss of yield.[75] Also, there is extensive concern in relation to the N that is not used by the plant, but lost by leaching of NO_3^- or de-nitrification from the soil and loss of ammonia to the atmosphere, all of which can have deleterious environmental effects and health hazards.[12] Furthermore, farmers are facing increasing economic pressures with the rising fossil-fuel costs required for the production of N fertilizers. To address both economic and ecological issues, plant breeders are identifying cultivars that can efficiently utilize N.

Burgeoning population of world needs crop genotypes responding to higher nitrogen and showing direct relationship to yield with use of nitrogen inputs, that is, high nitrogen-responsive genotypes. It is, therefore, imperative to identify the limiting steps in the control of nitrogen-use efficiency (NUE): N uptake and N assimilation during growth and development.

10.3 EFFECTS OF NITROGEN FERTILIZATION ON WHEAT

Wheat life cycle consists of two phases[66]: In the first phase (vegetative phase), young developing roots and leaves act as sink organs for the assimilation of inorganic N into amino acids.[65] These amino acids are further used for the synthesis of enzymes and proteins that help in building up plant architecture and the different components of the photosynthetic machinery. At later stage of plant development, generally starting after flowering, the remobilization of the N accumulated by these parts behave as a source of N by providing amino acids released from protein hydrolysis that are subsequently exported to reproductive and storage organs. However, it is still not clear whether it is the plant N availability or storage protein synthesis that limits the determination of grain yield in general and grain deposition.[103] In general, whole plant physiological studies[67] combined with ^{15}N-labeled experiments preferably performed in the field should be undertaken.[56] The results of the experiments will allow the identification of some of the key molecular and biochemical traits and NUE components that govern the adaptation to N-depleted environments before and after the grain filling in lines or hybrids exhibiting variable capacities to take up and utilize N.[83,101] The traditional concept of N-cycling in plant–soil systems considers that organic N must first be converted into inorganic N forms by soil microorganisms, before becoming available to a plant via its roots.[165,166] Thus, the assimilation and metabolism of inorganic N in plants is a complex trait, playing a pivotal role as N transport compounds in plants.[93]

For instance, NO_3^--use efficiency depends on uptake, reduction, translocation, temporary storage in vacuoles, energy availability, and so on. Plants are able to accumulate NO_3^- to high concentration, the majority of which is present in the vacuole by NO_3^{2-}/H^+ exchanger.[36] Since the assimilation of inorganic nitrogen into organic form has marked effect on plant productivity, biomass, and crop yield, plants have the capacity to grow successfully in a wide range of available nitrogen application. Several studies depicted the yield of an agricultural crop that strongly depends on the supply of mineral nutrients, particularly, nitrogen, which is a constituent of amino acids, which are required to synthesize proteins and other related compounds.[140]

Thus, N management at the whole plant level, the arbitrary separation of the plant life cycle into two phases,[106] remains obscure, since it is well known that N recycling can occur before flowering for the synthesis of new proteins in developing organs.[91] In addition, during the assimilatory phase when NH^{4+} incorporates into free amino acids, it is subjected to a high turnover as a result of photo-respiratory activity and it needs to be immediately re-assimilated into glutamine and glutamate.[123] Several past investigators have revealed that there is a significant genetic variability in several steps of nitrogen metabolism including nitrogen absorption, assimilation, and recycling in crops.[105]

10.3.1 NITROGEN SOURCES AVAILABLE TO CROPS

Nitrogen is an important nutrient required for crop development and is a major yield-determining factor.[49,50] Inadequate supply of available N frequently results in slow growth, depressed protein levels, poor yield of low-quality produce, and inefficient water use. However, its conversion to usable form by plant is an expensive process. Plants are able to take up N in the form of nitrate, ammonium ions, and amino acids[92] or obtain it through symbiotic fixation of nitrogen-fixing bacteria.[11] The major sink for N is ultimately proteins, which are available for human consumption in the form of seeds.[90] The productivity of seed in the form of grain yield can be enhanced if N uptake and its utilization efficiency are increased.

Nitrogen fertilizers are highly soluble and once applied to the soil may be lost from the soil–plant system or become unavailable to the plants due to the processes of leaching, NH_4^+ volatilization, denitrification, and immobilization. Thus, nitrogen shortage is one of the main constraints that limit the productivity of wheat and other cereals.[11] Therefore, the preferred form of N is taken up that depends on plant adaptation to soil conditions. Generally, plant adapted to low pH and reducing soils tend to take up NH_4^+ or amino acids, whereas plants adapted to higher pH and more aerobic soils prefer NO_3^-.[100] Regardless of these forms, NO_3^- and NH_4^+ in which N is supplied to plants, the microbial process of de-nitrification ensures the conversion of NH_4 into NO_3^- is the most abundant form of nitrogen available to plants.[35] Nitrogen fertilizers are effectively implemented in agriculture, thus stimulating vital processes in plants. Thus, nitrogen availability must be managed throughout the growing period of crop.

10.3.2 CARBON–NITROGEN METABOLIZING ENZYMES

The N in asparagine and glutamine can be transferred into a wide variety of amino acids, nucleic acid, ureides (R–CO–NH–CO–NH_2 or R–CO–NH–CO–NH–CO–R) and polyamines.[9,52] Key reactions in the metabolism of glutamate are the transfer of the α-amino group to a range of 2-oxo acid acceptors to form amino acids. The reversible reactions are carried out by pyridoxal phosphate-containing enzymes termed aminotransferases also known as transaminases, which can exhibit wide preferences for both the amino acid and oxo-acid substrates. Thus, nitrogen assimilation is a vital process controlling plant growth and development through inorganic nitrogen which is assimilated into amino acids. GOT and GPT might play an important role in the synthesis of amino acids and proteins.[84] Lopes et al.[96] studied the wheat nitrogen metabolism during grain filling, comparative roles of glumes, flag leaf, and grains in all three organs and no decrease in transaminase was detected. In literature, activity of GOT and GPT drastically appear to play a significant role in the de novo synthesis of amino acids and to facilitate the transfer of C from amino acids into carbohydrates.

10.3.2.1 GLUTAMATE OXALOACETATE TRANSAMINASE

GOT is a pyridoxal phosphate-dependent enzyme present in cytoplasmic and mitochondrial forms. GOT is the main enzyme involved in NH_4^+ assimilation.[104] This enzyme is also known as aspartate aminotransferase and is one of the most active enzymes in the cell. It exists in mitochondrial and cytosolic variants. The metabolic importance of this enzyme is that it brings about a free exchange of amino groups between glutamate (which is the most common amino acid) and aspartate which is a second major amino acid pool. Glutamate and aspartate are each required for separate but essential steps in the urea cycle. The free movement of nitrogen between the glutamate and aspartate pools is an important balancing process that is vital for normal cellular metabolism. The urea cycle consists of five reactions: two mitochondrial and three cytosolic (Table 10.1). The cycle converts two amino groups, one from NH_4^+ and one from ASP, and a carbon atom from HCO_3^-, into the relatively non-toxic excretion product urea at the cost of four "high-energy" phosphate bonds (three ATP hydrolyzed to two ADP and one AMP). Ornithine is the carrier of these carbon and nitrogen atoms.

This reaction is close to equilibrium in both the cytosol and the mitochondrial compartments (Fig. 10.1). It forms an integral part of the malate–aspartate shuttle which is effectively responsible for the "transport" of nicotinamide adenine dinucleotide (NADH) across the inner mitochondrial membrane. GOT might be one of the key enzymes in N-metabolism and participate in the consumption of glutamic acid for metabolism. It has been reported that GOT supplies a sole source of glutamate as the donor of amino group and this enzyme further catalyzes transfer of the amide group to oxalo-glutarate to form glutamate.[47,51,163] Earlier reports in crop plants have shown that this enzyme is responsive to N supply.[148] The activity of GOT enzyme was increased with increasing nitrogen doses in different crops such as maize, tomato, and wheat.[137]

TABLE 10.1 Reactions of Urea Cycle.

Step	Reactants	Products	Catalyzed by	Location
1	$NH_3 + HCO_3^- + 2ATP$	Carbamoyl phosphate + 2ADP + P_i	CPS1	Mitochondria
2	Carbamoyl phosphate + ornithine	Citrulline + P_i	OTC	Mitochondria
3	Citrulline + aspartate + ATP	Argininosuccinate + AMP + PP_i	ASS	Cytosol
4	Argininosuccinate	Arg + fumarate	ASL	Cytosol
5	Arg + H_2O	Ornithine + urea	ARG1	Cytosol

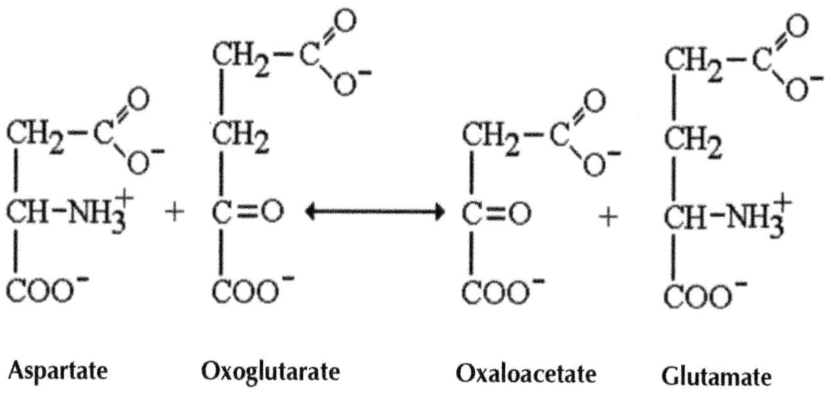

Aspartate + Oxoglutarate ⟷ Oxaloacetate + Glutamate

FIGURE 10.1 The reaction in cytosol and the mitochondrial compartments.

10.3.2.2 GLUTAMATE PYRUVATE TRANSAMINASE

GPT enzyme is also invovled in nitrogen metabolism. This active enzyme is also known as alanine amino transferases (AlaAT). GPT also occupies a critical position at the junction between carbon (2-oxoglutarate) and N (glutamate) metabolism and participates in balancing of the cellular levels of three major components: the NH_4^+ ions, 2-oxoglutarate, and glutamate.[13,41,110] Numerous studies have been carried out to indicate the role of GPT enzyme in $NH4^+$ assimilation and remobilization, which are tightly interrelated processes during the plant growth and development.[110] The possible relationship of the GPT activity with the rest of the enzyme involved in NH_4^+ assimilation clearly indicates its importance in determining N status of the plant.[79] In the case of roots and internodes, the response of GPT to N application was quite similar to GOT. GPT activity also increases at 15 DPA and thereafter decreases during 40 DPA. These findings were supported by ^{15}N-labeling experiments performed in the field, where GPT activity was decreased with ageing of plant.[83]

GPT maintains a nitrogen–carbon balance through the interaction between AlaAT and pyruvate (Fig. 10.2). The rapid decay of GPT is an important feature of senescence of leaves in dark. GPT is an important regulatory enzyme. It is logical that its activity may be affected more than that of GOT. Since the activity of transaminases is stimulated under conditions that are deleterious to protein synthesis, it was postulated that these enzymes may be acting in the direction of deamination to provide amino acids, especially glutamic acid or aspartic acid for the common N-pool. In addition, all NH_4^+ assimilation enzyme activities were higher in roots than in shoots, suggesting that their main role was photorespiration, which constitutes the main source of inorganic-N in the leaves. On the other hand, it also indicated that amino acids are first transaminated in roots, then transformed into other amino acids, and finally synthesized into proteins after transport into shoots.[99] Mo et al.[113] showed that NH_4 at high concentrations significantly increased the GPT activity of roots.

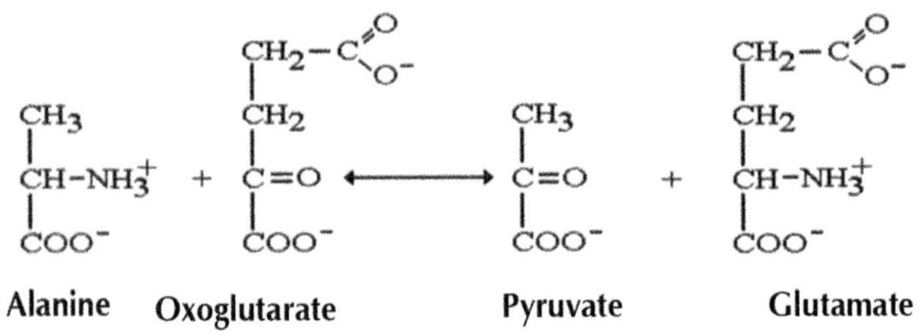

Alanine Oxoglutarate Pyruvate Glutamate

FIGURE 10.2 Interaction between AlaAT and pyruvate.

GPT play an important role in the synthesis of amino acids and proteins.[84] It is therefore probable that the resulting amino acids were substrates for GPT, thus inducing its enzymatic activities. The major role of GPT enzymes in leaves during ammonia assimilation is to replenish organic nitrogen which may get lost through photorespiration indicating GPT maximum activity was observed in leaves.[99]

10.3.2.3 ALKALINE INORGANIC PYROPHOSPHATASE

Alkaline inorganic pyrophosphatase (AIP) is another enzyme in the cell to provide control mechanism of starch synthesis whereby reactions producing pyrophosphatases as a product are involved in many biosynthetic reactions.[163] AIP enzyme is involved in many essential biosynthetic reactions producing inorganic pyrophosphatases. This enzyme is mainly distributed between mitochondria and glyoxymes. It is a monomer with a molecular mass of 55 kDa. It has a potential role in utilization of phosphomonoesters as the source of Pi (inorganic phosphate) required for maintenance of cellular metabolism.[125] AIP also catalyzes pyrophosphate hydrolysis, synthesis, and enriches the phosphate pool in plants by hydrolyzing inorganic pyrophosphate (IPP) to two molecules of Pi.[21,33] IPP is a byproduct of number of biosynthetic reactions and is essential for the regulation of many biochemical reactions in plant cells. AIP enzymes have extremely high affinity for pyrophosphate, which is associated with the anabolic process in the leaf. With plant development, AIP followed similar trend as shown by GOT and GPT activity. Both N levels and genotypes showed significant variation in AIP activity and their interaction was significant during both stages.

10.3.4 NITROGEN METABOLITES

10.3.4.1 AMINO ACID CONTENT

Amino acid composition is an important feature in determining the nutritional value of wheat grain for human and animal diets. The amino acid composition of wheat varies genotypically and environment, including rate and time of nitrogen fertilization, water availability, and temperature during grain filling period.[136] The cycling of amino acids between shoots and roots is one component of the signaling of whole-plant N status, enabling roots to regulate N-uptake accordingly.[33,73,86,87]

Previous studies indicated that proportions of certain amino acids in wheat protein may depend upon the total nitrogen content of the wheat. The traditional concept of N-cycling in plant–soil systems considers that organic N must first be converted into inorganic N forms by the soil microorganisms, before becoming available to a plant via its roots.[119,130] Further, inorganic N is assimilated into amino acids namely glutamate, glutamine, and asparagine which are then passed through amino acid-specific H^+ co-transporters along their entire root length and that they have the potential to outcompete microorganisms, in some situations. This suggests that this N-uptake pathway may have significance.[106] Hence, it was concluded that major transport of reduced N are amino acids, which are transported predominately in the phloem.

In maize under low N conditions, the amino acid content was decreased significantly and continually during leaf development. Slight decrease in amino acid content under low N supply could be interpreted as the dilution of a stable organic N pool by an increasing leaf volume. Higher N supply caused an overall increase in amino acid content and the large part of is stored in vacuole.[98] Thus, N fertilization can be used for nutritional improvement of cereals by increasing and maintaining protein, essential amino acid content.[154]

10.3.4.2 PROTEIN CONTENT

Nitrogen is an essential element in crop growth and makes an important constituent of proteins. Determining the total soluble proteins or total N appears to be an alternative for monitoring the N status of the plant. Thus, total soluble proteins and free amino acids are N-remobilizing pools.

The protein concentration in cereal grains is a genetic characteristic that can be modified by environmental conditions. Nitrogen in the form of amino acids is required to synthesize proteins and other related compounds and plays a role in almost all plant metabolic processes.[114] The protein content depends critically on the availability of all amino acids in requisite proportions, which in turn depends on the balance between assimilation of NO_3^- and NH_4^+ and the incorporation of the latter into amino acids.[3] Hence, whole plant processes such as N acquisition, translocation, and mobilization of carbon and nitrogen are important for determining protein concentration.

N fertilization has tremendous effect on grain quality of protein.[59,138] Wheat grain contains 8–20% protein, including the gluten storage proteins that are enriched in proline and glutamine. The abundant gluten protein constitutes up to 80% of total flour protein and confers properties of elasticity and extensibility that are essential for functionality of wheat flour.[146] The gluten protein consists of monomeric gliadins and polymeric glutenins. The gliadins constitute 30–40% of total flour protein and are polymorphic mixture of proteins soluble in 70% alcohol. They range in size from about 30 to 60 kDa and can be separated into a, g, and v subgroups, each containing many closely related proteins. The glutenin polymers consist of low-molecular-weight glutenin subunits (LMW-GS) of about 40 kDa linked by interchain disulfide bonds to high-molecular-weight glutenin subunits (HMW-GS) of about 90 kDa. The LMW-GS most closely resemble g-gliadins in sequence and comprise about 20–30% of the total protein while the HMW-GS accounts for about 5–10% of the total protein. The roles of individual gluten components in dough functionality are complex. Although HMW-GS constitute no more than 10% of total flour protein, yet they may be the most important determinants of bread-making quality because of their importance in the forming glutenin polymers.

Water-soluble albumins and salt-soluble globulins constitute 10–22% of total flour protein. Generally, albumin and globulin are not thought to play a critical role in flour quality, although the ratio of albumin to globulin was reported to correlate with bread-making quality in one of the studies to make such determination.[131] Increase in grain protein content and in gliadin to glutenin and HMW-GS to LMW-GS ratios were observed with increased nitrogen fertilizers. Wieser and Seilmeier[164] used reversed-phase high-performance liquid chromatography (RP-HPLC) to conduct a detailed quantitative study of the effects of nitrogen fertilizer on individual gliadin and glutenin components in 13 varieties of hexaploid wheat. With increased fertilizer, amount of protein per mg of flour was increased by 44–68%. There were 2–3-fold increase in the amount of v-gliadin per mg flour and increase of 56–101% in HMW-GS, whereas little changes were observed in a- and g-gliadin and LMW-GS as proportion of total flour protein.

Albumins and globulin were controlled more by genotypes than nitrogen treatment, whereas prolamin and glutenin were largely determined by nitrogen.[54] Protein content and level of v-gliadin were higher for flour from plants grown under the high fertilizer.[42] It has been indicated that different N rates can cause a change in the total amount of different grain proteins. In fact, the composition of the protein fractions has been more affected by N management than the temperature.[156,157]

It was observed in mustard that highest recovery of applied N in the form of protein was under the influence of 30, 60, and 90 kg/N ha, respectively, compared to control.[108] In the case of rice, protein content was increased with increasing level of nitrogen application up to 30 kg/N ha and further increase in application level caused slight decrease in protein content. Leaf protein was not influenced by N treatments.[1] The application of a nitrogen fertilizer in flag leaf stage would enrich the nitrogen content of the vegetative tissues. As a consequence, more nitrogen would be available to be translocated later to the grain, leading to increase in the final protein concentration reported by Scott et al.[143] and Kato.[78]

Studies revealed that an increase in the nitrogen fertilization rate has a favorable effect on grain quality by producing an increase in protein contents.[57] In other words, a greater nitrogen supply increased grain protein concentration linearly, while grain yield response to added nitrogen had a diminishing return relationship. The authors further indicated that when nitrogen was very limiting, small nitrogen addition resulted in greater grain yield with decreased protein concentration caused by dilution of the plant nitrogen. Thus, nitrogen nutrition is the main factor affecting storage proteins as well as the technological quality of the grain.

Wieser and Seilmeier[164] reported that protein content was significantly influenced by nitrogen. Sip et al.[149] reported that application of nitrogen in split doses increased grain protein content by 1.55%. Triboi et al.[158] also reported that nitrogen supply is the most important environmental factor affecting protein content and composition.

Earlier reports of Diaz et al.[38] indicated that both amino acids and soluble protein concentration decreased with leaf ageing. The progressive decrease in leaf N during grain-filling period corresponded to the degradation of leaf proteins.[82]

Reports in the literature also indicated that regardless of the N fertilization regime, larger amount of protein present in the flag leaf is due to photosynthetic apparatus that maximizes carbon assimilation and N availability.[58] The increase in total soluble proteins may be the result of enhanced amino acid formation during N assimilation. Garrido-Lestache et al.[57] reported that increase in N fertilization rate has a favorable effect on grain quality by producing an increase in protein concentration. Khan et al.[81] observed the maximum increase in seed protein contents in maize by the application of highest 300 kg N/ha rather than 75 kg N/ha.

10.3.4.3 SUGAR CONTENT

Sugars are essential to plant growth and metabolism, both as energy source and structural components. The interaction of nitrogen supply and carbohydrate availability has been interpreted as the regulatory mechanism of the symbiosis. Application of nitrogen had also significant effect on sugar content in cereals such as wheat, rice, maize. Very few studies have examined the effect of external nitrogen supply on the internal sugar pool.

Nitrogen assimilation needs carbon which is provided by carbohydrates. In wheat, the increase in dry matter is due to the increased light intercepting area, resulting in more assimilation of photosynthates.[162] Almodares et al.[6–8] reported that N fertilizer increased soluble carbohydrates content (sucrose) in sweet sorghum. This relationship between N fertilizer and consumption of soluble carbohydrates in plants may be due to N assimilation. Monreala et al.[115] stated that raising nitrogen fertilizer rates up to 90 kg N/fed increased sugar yields significantly.

Recent findings have demonstrated that nitrogen translocation facilitates kernel utilization of sugars.[171] Similar studies in *Allium porrum* reported that when nitrogen levels are limiting, photosynthesis is not fully used in the synthesis of organic compounds and then sugars are accumulated.[77,109] The findings of Delden[37] and Zhao et al.[171] indicated that raising nitrogen fertilizer rates up to 120 kg N/fed increased sugar yields significantly.

10.3.4.4 STARCH CONTENT

Starch is the major storage nutrient of plant. Grain dry matter contains 55–75% (w/w) of starch.[88] Native starch is a mixture of two homopolysaccharides composed of α-d-glucopyranosyl units-amylose and amylopectin. Starch in the endosperm of wheat (*T. aestivum*) is a major form of carbon reserves and comprises 65–75% of the final dry weight of the grain.[43,44,153] Grain filling is therefore mainly a process of starch biosynthesis and accumulation.[167] It is predominantly composed of two α-glucan types: amylose and amylopectin. They usually occur in 1:3 weight ratios. Amylose is essentially a linear form composed of alpha-1,4-glucosidic linkages, while amylopectin has a much larger degree of polymerization (dp) in the range 50,000–500,000 and is highly branched (4–5%) and has average chain lengths of 20–25 glucose residues. Amylose/amylopectin ratio is a useful descriptor of starch composition and functionality because starch properties like gelatinization, pasting, starch breakdown, starch components solubility, gelation, retro-gradation characteristics, and extrusion cooked pasta characters are influenced by N.[26,85,139,150,169] In rice, increase in nitrogen level decreases starch and amylose content significantly. In endosperm during grain filling of cereal plants, starch debranching enzyme (DBE) probably plays a crucial role in the starch synthesis.[39] Similar studies indicated that DBE activity under high-nitrogen condition remained at a low level resulting in the decrease of starch content.

The effect of nitrogen supply was studied on flag leaf photosynthesis and grain starch accumulation of wheat from its anthesis and maturity under drought or waterlogging. The results revealed that total soluble sugar content in the grain was reduced under both drought and water-logging, while that in leaf was decreased under water-logging but increased under drought. It was found that both leaf photosynthesis and grain starch accumulation could be regulated by nitrogen supply under stress of soil drought or water-logging from anthesis to maturity of wheat.

Labuschagne et al.[89] reported that starch content was significantly affected by higher N doses. Moreover, it is known that in developed kernels proteins occur first and thereafter starch is produced.[151] Therefore, high protein contents were accompanied by lowering of starch content in wheat varieties. Protein and starch contents were found to be inversely correlated, which is in agreement with results by Gooding and Davies[61] and Viswanathan and Khanna-Chopra.[160]

10.3.5 EFFECTS OF NITROGEN ON YIELD AND ITS ATTRIBUTES

Fertilizer management is an important part of the overall management package target toward realizing higher yield.[19] Grain yield is the main target of crop production. Number of spikes, grains per spike, and 1000-grain weight (TGW) were also considered the most important yield components of wheat, which responded differently to N levels.[134] Usually the number of grains/spike is determined at panicle primordial formation stage which is strongly dependent on genetic factors rather than management factors.[142]

The enhancement of seed yield per plant may be under optimal N nutrition and CO_2 assimilation is favorably upregulated[144] resulting in an adequate supply of photo-assimilates to the developing meristems which maintain their growth. Thus, more reproductive structures are produced per plant such as grains. Bakht et al.[16] reported the addition of different levels of nitrogen to different varieties to enhance the number of grains per spike.

In cereals, leaves remain green for a long time during period of ear emergence, thus increasing the photosynthetic activity.[126,132] More leaves/plant imply higher yield. Approximately 70–90% of the final grain yield is derived from photosynthates (products of photosynthesis) produced by the plant during grain filling. The flag leaf and head usually contribute most, but certainly not all, to the photosynthetic material and thus to the grain.[168]

Karic et al.[77] applied four nitrogen levels (0, 50, 100, and 200 kg N ha^{-1}) to leek culture and reported that application of 200 kg N ha^{-1} resulted in a maximum number of leaves per plant (14.4), but no effect was observed on the number of leaves up to 100 kg N ha^{-1}. Similar studies carried out by Shahbazi[145] showed that there was a significant difference in the number of leaves among nitrogen levels (0, 50, 100, 150, and 200 kg N ha^{-1}) and that the highest leaf number was obtained with 200 kg N ha^{-1}.

Higher levels of nitrogen are reported to enhance the grain yield in correspondence to increased number of grains in cereals and others crops. With the application of high rates of N fertilizer, grain yield was increased in *Brassica juncea*,[147,155], corn,[70,71] and barley.[20] Application of 120 kg N/ha fertilizer increased yield up to 40% for amaranth and 94% for quinoa as compared to control.[170] Increased grain yield due to increased N application could be ascribed to increased biomass production, improved harvest index, and increased seed-set with N fertilizer.[111] Positive effect of N on grain yield and yield attributes of sweet sorghum were reported by Huger et al.[72] N fertilizer not only promoted mineral elements in the soil, but also is usually essential for growth enhancement and maximizing yield.[14]

Plant height and number of grains per ear were increased with increasing N levels in corn.[122] Increase in grains per ear at higher nitrogen level might be due to lower competition for nutrient that allows the plants to accumulate more biomass with higher capacity to convert more photosynthesis into sink resulting in more grains per year. These results are in agreement with Zeidan et al.[170] Aynehband et al.[14] reported that higher concentration of soil N decreased and lower concentration increased the biomass in *T. aestivum*.

Hasanzadeh et al.[63] showed positive relationship between N level and grain yield. Numbers of grains per ear also play an important to determine the grain yield. Increase in N application from 60 to 180 kg N/ha significantly led to increase in TGW.[71] Thus, increase in TGW may be due to increased photosynthetic activity that increases accumulation of metabolites, with direct impact on seed weight. Highest number of tillers were found at 150% N being followed by 100% N and 50% N. The increase in the number of fertile tillers with increase in N levels can be attributed to reduction in mortality of tillers and enabling the production of more tillers from the main stem.[46,102,117,162] In wheat, the role of N in encouraging metabolic processes and hence consequently their growth, spike initiation, and grain filling is responsible for the increase of spike length, number of spikelets and grains/spike, TGW, and ultimately grain yield.[45,141] As expected, modern cultivars generally out-yield the older ones across a range of fertility and N levels.[34] Grain yield is associated with an increase in the C/N ratio as a result of an improvement of N-uptake during the grain-filling period.[23,24,68,107]

10.3.6 EFFECTS OF NITROGEN DOSE ON NUE

NUE is defined as units of economic yield per unit N in soil.[63] In fact, NUE is complex set of process and the ways of estimating NUE depend on the crop and the harvestable product. NUE is actually the product of both nitrogen uptake efficiency (NUpE) which is a root-associated trait, that is, the ability of the plant to remove N from the soil as nitrate (NO_3^-) and ammonium (NH_4^+) ions and nitrogen utilization efficiency (NUtE), which is a function of canopy activity, that is, the ability to use N to produce grain yield.[20,116] Total NUpE is the total dry matter, that is, yield (grain plus straw) divided by the entire N in the aboveground parts of the crop maturity. NUtE is an important part of NUE under high N conditions. Therefore, NUE at the plant level is its ability to utilize the available nitrogen (N) resources to optimize its productivity. This includes nitrogen uptake and assimilatory processes, redistribution within the cell, and balance between storage and current use at the cellular and whole-plant level.[2] NUE for crop plants is of great concern throughout the world.

In simple terms, fertilizer efficiency is the ratio of output (economic yield) to input (fertilizer) for a process or complex system. Thus, obtaining high NUE is very important in actual crop production. NUE can show variation in genotypes. Limon-Ortega et al.[95] and Zhao et al.[172] indicated that a decrease in NUE with increasing fertilizer rates may be due to the increase in grain yield which is less than the N supply in soil and fertilizer.

At low N availability, C_3 plants have greater NUE than C_4 plants, while at high availability the opposite is true.[66] The large variation in cultivars NUE reported by Bellido et al.[22] studies may be due to difference in climate, cultivar, and management practices. Splitting nitrogen fertilizer to many doses increases efficiency of the fertilizers used by decreasing leaching to large extent and increasing both yield and quality of crop in favor of the highest number of splitting as indicated by Ali.[4]

Field-based studies have shown differences in the NUE of barley genotypes by Anbessa et al.[10] For maize, genotype comparison under low N and high N inputs revealed that the genotypes, which were adapted to low N supply, were different from those adapted to high N supply.[55]

Similarly, in Indian mustard genotypes, NUE indicated that plants with high NUpE and high physiological NUtE are not only able to take up N efficiently, but also utilize it efficiently.[25,127] Genotypes with high NUpE accumulated higher N content than those with low NUpE under limited conditions. High physiological NUtE is essential for optimum seed yield, because these genotypes absorb N efficiently.

The target is to synchronize the availability of N with the time/stage of maximum N demand by the crop. Greater synchronization between crop demand and nutrient supply is necessary to improve NUE, especially for N. Split applications of N during the growing season, rather than a single, large application prior to planting, are known to be effective in increasing NUE.[31]

The efficiency of N applied in satisfying the N demand of the crop depends on the type of fertilizer, timing of fertilizer application, crop sequence, the supply of residual and mineralized N, and seasonal trends.[27,28] Therefore, numerous strategies such as use of N sources, slow release fertilizer, placement techniques, and nitrification inhibitors have been devised to reduce nitrogen losses and improve fertilizer-use efficiency.

Paponov et al.[128] observed at low N supply that differences among maize hybrids for NUE were largely due to variations in utilization of accumulated N, but with high N they were largely due to variation among genotypes and levels of N supplied. However, absence of significant interaction has been reported by Spanakakis.[152] In general, presence or absence of interactions strongly depended upon the behavior of genotypes.[5]

Enhancing NUE is especially relevant to cereal crops for which large amounts of N fertilizers are required to attain maximum yield and NUE is estimated to be far less than 50%.[118] Despite its importance, a clear understanding of the major mechanisms and inheritance of NUE is lacking[18] as NUE in plants is a complex trait and in addition to soil N availability, it can also depend on a number of internal and external factors such as photosynthetic carbon fixation to provide energy.[94] However, for most plant species, NUE mainly depends on how plants extract inorganic N from the soil, assimilate NO_3^- and NH_4^+, and recycle organic N.[106]

NUE was decreased with increasing dose of fertilizer,[80,121,161] which may be due to excessive N losses that decreased N utilization efficiency. Jamaati-s-Somarin et al.[74] and Kanampiu et al.[76] reported similar results in wheat and potato tuber. Lopez-Bellido[97] indicated that a decrease in NUE for wheat with increasing fertilizer rates is because grain yield rises less than N supply in soil and fertilizer. Higher NUE leads to maximum tillers, high number of grains per panicle, and high TGW in wheat[80] and maize.[62,120]

Many reports indicate that NUE decreases with increasing N rate. Previous results concluded that variations of NUE appeared to result from differences among genotypes and levels of N supplied. The application of high N may result in poor N uptake and low NUE due to excessive N losses and decreased N utilization efficiency (grain weight produced per unit plant N). Thus, both N capture (uptake) and N conversion (utilization) play an important role in improving NUE. It appears that the NUE of barley genotypes grown in field depends on the level of N supplied.[20,133] Experiments conducted with low N supply indicated genetic variation in NUE of maize which was related to N utilization efficiency, while at high N supply genetic variation in NUE was due to N uptake and N utilization efficiencies. So in order to improve NUE, both N uptake and N utilization need to be efficient.[116]

Nitrogen is the most limiting nutrient for crop production in many of the world's agricultural areas and its efficient use is important for the economic sustainability of cropping systems. Furthermore, the dynamic nature of N and its propensity for loss from soil–plant systems creates a unique and challenging environment for its efficient management. Recovery of N in crop plant is usually less than 50% worldwide. Low recovery of N in annual crop is associated with its loss by volatilization, leaching, surface runoff, de-nitrification, and plant canopy. Low recovery of N is not only responsible for higher cost of crop production, but also for environmental pollution. Hence, improving NUE is desirable to improve crop yields, reducing cost of production, and maintaining environmental quality. To improve N efficiency in agriculture, integrated N management strategies that take into consideration improved fertilizer along with soil and crop management practices are necessary. Including livestock production with cropping offers one of the best opportunities to improve NUE. Synchronization of N supply with crop demand is essential in order to ensure adequate quantity of uptake, utilization, and optimum yield. Good et al.[60] stated that NUE was more complex and the ways of estimating NUE depended on the crop and harvestable product. Thus, efficient N utilization is crucial for economic wheat production and protection of ground and surface water.[161] Better N utilization is an integration of better N assimilation and remobilization within the genotypes.[17,29,30,40,53,106,159] Uptake and partitioning between straw and grain are the two major components of N economy in plants. For environmental and economic reasons, nitrogen fertilizers should be utilized as efficiently as possible in agriculture.

The NUE of plant depends on several factors including application time, application rate of nitrogen fertilizer, cultivar, and climatic conditions.[124] The management of the time of nitrogen application is essential to ensure sustained nutrition at the end of vegetative growth. Therefore, the total amount of N should be divided into suitable fractions to be applied to best satisfy the requirement of the growing crop. Modifying the timing and the application method of N can also lead to an improvement in absorption efficiency. One of the main causes of low NUE in actual N management practices is the scarce synchrony between the N soil input and crop demand.[48,135]

10.4 SUMMARY

Wheat crop requires large quantities of nitrogenous fertilizers to attain maximal growth and productivity. Increased use of fertilizer in the form of nitrogen in agricultural production has raised concerns because the nitrogen surplus is at the risk of leaving the plant–soil system causing environmental contamination and also increased costs associated with the manufacture and distribution of nitrogen fertilizer. Wheat grain proteins play a crucial role in forming the strong, cohesive dough that will retain gas and produce light baked products. These properties make wheat suitable for the preparation of a great diversity of food products: breads, noodles, pasta, cookies, cakes, pastries, and many other foods. The mature wheat grains comprise of 18–20% proteins which are further classified into four categories on the basis of their solubility: albumin (water), globulin (salt), gluten (alkali), and prolamin (propanol).

Gluten are large complex proteins composed of glutenin and gliadins, which are important in baking quality because of their impact on water absorption capacity of the dough. Their elasticity and extensibility can affect wheat flour quality extensively. Albumin and globulin probably have critical role in flour quality, while they also have dual role as nutrient reserves for the germinating embryo. The distribution of nitrogen assimilation throughout the plants depends on the plant species, age, and environmental factors. Ammonium originates in the plant from nitrate reduction by nitrate reductase, direct absorption, photorespiration, or deamination of nitrogenous compounds. It is assimilated into organic molecules primarily by the combined catalytic action of two enzymes, glutamine synthetase, and glutamate synthase. GOT and GPT are responsible for the biosynthesis of glutamine to other amino acids, forming storage protein molecule, and thus play an important role in regulation of nitrogen metabolism in crop plants. Glutamate dehydrogenase catalyzes the amination of 2-oxoglutarate and the deamination of glutamate; the direction of the activity depends on specific environmental conditions.

Starch with varying pasting characteristics are of major concern for food processing because of their potential to modify the texture and quality of the end use-based products. It also serves as a source of carbon during yeast fermentation in bread making, in setting of the bread loaf, and in retrogradation during storage. Starch is divided into two broad categories: amylose and amylopectin. Amylose consists of glucose units having linear chain linked by α-1,4 linkage, while amylopectin has an additional α-1,6 linkages. The content of amylose varies from 20 to 30%, whereas amylopectin constitutes about 75% of cereal starches. High amylose content (>40%) in starch is used as thickeners and as strong gelling agents, while amylopectin content in starch improves homogeneity, stability, texture of gelled starch that enhances the stability of starch gel at frosting and defrosting of frozen foods. Albumins and globulin were controlled merely by genotypes than nitrogen treatment, whereas prolamin and glutenin were largely determined by nitrogen. Reports indicate that different N rates can cause changes in the total amount of different grain proteins.

In fact, the composition of the protein fractions has found to be much more affected by N management than temperature. Although N application greatly influences starch and protein composition, not enough information is available on the effect of nitrogen fertilizer on protein sub-fractions and starch components with little emphasis on quality characteristics. However, none of the earlier studies reported the effects of different doses of nitrogen application on grain quality parameters such as starch (amylose and amylopectin) and proteins (albumin, globulin, gluten, and prolamin) sub-components of wheat.

KEYWORDS

- **amylose/amylopectin ratio**
- **arbitrary separation**
- **asparagine**
- **aspartate**
- **aspartate aminotransferase**
- **aspartic acid**
- **cellular metabolism**
- **globulins**
- **low-molecular-weight glutenin subunits (LMW-GS)**
- **monomeric gliadins**
- **N assimilation**
- **nitrogen assimilation**
- **NO_3^- leaching**
- **N-pool**
- **nucleic acid**
- **N-uptake pathway**
- **photorespiratory activity**
- **photosynthates**

- **starch**
- **starch breakdown**
- ***Triticum aestivum***
- **water-soluble albumins**
- **yield determining factor**
- **α-amino group**

REFERENCES

1. Abedi, T.; Alemzadeh, A.; Kazemeini, S. A. Effect of Organic and Inorganic Fertilizers on Grain Yield and Protein Banding Pattern of Wheat. *Aust. J. Crop Sci.* **2010,** *4*, 384–389.
2. Abeledo, L. G.; Savin, R.; Slafer, G. A. Wheat Productivity in the Mediterranean Ebro Valley: Analyzing the Gap Between Attainable and Potential Yield with a Simulation Model. *Eur. J. Agron.* **2008,** *28*, 541–550.
3. Ali, A.; Jha, P.; Sandhu, K. S.; Raghuram, N. *Spirulina* Nitrate-assimilating Enzymes (NR, NiR, GS) have Higher Specific Activities and are More Stable Than Those of Rice. *Physiol. Mol. Biol. Plants* **2008,** *14*, 179–182.
4. Ali, E. A. Impact of Nitrogen Application Time on Grain and Protein Yields as well as Nitrogen Use Efficiency of Some Two-row Barley Cultivars in Sandy Soil. *Am.-Eurasian J. Agric. Environ. Sci.* **2011,** *10*, 425–433.
5. Alizadeh, K.; Ghaderi, J. Variation of Nitrogen Uptake Efficiency in Local Landraces of Wheat in Mahabad-Iran. *J. Agric. Social Sci.* **2006,** *3*, 122–124.
6. Almodares, A.; Jafarinia, M.; Hadi, M. R. The Effects of Nitrogen Fertilizer on Chemical Compositions in Corn and Sweet Sorghum. *J. Agric. Environ. Sci.* **2009,** *6*, 441–446.
7. Almodares, A.; Ranjbar, M.; Hadi, M. R. Effects of Nitrogen Treatments and Harvesting Stages on the Aconitic Acid, Invert Sugar and Fiber in Sweet Sorghum Cultivars. *J. Environ. Biol.* **2010,** *31*, 1001–1005.
8. Almodares, A.; Taheri, R.; Chung, M.; Fathi, M. The Effect of Nitrogen and Potassium Fertilizers on Growth Parameters and Carbohydrate Content of Sweet Sorghum Cultivars. *J. Environ. Biol.* **2008,** *29*, 849–852.
9. Amarante, L.; Lima, J. D.; Sodek, L. Growth and Stress Conditions Cause Similar Changes in Xylem Amino Acids for Different Legume Species. *Environ. Exp. Bot.* **2006,** *58*, 123–129.
10. Anbessa, Y.; Juskiw, P.; Good, A.; Nyachiro, J.; Helm, J. Genetic Variability in Nitrogen Use Efficiency of Spring Barley. *Crop Sci.* **2009,** *49*, 1259–1269.
11. Andrews, M.; Lea, P. J.; Raven, J. A.; Lindsay, K. Can Genetic Manipulation of Plant Nitrogen Assimilation Enzymes Result in Increased Crop Yield and Greater N-use Efficiency? An Assessment. *Ann. Appl. Biol.* **2004,** *145*, 25–40.
12. Anjana, K.; Kaushik, A.; Kiran, B.; Nisha, R. Biosorption of Cr(VI) by Immobilized Biomass of Two Indigenous Strains of Cyanobacteria Isolated from Metal Contaminated Soil. *J. Hazard. Mater.* **2007,** *48*, 383–386.

13. Aubert, S.; Bligny, R.; Douce, R.; Ratcliffe, R. G.; Roberts, J. K. M. Contribution of Glutamate Dehydrogenase to Mitochondrial Metabolism Studied by ^{13}C and ^{31}P Nuclear Magnetic Resonance. *J. Exp. Bot.* **2001,** *52*, 37–45.
14. Aynehband, A.; Moezi, A. A.; Sabet, M. Agronomic Assessment of Grain Yield and Nitrogen Loss and Gain of Old and Modern Wheat Cultivars Under Warm Climate. *Afr. J. Agric. Res.* **2010,** *5*, 222–229.
15. Azza, A. M. M.; Sahar, M. Z.; El Mesiry, T. Nitrogen Forms Effects on the Growth and Chemical Constituents of *Taxodium distichum* Grown Under Salt Conditions. *Aust. J. Basic Appl. Sci.* **2008,** *2*(3), 527–534.
16. Bakht, J.; Shafi, M.; Zubair, M.; Khan, M. A.; Shah, Z. Effect of Foliar vs. Soil Application of Nitrogen on Yield and Yield Components of Wheat Varieties. *Pak. J. Bot.* **2010,** *42*(4), 2737–2745.
17. Barbottin, A.; Lecomte, C.; Bouchard, C.; Jeuffroy, M. H. Nitrogen Remobilization During Grain Filling in Wheat: Genotypic and Environmental Effects. *Crop Sci.* **2005,** *45*, 1141–1150.
18. Basra, A. S.; Goyal, S. S. Mechanisms of Improved Nitrogen-Use Efficiency in Cereals. In *Quantitative Genetics, Genomics and Plant Breeding;* Kang, M. S., Ed.; CABI Publishing: London, UK; 2002; pp 269–288.
19. Bayoumi, T. Y.; El-Demardash, I. S. Influence of Nitrogen Application on Grain Yield and End Use Quality in Segregating Generations of Bread Wheat (*Triticum aestivum* L). *Afr. J. Biochem. Res.* **2008,** *2*(6), 132–140.
20. Beatty, P. H.; Anbessa, Y.; Juskiw, P.; Carroll, R. T.; Wang, J.; Good, A. G. Nitrogen Use Efficiencies of Spring Barley Grown Under Varying Nitrogen Conditions in the Field and Growth Chamber. *Ann. Bot.* **2010,** *105*, 1171–1182.
21. Beknazarow, B. O.; Valikhanov, M. N. Properties of Cotton Inorganic Pyrophosphatase. *Appl. Biochem. Microbiol.* **2007,** *43*, 153–158.
22. Bellido, L. R. J.; Bellido, L. L.; Benıtez-Vega, J.; Bellido, L. F. J. Tillage System, Preceding Crop and Nitrogen Fertilizer in Wheat Crop, I: Soil Water Content. *Agron. J.* **2007,** *99*, 59–65.
23. Bernard, S. M.; Habash, D. Z. The Importance of Cytosolic Glutamine Synthetase in Nitrogen Assimilation and Recycling. *New Phytol.* **2009,** *182*, 608–620.
24. Bertheloot, J.; Martre, P.; Andrieu, B. Dynamics of Light and Nitrogen Distribution during Grain Filling Within Wheat Canopy. *Plant Physiol.* **2008,** *148*, 1707–1720.
25. Bhat, M. A.; Singh, R.; Dash, D. Effect of INM on Uptake and Use Efficiency of N and S in Indian Mustard on an Inceptisol. *Crop Res.* **2005,** *30*, 23–25.
26. Biliaderis, C. G. The Structure and Interactions of Starch with Food Constituents. *Can. J. Physiol. Pharmacol.* **1991,** *69*(1), 60–78.
27. Blankenau, K.; Olfs, H. W.; Kuhlmann, H. Strategies to Improve the Use Efficiency of Mineral Fertilizer Nitrogen Applied to Winter Wheat. *J. Agron. Crop Sci.* **2002,** *188*, 146–154.
28. Borghi, B. Nitrogen as Determinant of Wheat Growth and Yield. In *Wheat Ecology and Physiology of Yield Determination;* Satorre, E. H., Slafer, G. A., Eds.; Food Products Press: New York, 2000; pp 67–48.
29. Borras, L.; Slafer, G. A.; Otegui, M. E. Seed Dry Weight Response to Source–Sink Manipulations in Wheat, Maize and Soybean: A Quantitative Reappraisal. *Field Crops Res.* **2004,** *86*, 131–146.

30. Calderini, D. F.; Reynolds, M. P.; Slafer, G. A. Source–Sink Effects on Grain Weight of Bread Wheat, Durum Wheat, and Triticale at Different Locations. *Aust. J. Agric. Res.* **2006,** *57*, 227–233.
31. Cassman, K. G.; Dobermann, A.; Walters, D. Agroecosystems, Nitrogen-Use Efficiency, and Nitrogen Management. *AMBIO* **2002,** *31*, 132–140.
32. Cathcart, R. J.; Swanton, C. J. Nitrogen Management will Influence Threshold Values of Green Foxtail (*Setaria viridis*) in Corn. *Weed Sci.* **2003,** *51*, 975–986.
33. Cooper, H. D.; Clarkson, D. T. Cycling of Amino-Nitrogen and Other Nutrients Between Shoots and Roots in Cereals—A Possible Mechanism Integrating Shoot and Root in the Regulation of Nutrient Uptake. *J. Exp. Bot.* **1989,** *40*, 753–762.
34. Coque, M.; Gallais, A. Genetic Variation for N-remobilization and Post Silking N-uptake in a Set of Recombinant Inbred Lines. 1. Evaluation by ^{15}N Labeling, Heritabilities and Correlation Among Traits for Testcross Performance. *Crop Sci.* **2007,** *47*, 1787–1796.
35. Davidson, E. A.; Seitzinger, S. The Enigma of Progress in Denitrification Research. *Ecol. Appl.* **2006,** *16*, 2057–2063.
36. De Angeli, A.; Monachello, D.; Ephritikhine, G.; Frachisse, J. M.; Thomine, S.; Gambale, F.; Barbier-Brygoo, H. *The Nitrate/Proton Antiporter AtCLCa Mediates Nitrate Accumulation in Plant Vacuoles. Nature* **2006,** *442, 939–942.*
37. Delden, A. V. Yield and Growth Components of Potato and Wheat under Organic Nitrogen Management. *Agron. J.* **2001,** *93*, 1370–1385.
38. Diaz, C.; Lemaître, T.; Christ, C.; Azzopardi, M.; Kato, Y.; Sato, F.; Morot-Gaudry, J. F.; Dily, F. L.; Masclaux-Daubresse, C. Nitrogen Recycling and Remobilization are Differentially Controlled by Leaf Senescence and Development Stage in *Arabidopsis* Under Low Nitrogen Nutrition. *Plant Physiol.* **2008,** *147*, 1437–1449.
39. Dinges, J. R.; Colleoni, C.; James, M. G.; Myers, A. M. Mutational Analysis of the Pullulanase Type Debranching Enzyme of Maize Indicates Multiple Functions in Starch Metabolism. *Plant Cell* **2003,** *15*, 666–680.
40. Dreccer, M. F.; Grashoff, C.; Rabbinge, R. Source–Sink Ratio in Barley (*Hordeum vulgare* L.) During Grain Filling: Effects on Senescence and Grain Protein Concentration. *Field Crops Res.* **1997,** *49*, 269–277.
41. Dubois, F.; Tercé-Laforgue, T.; Gonzalez-Moro, M. B.; Estavillo, J. M.; Sangwan, R.; Gallais, A.; Hirel, B. Glutamate Dehydrogenase in Plants: Is There a New Story for an Old Enzyme? *Plant Physiol. Biochem.* **2003,** *41*, 565–576.
42. Dupont, F. M.; Altenbach, S. B. Molecular and Biochemical Impacts of Environmental Factors on Wheat Grain Development and Protein Synthesis. *J. Cereal Sci.* **2003,** *38*, 133–146.
43. Edwards, M. A.; Osborne, B. G.; Henry, R. J. Effect of Endosperm Starch Granule Size Distribution on Milling Yield in Hard Wheat. *J. Cereal Sci.* **2008,** *48*, 180–192.
44. Edwards, M. A.; Osborne, B. G.; Henry, R. J. Puroindoline Genotype, Starch Granule Size Distribution and Milling Quality. *J. Cereal Sci.* **2010,** *52*, 314–320.
45. El-Gizawy, N. K. B. In *Yield and Nitrogen Use Efficiency as Influenced by Rates and Sources of Nitrogen Fertilizers of Some Wheat Varieties*, The 11th Conference of Agronomy Dept., Fac. Agric, Assiut Univ., 2005, pp 51–64.
46. Ersin, C.; Nafiz, C.; Rustu, H.; Saban, Y.; Suleyman, A. The Effects of Nitrogen and Phosphorus Fertilization on the Plant Characteristics of Turkish Yellow Bluestem (*Bothriochloa ischaemum* L.). *Int. J. Agric. Biol.* **2006,** *8*, 154–156.
47. Esposito, S.; Guerriero, G.; Vona, V.; Di Martino Rigano, V.; Carfagna, S.; Rigano, C. Glutamate Synthase Activities and Protein Changes in Relation to Nitrogen Nutrition in

Barley: The Dependence on Different Plastidic Glucose-6P Dehydrogenase Isoforms. *J. Exp. Bot.* **2005,** *56*, 55–64.
48. Fageria, N. K.; Baligar, V. C. Enhancing Nitrogen Use Efficiency in Crop Plants. *Adv. Agron.* **2005,** *88*, 97–185.
49. Feng, H.; Yan, M.; Fan, X., Li, B.; Shen, Q.; Miller, A. J.; Xu, G. Spatial Expression and Regulation of Rice High-Affinity Nitrate Transporters by Nitrogen and Carbon Status. *J. Exp. Bot.* **2011,** *62*, 2319–2332.
50. Feng, Y.; Cao, L. Y.; Wu, W. M.; Shen, X. H.; Zhan, X. D.; Zhai, R. R.; Wang, R. C.; Chen, D. B.; Cheng, S. H. Mapping QTLs for Nitrogen-Deficiency Tolerance at Seedling Stage in Rice (*Oryza sativa* L.). *Plant Breed.* **2010,** *129*, 652–656.
51. Ferrario-Méry, S.; Valadier, M. H.; Godefroy, N.; Miallier, D.; Hirel, B.; Foyer, C. H.; Suzuki, A. Diurnal Changes in Ammonium Assimilation in Transformed Tobacco Plants Expressing Ferredoxin-dependent Glutamate Synthase mRNA in the Antisense Orientation. *Plant Sci.* **2002,** *163*, 59–67.
52. Forde, B. G.; Lea, P. J. Glutamate in Plants: Metabolism, Regulation and Signalling. *J. Exp. Bot.* **2007,** *58*, 2339–2358.
53. Foulkes, M. J.; Hawkesford, M. J.; Barraclough, P. B.; Holdsworth, M. J.; Kerr, S.; Kightley, S.; Shewry, P. R. Identifying Traits to Improve the Nitrogen Economy of Wheat: Recent Advances and Future Prospects. *Field Crops Res.* **2009,** *114*, 329–342.
54. Fuertes-Mendizábal, T.; Aizpurua, A.; González-Moro, M. B.; Estavillo, J. M. Improving Wheat Bread Making Quality by Splitting the N Fertilizer Rate. *Eur. J. Agron.* **2010,** *33*, 52–61.
55. Gallais, A.; Coque, M. Genetic Variation and Selection for Nitrogen Use Efficiency in Maize: A Synthesis. *Maydica* **2005,** *50*, 531–547.
56. Gallais, A.; Coque, M.; Quilleré, I.; Prioul, J. L.; Hirel, B. Modelling Post-silking N-fluxes in Maize Using ^{15}N-labelling Field Experiments. *New Phytol.* **2006,** *172,* 696–707.
57. Garrido-Lestache, E. L.; Lopez-Bellido, R. J.; Lopez-Bellido, L. Durum Wheat Quality Under Mediterranean Conditions as Affected by N Rate, Timing and Splitting, N Form and S Fertilization. *Eur. J. Agron.* **2005,** *23*, 265–278.
58. Gastal, F.; Lemaire, G. N Uptake and Distribution in Crops: An Agronomical and Ecophysiological Perspective. *J. Exp. Bot.* **2002,** *53*, 789–799.
59. Gerba, L.; Getachew, B.; Walelign, W. Nitrogen Fertilization Effects on Grain Quality of Durum Wheat (*Triticum turgidum* L. var. Durum) Varieties in Central Ethiopia. *Agric. Sci.* **2013,** *4*, 123–130.
60. Good, A. G.; Shrawat, A. K.; Muench, D. G. Can Less Yield More? Is Reducing Nutrient Input into the Environment Compatible with Maintaining Crop Production? *Trends Plant Sci.* **2004,** *9,* 597–605.
61. Gooding, M. J.; Davies, W. P. *Wheat Production and Utilization*; CAB International: Wallingford, UK, 1997; 323 p.
62. Hanan, S. S.; Mona, G.; El-Kadar, A.; El-Alia, H. I. Yield and Yield Components of Maize as Affected by Different Sources and Application Rates of Nitrogen Fertilizer. *Res. J. Agric. Biol. Sci.* **2008,** *4*, 399–412.
63. Hasanzadeh, G. A.; Fathollahzadeh, A.; Nasrollahzadeh, A. A.; Akhondi, N. Agronomic Nitrogen Efficiency in Different Wheat Genotypes in West Azerbaijan Province. *Electron. J. Crop Prod.* **2008,** *1*(1), 82–100.
64. Hawkesford, M. J. Reducing the Reliance on Nitrogen Fertilizer for Wheat Production. *J. Cereal Sci.* **2014,** *59*(3), 276–283.

65. Hirel, B.; Lea, P. Ammonia Assimilation. In *Plant Nitrogen;* Lea, P. J., Morot-Gaudry, J. F., Eds.; Springer-Verlag: Berlin, 2001; pp 79–100.
66. Hirel, B.; Gouis, J. L.; Ney, B.; Gallais, A. The Challenge of Improving Nitrogen Use Efficiency in Crop Plants: Towards a More Central Role for Genetic Variability and Quantitative Genetics Within Integrated Approaches. *J. Exp. Bot.* **2007,** *58,* 2369–2387.
67. Hirel, B.; Martin, A.; Tercé-Laforgue, T.; Gonzalez-Moro, M. B.; Estavillo, J. M. Physiology of Maize I: A Comprehensive and Integrated View of Nitrogen Metabolism in a C4 Plant. *Physiol. Plant.* **2005,** *124,* 167–177.
68. Hirel, B.; Bertin, P.; Quillere, I.; Bourdoncle, W.; Attagnant, C.; Dellay, C.; Gouy, A.; Cadiou, S.; Retailliau, C.; Falque, M.; Gallais, A. Towards a Better Understanding of the Genetic and Physiological Basis for Nitrogen Use Efficiency in Maize. *Plant Physiol.* **2001,** *125,* 1258–1270.
69. Hirel, B.; Tétu, T.; Lea, P. J.; Dubois, J. Improving Nitrogen Use Efficiency in Crops for Sustainable Agriculture. *Sustainability* **2011,** *3,* 1452–1485.
70. Hokmalipour, S.; Darbandi, M. H. Effects of Nitrogen Fertilizer on Chlorophyll Content and Other Leaf Indicate in Three Cultivars of Maize (*Zea mays* L.). *World Appl. Sci. J.* **2011,** *15,* 1780–1785.
71. Hokmalipour, S.; Seyedsharifi, R.; Jamaati-e-Somarin, S. H.; Hassanzadeh, M.; Shiri-e-Janagard, M.; Zabihi-e-Mahmoodabad, R. Evaluation of Plant Density and Nitrogen Fertilizer on Yield, Yield Components and Growth of Maize. *World Appl. Sci. J.* **2010,** *8,* 1157–1162.
72. Hugar, A. Y.; Jayadeva, H. M.; Rangaswamy, B. R.; Shivanna, S.; Chandrappa, H. Assessing the Effect of Nitrogen and Harvesting Stages on Yield and Yield Attributes of Sweet Sorghum Genotypes. *Agric. Sci. Digest* **2010,** *30,* 139–141.
73. Imsande, J.; Touraine, B. N Demand and the Regulation of Nitrate Uptake. *Plant Physiol.* **1994,** *105,* 3–7.
74. Jamaati-e-Somarin, S.; Tobeh, A.; Hassanzadeh, M.; Saeidi, M.; Gholizadeh, A.; Zabihi-e-Mahmoodabad, R. Effects of Different Plant Density and Nitrogen Application Rate on Nitrogen Use Efficiency of Potato Tuber. *Pak. J. Biol. Sci.* **2008,** *11,* 1949–1952.
75. Ju, X. T.; Xing, G. X.; Chen, X. P.; Zhang, S. L.; Zhang, L. J.; Liu, X. J.; Zhen, C. L.; Yin, B.; Christie, P.; Zhu, Z. L.; Zhang, F. S. Reducing Environmental Risk by Improving N Management in Intensive Chinese Agricultural Systems. *Proc. Natl. Acad. Sci. USA* **2009,** *106*(9), 3041–3046.
76. Kanampiu, F. K.; Raun, W. R.; Johnson, G. V. Effect of Nitrogen Rate on Plant Nitrogen Loss in Winter Wheat Varieties. *J. Plant Nutr.* **1997,** *20,* 389–404.
77. Karic, L.; Vukasinovic, S.; Znidarcic, D. Response of Leek (*Allium porrum* L.) to Different Levels of Nitrogen Dose Under Agro-climate Conditions of Bosnia and Herzegovina. *Acta Agric. Slovenica* **2005,** *85,* 219–226.
78. Kato, Y. Grain Nitrogen Concentration in Wheat Grown Under Intensive Organic Manure Application on Andosols in Central Japan. *Plant Prod. Sci.* **2012,** *15*(1), 40–47.
79. Keresi, S. T.; Malencic, D. R.; Popovic, M. T.; Kraljevic-Balalic, M. J. A.; Llic, A. D. Nitrogen Metabolism Enzymes, Soluble Protein and Free Proline Content in Soybean Genotypes and their F1 Hybrids. *Proc. Natl. Acad. Sci. USA* **2008,** *115,* 21–26.
80. Khaleque, M. A.; Paul, N. K.; Meisner, C. A. Yield and N Use Efficiency of Wheat as Influenced by Bed Planting and Application. *Bangladesh J. Agric. Res.* **2008,** *33,* 439–448.
81. Khan, H. Z.; Iqbal, S.; Iqbal, A.; Akbar, N.; Jones, D. L. Response of Maize (*Zea mays* L.) Varieties to Different Levels of Nitrogen. *Crop Environ.* **2011,** *2*(2), 15–19.

82. Kichey, T.; Heumez, E.; Pocholle, D.; Pageau, K.; Vanacker, H.; Dubois, F.; Le Gouis, J.; Hirel, B. Combined Agronomic and Physiological Aspects of Nitrogen Management in Wheat Highlight a Central Role for Glutamine Synthetase. *New Phytol.* **2006,** *169,* 265–278.
83. Kichey, T.; Hirel, B.; Heumez, E.; Dubois, F.; Le Gouis, J. Wheat Genetic Variability for Post-Anthesis Nitrogen Absorption and Remobilization Revealed by ^{15}N Labelling and Correlations with Agronomic Traits and Nitrogen Physiological Markers. *Field Crops Res.* **2007,** *102,* 22–32.
84. Kikuchi, H.; Hirose, S.; Toki, S.; Akama, K.; Takaiwa, F. Molecular Characterization of a Gene for Alanine Aminotransferase from Rice (*Oryza sativa*). *Plant Mol. Biol.* **1999,** *39,* 149–159.
85. Kim, H. S.; Huber, K. C. Impact of A/B-type Granule Ratio on Reactivity, Swelling, Gelatinization, and Pasting Properties of Modified Wheat Starch. Part I: Hydro Xypropylation. *Carbohyd. Polym.* **2010,** *80*(1), 94–104.
86. Krapp, A.; Fraisier, V.; Scheible, W.; Quesada, A.; Gojon, A.; Stitt, M.; Caboche, M.; Vedele, D. F. Expression Studies of Nrt2:1Np, a Putative High-Affinity Nitrate Transporter: Evidence for Its Role in Nitrate Uptake. *Plant J.* **1998,** *14,* 723–731.
87. Krapp, A.; Saliba-Colombani, V.; Daniel-Vedele, F. Analysis of C and N Metabolisms and of C/N Interactions using Quantitative Genetics. *Photosynth. Res.* **2005,** *83,* 251–263.
88. Kulp, K.; Ponte, J. G. *Handbook of Cereal Science and Technology*, 2nd Edition (Revised and Expanded); Marcel Dekker: New York, 2000; 790 p.
89. Labuschagne, M. T.; Geleta, N.; Osthoff, G. The Influence of Environment on Starch Content and Amylose to Amylopectin Ratio in Wheat. *Starch-Starke* **2007,** *59,* 234–238.
90. Landry, J.; Delhaye, S. Influence of Genotype and Texture on Zein Content in Endosperm of Maize Grains. *Ann. Appl. Biol.* 2007, *151,* 349–356.
91. Lattanzi, F. A.; Schnyder, H.; Thornton, B. The Sources of Carbon and Nitrogen Supplying Leaf Growth. Assessment of the Role of Stores with Compartmental Models. *Plant Physiol.* **2005,** *137,* 383–395.
92. Lea, P. J.; Azevedo, R. A. Nitrogen Use Efficiency, 1: Uptake of Nitrogen from the Soil. *Ann. Appl. Biol.* 2006, *149,* 243–247.
93. Lea, P. J.; Miflin, B. J. Glutamate Synthase and the Synthesis of Glutamate in Plants. *Plant Physiol. Biochem.* **2003,** *41,* 555–564.
94. Lewis, J. D.; McKane, R. B.; Tingey, D. T.; Beedlow, P. A. Vertical Gradients in Photosynthetic Light Response within an Old-growth Douglas-fir and Western Hemlock Canopy. *Tree Physiol.* **2000,** *20,* 447–456.
95. Limon-Ortega, A.; Sayre, K. D.; Francis, C. A. Wheat Nitrogen Use Efficiency in a Bed Planting System in Northwest Mexico. *Agron. J.* **2000,** *92,* 303–308.
96. Lopes, M. S.; Cortadellas, N.; Kichey, T.; Dubois, F.; Habash, D. Z.; Araus, J. L. Wheat Nitrogen Metabolism During Grain Filling: Comparative Role of Glumes and the Flag Leaf. *Planta* **2006,** *225*(1), 165–181.
97. Lopez-Bellido, R. J.; Lopez-Bellido, L. Efficiency of Nitrogen in Wheat Under Mediterranean Condition: Effect of Tillage, Crop Rotation and N Fertilization. *Field Crop Res.* **2001,** *71,* 31–64.
98. Lošák, T.; Vollmannn, J.; Hlušek, J.; Peterka, J.; Filipík, R.; Prášková, L. Influence of Combined Nitrogen and Sulphur Fertilization on False Flax (*Camelina sativa* [L.] Crtz.) Yield and Quality. *Acta Aliment* **2010,** *39,* 431–444.

99. Ma, L. Absorption and Utilization of Amino Acids by Plant. *J. Southwest Univ. Sci. Technol.* **2004,** *3*, 102–107 (in Chinese).
100. Maathuis, F. J. M. Physiological Functions of Mineral Macronutrients. *Curr. Opin. Plant Biol.* **2009,** *12*, 250–258.
101. Martin, A.; Belastegui-Macadam, X.; Quillere, I.; Floriot, M.; Valadier, M. H.; Pommel, B.; Andrieu, B.; Donnison, I.; Hirel, B. Nitrogen Management and Senescence in Two Maize Hybrids Differing in the Persistence of Leaf Greenness. Agronomic, Physiological and Molecular Aspects. *New Phytol.* **2005,** *167*, 483–492.
102. Martin, R. J.; Sutton, H. K.; Muyle, T. N.; Gillespie, R. N. Effect of Nitrogen Fertilizer on the Yield and Quality of Six Cultivars of Autumn Sown Wheat. *N. Z. J. Crop Hort. Sci.* **1992,** *20*, 273–282.
103. Martre, P.; Porter, J. R.; Jamieson, P. D.; Triboi, E. Modeling Grain Nitrogen Accumulation and Protein Composition to Understand Sink/Source Regulations of Nitrogen Remobilization for Wheat. *Plant Physiol.* **2003,** *133*, 1959–1967.
104. Masclaux-Daubresse, C.; Reisdorf-Cren, M.; Pageau, K.; Lelandais, M.; Grandjean, O.; Kronenberger, J.; Valadier, M. H.; Feraud, M.; Jouglet, T.; Suzuki, A. Glutamine Synthetase-Glutamate Synthase Pathway and Glutamate Dehydrogenase Play Distinct Roles in the Sink–Source Nitrogen Cycle in Tobacco. *Plant Physiol.* **2006,** *140*, 444–456.
105. Masclaux-Daubresse, C.; Valadier, M. H.; Brugiere, N.; Morot-Gaudry, J. F.; Hirel, B. Characterization of the Sink/Source Transition in Tobacco (*Nicotiana tabacum* L.) Shoots in Relation to Nitrogen Management and Leaf Senescence. *Planta* **2000,** *211*, 510–518.
106. Masclaux-Daubresse, C.; Daniel-Vedele, F.; Dechorgnat, J.; Chardon, F.; Gaufichon, L.; Suzuki, A. Nitrogen Uptake, Assimilation and Remobilization in Plants: Challenges for Sustainable and Productive Agriculture. *Ann. Bot.* **2010,** *105*, 1141–1157.
107. Masclaux-Daubresse, C.; Reisdorf-Cren, M.; Orsel, M. Leaf Nitrogen Remobilization for Plant Development and Grain Filling. *Plant Biol.* **2008,** *10*(Suppl. 1), 23–36.
108. Meena, B. S.; Sumeriya, H. K. Influence of Nitrogen Levels, Irrigation and Interculture on Oil and Protein Content, Soil Moisture Status and Interaction Effects of Mustard (*Brassica juncea*). *Crop Res.* **2003,** *26*, 409–413.
109. Mengel, K.; Kirkby, E. A. Nutrition and Plant Growth. In *Principles of Plant Nutrition. I*; International Potash Institute: Bern, Switzerland, 1978; pp 211–256.
110. Miflin, B. J.; Habash, D. Z. The Role of Glutamine Synthetase and Glutamate Dehydrogenase in Nitrogen Assimilation and Possibilities for Improvement in the Nitrogen Utilization of Crops. *J. Exp. Bot.* **2002,** *53*, 979–987.
111. Miri, K.; Rana, D. S.; Rana, K. S.; Kumar, A. Productivity, Nitrogen-Use Efficiency and Economics of Sweet Sorghum (*Sorghum bicolor*) Genotypes as Influenced by Different Levels of Nitrogen. *Indian J. Agron.* **2012,** *57*, 49–54.
112. Mishra, B. *Project Director Report (2007–2008)*; Issued by the Project Director, Directorate of Wheat Research, Karnal. Presented at the 47th All India Wheat and Barley Research Workers' Meet, Hisar (India), August 17–20, 2008, 243 p.
113. Mo, L.; Wu, L.; Tao, Q. Effects of Amino Acid–N and Ammonium–N on Wheat Seedlings under Sterile Culture. *Ying Yong Sheng Tai Xue Bao* **2003,** *14*(2), 184–186 (in Chinese).
114. Mokhele, B.; Zhan, X.; Yang, G.; Zhang, X. Review: Nitrogen Assimilation in Crop Plants and Its Affecting Factors. *Can. J. Plant Sci.* **2012,** *92*, 399–405.

115. Monreala, J. A.; Jiméneza, E. T.; Remesala, E.; MorilloVelardeb, R.; García-Mauriñoa, S.; Echevarríaa, C. Proline Content of Sugar Beet Storage Roots: Response to Water Deficit and Nitrogen Fertilization at Field Conditions. *Environ. Exp. Bot.* **2007,** *60*(2), 257–267.
116. Moose, S.; Below F. E. Biotechnology Approaches to Improving Maize Nitrogen Use Efficiency. In *Molecular Genetic Approaches to Maize Improvement, Biotechnology in Agriculture and Forestry;* Kriz, A. L., Larkins, B. A., Eds.; Springer-Verlag: Berlin, 2009; pp 65–77.
117. Muneshwar, S.; Sing, V. P.; Reddy, K. S.; Sing, M. Effect of Integrated Use of Fertilizer Nitrogen Farmyard Manure or Green Manure on Transformation of N, K and S and Productivity of Rice–Wheat System on a Vertisol. *J. Indian Soc. Soil Sci.* **2001,** *49*, 430–434.
118. Murphy, K. M.; Campbell, K. G.; Lyon, S. R.; Jones, S. S. Evidence of Varietal Adaptation to Organic Farming Systems. *Field Crops Res.* **2007,** *102*, 172–177.
119. Näsholm, T.; Kielland, K.; Ganeteg, U. Uptake of Organic Nitrogen by Plants. *New Phytol.* **2009,** *182*, 31–48.
120. Nasser, K. H.; El-Gizawy, B. Effects of Nitrogen Rate and Planting Density on Agronomic Nitrogen Efficiency and Maize Yields Following Wheat and Faba Bean. *Am.-Eurasian J. Agric. Environ. Sci.* **2009,** *5*, 378–386.
121. Nasseri, A.; Fallahi, A. H.; Siadat, A.; Eslami-Gumush, T. K. Protein and N-use Efficiency of Rainfed Wheat Responses to Supplemental Irrigation and Nitrogen Fertilization. *Arch. Agron. Soil Sci.* **2009,** *55*, 315–325.
122. Nemati, A. R.; Sharifi, R. S. Effects of Rates and Nitrogen Application Timing on Yield, Agronomic Characteristics and Nitrogen Use Efficiency in Corn. *Int. J. Agric. Crop Sci.* **2012,** *4*, 534–539.
123. Novitskaya, L.; Trevanion, S. J.; Driscoll, S.; Foyer, C. H.; Noctor, G. How Does Photorespiration Modulate Leaf Amino Acid Contents? A Dual Approach through Modelling and Metabolite Analysis. *Plant Cell Environ.* **2002,** *25*, 821–835.
124. Okamoto, M.; Okada, K. Differential Responses of Growth and Nitrogen Uptake to Organic Nitrogen in Four Gramineous Crops. *J. Exp. Bot.* **2004,** *55*, 1577–1185.
125. Orhanovic, S.; Pavela-Vrancic, M. Alkaline Phosphatase Activity in Sea Water: Influence of Reaction Conditions on the Kinetic Parameters of ALP. *Croatica Chem. Acta* **2000,** *73*, 819–830.
126. Ozen, H. C.; Onay, A. *Plant Physiology*; Nobel Printing Press: Mumbai, India 2007; 1220 p.
127. Panda, B. B.; Bandyopadhyay, S. K.; Shivay, Y. S. Effect of Irrigation Level, Sowing Dates and Varieties on Yield Attributes, Yield, Consumptive Water Use and Water-Use Efficiency of Indian Mustard. *Indian J. Agric. Sci.* **2004,** *74*(6), 339–342.
128. Paponov, I. A.; Sambo, P.; Erley, G.; Presterl, T.; Geiger, H. H.; Engels, C. Grain Yield and Kernel Weight of Two Maize Genotypes Differing in Nitrogen Use Efficiency at Various Levels of Nitrogen and Carbohydrate Availability During Flowering and Grain Filling. *Plant Soil* **2005,** *272*, 111–123.
129. Pathak, R. R.; Lochab, S.; Raghuram, N. Plant Systems: Improving Plant Nitrogen-Use Efficiency. In *Comprehensive Biotechnology*, 2nd Edition; Moo-Young, M., Ed.; Elsevier: The Netherlands; 2011; Vol. 4, pp 209–218.
130. Paungfoo-Lonhienne, C.; Visser, J.; Lonhienne, T. A.; Schmidt, S. Past, Present and Future of Organic Nutrients. *Plant Soil* **2012,** 359, 1–18.

131. Persson, J.; Nasholm, T. Regulation of Amino Acid Uptake by Carbon and Nitrogen in *Pinus sylvestris*. *Planta* **2003,** *217*, 309–315.
132. Pollmer, W. G.; Eberhard, D.; Klein, D.; Dhillon, B. S. Genetic Control of Nitrogen Uptake and Translocation Maize. *Crop Sci.* **1979,** *19*, 82–86.
133. Rahimizadeh, M.; Kashani, A.; Zare-Feizabadi, A.; Koocheki, A.; Nassiri-Mahallati, M. Nitrogen Use Efficiency of Wheat as Affected by Preceding Crop, Application Rate of Nitrogen and Crop Residues. *Aust. J. Crop Sci.* **2010,** *4*, 363–368.
134. Rahman, M. A.; Sarker, M. A. Z.; Amin, M. F.; Jahan, A. H. S.; Akhter, M. M. Yield Response and Nitrogen Use Efficiency of Wheat Under Different Doses and Split Application of Nitrogen Fertilizer. *Bangladesh J. Agric. Res.* **2011,** *36*, 231–240.
135. Raun, W. R.; Johnson, G. V. Improving Nitrogen Use Efficiency for Cereal Production. *Agron. J.* **1999,** *91*, 357–363.
136. Rharrabti, Y.; Elhani, S.; Martos-N˙Òez, V.; Garcİa del Moral, L. F. Environmentally Induced Changes in Amino Acid Composition in the Grain of Durum Wheat Grown Under Different Water and Temperature Regimes in a Mediterranean Environment. *J. Agric. Food Chem.* **2001,** *49*, 3802–3807.
137. Rosales, E. P. M. F.; Iannone, M.; Groppa, D. M.; Benavides, P. Nitric Oxide Inhibits Nitrate Reductase Activity in Wheat Leaves. *Plant Physiol. Biochem.* **2011,** *49*, 124–130.
138. Saeed, B.; Gul, H.; Ali, F.; Khan, A., Anwar, S. N.; Alam, S.; Khalid, S.; Naz, A.; Fayyaz, H. A. Contribution of Soil and Foliar Fertilization of Nitrogen and Sulfur on Physiological and Quality Assessment of Wheat (*Triticum aestivum* L.). *Nat. Sci.* **2013,** *5,* 1012–1018.
139. Samaan, J.; El-Khayat, G. H.; Manthey, F. A.; Fuller, M. P.; Brennan, C. S. Durum Wheat Quality: II. The Relationship of Kernel Physicochemical Composition to Semolina Quality and End Product Utilization. *Int. J. Food Sci. Technol.* **2006,** *41*, 47–55.
140. Sawan, Z. M.; Mahmoud, M. H.; El-Guibali, A. H. Response of Yield, Yield Components, and Fiber Properties of Egyptian Cotton (*Gossypium barbadense* L.) to Nitrogen Fertilization and Foliar-Applied Potassium and Mepiquat Chloride. *J. Cotton Sci.* **2006,** *10*, 224–234.
141. Sawires, E. S. Yield and Yield Attributes of Wheat in Relation to Nitrogen Fertilization and Withholding of Irrigation to Different Stages of Growth. *Ann. Agric. Sci.* **2000,** *45*, 439–452.
142. Schwarte, A. J.; Gibson, L. R.; Karlen, D. L.; Dixon, P. M.; Lieman, M.; Jannink, J. L. Planting Date Effects on Winter Triticale Yield and Yield Components. *Crop Sci.* **2006,** *46*, 1218–1224.
143. Scott, W. R.; Martin, R. J.; Stevenson, K. R. Soil Fertility Limitations to Wheat Yield and Quality. In *Wheat Symposium: Limits to Production and Quality;* Griffin, W. B., Pollock, K. M., Bezar, H. J., Eds.; Agronomy Society of New Zealand Special Publication 8. Agronomy Society of New Zealand: Christchurch, 1992; pp 47–56.
144. Shah, S. H. Effects of Nitrogen Fertilization on Nitrate Reductase Activity, Protein, Oil Yields of *Nigella sativa* L. as Affected by Foliar GA_3 Application. *Turk. J. Bot.* **2008,** *32*, 165–170.
145. Shahbazi, M. Effects of Different Nitrogen Levels on the Yield and Nitrate Accumulation in the Four of Lettuce Cultivars. MSc Thesis, Department of Horticulture, Science and Research Branch, Islamic Azad University, Tehran, Iran, 2005, 99 p.
146. Shewry, P. R.; Field, J. M.; Faulks, A. J.; Parmar, S.; Miflin, B. J.; Dietler, M. D.; Lew, E. J. L.; Kasarda, D. D. The Purification and N-terminal Amino Acid Sequence Analysis

of the High Molecular Weight Gluten Polypeptides of Wheat. *Biophys. Acta* **1984,** *788*(1), 23–34.
147. Singh, B.; Singh, Y.; Ladha, J. K.; Bronson, K. F.; Balasubramanian, V.; Singh, J.; Khind, C. S. Chlorophyll-meter and Leaf Color Chart-based Nitrogen Management for Rice and Wheat in Northwestern India. *Agron. J.* **2002,** *94,* 821–829.
148. Singletary, G. W.; Doehlert, D. C.; Wilson, C. M.; Muhitch, M. J.; Below, F. E. Response of Enzymes and Storage Proteins of Maize Endosperm to Nitrogen Supply. *Plant Physiol.* **1990,** *94*(3), 858–864.
149. Sip, V.; Skorpik, M.; Chrpova, J.; Sonttaikova, U.; Bartova, S. Effect of Cultivar and Cultural Practices on Grain Yield and Bread Quality of Winter Wheat. *Rostinna Vyroba* **2000,** *46*(4), 159–167.
150. Sissons, M. J.; Batey, I. L. Protein and Starch Properties of Some Tetraploid Wheats. *Cereal Chem.* **2003,** *80*, 468–475.
151. Sowers, K. E.; Miller, B. C.; Pan, W. L. Optimizing Grain Yield in Soft White Winter Wheat with Split Nitrogen Applications. *Agron. J.* **1994,** *86*, 1020–1025.
152. Spanakakis, A. Breeding of Winter Wheat with Improved Efficiency (Züchtung von Winterweizen mit verbesserter Neffizienz). In *Nitrogen Efficiency of Crops (Stickstoff-Effizienz Landwirtschaftlicher Kulturpflanzen);* Millers, C., Ed.; Erich Schmidt Verlag: Germany, 2000; pp 110–115.
153. Stone, B.; Morell, M. K. Carbohydrates. In *Wheat: Chemistry and Technology*, 4th Edition; Khan, K., Shewry, P. R., Eds.; AACC International, Inc.: USA, 2009; pp 299–362.
154. Thanapornpoonpong, S. N.; Vearasilp, S.; Pawelzik, E.; Gorinstein, S. Influence of Various Nitrogen Applications on Protein and Amino Acid Profiles of Amaranth and Quinoa. *J. Agric. Food Chem.* **2008,** *56*, 11464–11470.
155. Thanki, J. D.; Patel, A. M.; Patel, M. P. Effect of Date of Sowing, Phosphorus and Biofertilizer on Growth, Yield and Quality of Summer Sesame (*Sesamum indicum* L.). *J. Oilseeds Res.* **2004,** *21*, 301–302.
156. Triboi, E.; Triboi-Blondel, A. M. Cropping Systems Self-sufficient in Nitrogen and Energy, Reality or Utopia? *Alter Agric.* **2008,** *89,* 17–18 (French).
157. Triboi, E.; Triboi-Blondel, A. M. In *Cropping Systems Self-sufficient in Nitrogen and Energy, Reality or Utopia?* International Conference on Organic Agriculture and Climate Change, (French) ENITA, France, Apr 17–18, 2008, pp 115–120.
158. Triboi, E.; Abad, A.; Michelena, A.; Lloveras, J.; Ollier, J. L.; Daniel, C. Environmental Effects on the Quality of Two Wheat Genotypes: 1. Quantitative and Qualitative Variation of Storage Proteins. *Eur. J. Agron.* **2000,** *13*, 47–64.
159. Van Oosterom, E. J.; Chapman, S. C.; Borrel, A. K.; Broad, I. J.; Hammer, G. L. Functional Dynamics of the Nitrogen Balance of Sorghum, II: Grain Filling Period. *Field Crops Res.* **2010,** *115*, 29–38.
160. Viswanathan, C.; Khanna-Chopra, R. Effect of Heat Stress on Grain Growth, Starch Synthesis and Protein Synthesis in Grains of Wheat (*Triticum aestivum* L.) Varieties Differing in Grain Weight Stability. *J. Agron. Crop Sci.* **2001,** *186*, 1–7.
161. Vukovic, I.; Mesic, M.; Zgorelec, Z.; Jurisic, A. Nitrogen Use Efficiency in Winter Wheat. *Cereal Res. Commun.* **2008,** *36*, 1199–1202.
162. Warraich, E. A.; Ahmad, N.; Basra, S. M. A.; Afzal, I. Effect of Nitrogen on Source–Sink Relationship in Wheat. *Int. J. Agric. Biol.* **2002,** *4*, 300–302.
163. Wickert, S.; Marcondes, J.; Lemos, M. V.; Lemos, E. G. M. Nitrogen Assimilation in Citrus Based on CitEST Data Mining. *Genet. Mol. Biol.* **2007,** *30*, 810–818.

164. Wieser, H.; Seilmeier, W. The Influence of Nitrogen Fertilization on Quantities and Proportions of Different Protein Types in Wheat Flour. *J. Sci. Food Agric.* **1998,** *76,* 127–131.
165. Xu, G.; Fan, X.; Miller, A. J. Plant Nitrogen Assimilation and Use Efficiency. *Annu. Rev. Plant Biol.* **2012,** *63,* 153–182.
166. Xu, X. L.; Hua, O. Y.; Cao, G. M.; Pei Z. Y.; Zhou C. P. Uptake of Organic Nitrogen by Eight Dominant Plant Species in Kobresia Meadows. *Nutrient Cycl. Agroecosyst.* **2004,** *69,* 5–10.
167. Yang, W. Y.; Xiang, Z. F.; Ren, W. J.; Wang, X. C. Effect of S-3307 on Nitrogen Metabolism and Grain Protein Content in Rice. *Chin. J. Rice Sci.* **2005,** *19,* 63–67.
168. Yildirim, M.; Akinci, C.; Koc, M.; Barutcular, C. Applicability of Canopy Temperature Depression and Chlorophyll Content in Durum Wheat Breeding. *Anadolu J. Agric. Sci.* **2009,** *24*(3), 158–166.
169. Yue, P.; Rayas-Duarte, P.; Elias, E. Effect of Drying Temperature on Physicochemical Properties of Starch Isolated from Pasta. *Cereal Chem.* **1999,** *76*(4), 541–547.
170. Zeidan, M. S.; Amany, A.; Bahr El-Kramany, M. F. Effect of N-fertilizer and Plant Density on Yield and Quality of Maize in Sandy Soil. *Res. J. Agric. Biol. Sci.* **2006,** *2,* 156–161.
171. Zhao, G. Q.; Ma, B. L.; Ren, C. Z. Response of Nitrogen Uptake and Partitioning to Critical Nitrogen Supply in Oat Cultivars. *Crop Sci.* **2009,** *49,* 1040–1048.
172. Zhao, R. F.; Chen, X. P.; Zhang, F. S.; Zhang, H.; Schroder, J.; Romheld, V. Fertilization and Nitrogen Balance in a Wheat–Maize Rotation System in North China. *Agron. J.* **2006,** *98,* 935–945.

CHAPTER 11

SYMPLOCOS PANICULATA (SAPPHIRE BERRY): A WOODY AND ENERGY-EFFICIENT OIL PLANT

QIANG LIU[1], YOUPING SUN[2,*], JINGZHENG CHEN[3], PEIWANG LI[3], GENHUA NIU[2], CHANGZHU LI[1,*], and LIJUAN JIANG[3,*]

[1]*Central South University of Forestry and Technology, 498 South Shaoshan Road, Changsha, Hunan 410004, China*

[2]*Assistant Professor, Department of Plants, Soils and Climate Utah State Uni 4820 Old Main Hill Logan UT 84322, USA*

[3]*Hunan Academy of Forestry, 658 South Shaoshan Road, Changsha, Hunan 410004, China*

[*]*Corresponding author. E-mail: youping.sun@usu.edu; znljiang2542@163.com; lichangzhu2013@aliyun.com*

CONTENTS

ABSTRACT

Research activities on the production of biodiesel from the fruit oil are summarized. This chapter also documents the treasures of indigenous knowledge of other multiple uses such as ornamental and medicinal uses, natural dyes production, and livestock feed.

11.1 INTRODUCTION

Biofuel as an alternative to petrodiesel has received considerable attention due to diminishing availability of discoverable fossil fuel reserves and environmental consequences of exhaust gases from fossil fuels. The U.S. Energy Information Administration[29] has predicted that renewable energy except hydropower would account for 28% of the overall growth in electricity generation during 2012–2040. Biofuel is gaining popularity among the most important types of renewable energy, and the global demand for liquid biofuel has tripled between 2000 and 2007.[11] Bioethanol is produced from agricultural feedstocks such as corn, miscanthus, sweet potato, sugar cane, sorghum, and switchgrass,[7,26,28] whereas biodiesel, alkyl esters of fatty acids, is made from edible animal fats and vegetable oils derived from coconut, linseed, palm seed, rapeseed, soybean, and sunflower.[14] A majority of biofuel feedstocks are food crops. As a consequence, biofuel feedstock production may compete in the long run with food supply and/or with food crops for arable land.

Production of biofuel from edible oil is not feasible in China since there is a huge population of 1.35 billion and relatively inadequate arable land resources per capita. Oil-bearing plants that produce nonedible oils in appreciable quantity and adapt to noncropped marginal lands and wastelands would be better feedstock of choice for biofuel production. Fortunately, there are more than 4000 species of plants in China with potential for bioenergy production.[16] A total of 38 oilseed crops have been identified as potential energy plants in China and are mainly distributed in tropical and subtropical areas.[15] *Symplocos paniculata* (Thunb.) Miq (sapphire berry or Asiatic sweet leaf) is one of the candidate bioenergy plants.[16]

This chapter explores potential and adaptability of *S. paniculata* (sapphire berry or Asiatic sweet leaf) cultivation for ornamental and medicinal uses, natural dyes production, and livestock feedstock.

11.2 NATURAL DISTRIBUTION AND MORPHOLOGY

S. paniculata belongs to *Symplocaceae*. It is comprised of more than 300 species widely distributed in tropical and subtropical areas of the world. Among them, a total of 77 species are located in the mountainous regions of the eastern China.[30] Species in this family are usually found in the open area in the forest at an elevation of 100–1600 m. Along Daweishan Mountain (Liuyang, Hunan) from bottom to top, one can easily locate *S. chinensis*, *S. paniculata*, and *S. tanakana*.[19] Genetic analysis showed that the geographical elevation is greatly correlated with the genetic differentiation of *S. paniculata*.[22]

S. paniculata is widely distributed in Japan, Korea, Bhutan, India, Laos, Myanmar, and Vietnam[27] and most places in China.[12] It has a high adaptability to different temperature zones and varying soil conditions. It grows well in barren, salty, and severe drought soil like marginal land and hyper arid areas. Because of its developed root system with large active absorption root surface and high tolerance of disease and insect, this species plays an important role in maintaining ecosystem function and eliminating desertification and erosion.[9] In addition, *S. paniculata* is a national second-level fire-resistant species.[34]

S. paniculata is a deciduous shrub or small tree native to China with ecological and economic importance.[21] It typically grows 1–5 m tall, with spreading habit and taupe barks. The dark green leaves are ovate to obovate, finely toothed, and about 3-in. long (Fig. 11.1). It features 2–3-in. long panicles of creamy white, fragrant flowers in late spring followed by clusters of sapphire-blue berries (~1/3-long drupes) in fall. It usually blooms in April,[33] and its fruits become mature at the end of September and early October.[21]

FIGURE 11.1 *Symplocos paniculata* tree grows in the wild (A), with green leaves (B), sapphire-blue berries (C), white flowers (D), and taupe barks (E).

A mature tree can yield up to 20 kg of fruit.[32] The whole fruit contains 36.6% oil,[20] of which 79.8% is unsaturated fatty acid. Oil is mainly located in the pericarp of fruit (Fig. 11.2). Due to high fruit yield and oil content, *S. paniculata* serves as an ideal biodiesel feedstock[18] and edible oil plant.[10]

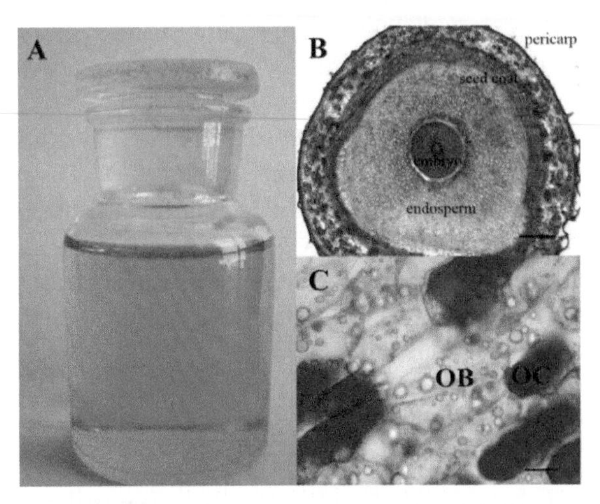

FIGURE 11.2 The oil produced from *Symplocos paniculata* fruit (A); transverse section of whole fruit (B, bar = 1 mm) showing the spatial distribution of oil cells (OC); and oil bodies (OB) (C, bar = 10 μm).

11.3 USE OF *S. PANICULATA* OIL

11.3.1 OIL EXTRACTION FROM *S. PANICULATA* FRUIT

Owing to the high oil content and better fatty acid composition,[20,36] the fruit oil of *S. paniculata* is characterized as a potential source for cooking oil, biodiesel production, and other industrial uses such as ink surfactants, lubricants, and soap. In the *S. paniculata*, palmitic acid, stearic acid, oleic acid, linoleic acid, and linoleic constitute 13.9, 1.6, 53.9, 31.8, and 0.67% in the seed oil compared to 18.3, 1.3, 50.6, 26.7, and 2.5% in the fruit oil, respectively.

11.3.2 COOKING OIL

S. paniculata has long been used as a cooking oil.[10] Zuo et al.[36] reported that oil extracted from *S. paniculata* seeds had high contents of oleic and linoleic acid and low contents of stearic acid. The transparency, smell, and color of the oil produced from *S. paniculata* seeds are similar to that of ordinary cooking oil such as peanut oil.[17] In *S. paniculata*, acid value, iodine value, saponification, and refractive index have been reported to be 18.6, 71.9, 152.2, and 1.4718, respectively.[17] But the acid value is higher than peanut oil, and iodine value and saponification value are lower than that of peanut oil. Refinery processes are needed to improve the oil quality for use as cooking oil. Furthermore, a toxicity test was conducted by treating mice with *S. paniculata* oil at a moderate dose of 0.5–1 mL according to the maximum gastric volume (50 mL · kg^{-1}), and no acute toxic symptom have been observed.[36] Increased motor activity, sedation, acute convulsion, coma, and death were not found during the observation at regular intervals for 24 h up to 7 days.[36]

11.3.3 BIODIESEL PRODUCTION

The physicochemical properties are identification indexes of oil quality and they play a vital role in biodiesel production[17]. The physicochemical properties of several woody oil plant species have been identified in Table 11.2.[17] Acid value is a measure of the amount of carboxylic acid groups in fatty acid. The acid value of *S. paniculata* oil is 18.59 mg of KOH per gram, which is higher than that of other oil plant species listed in the table. The iodine value is an index for the amount of unsaturation in fatty acids. The iodine value of *S. paniculata* oil is 71.93 g per 100 g, which is 25% lower than peanut oil. Saponification value is a measure of the average molecular weight (or chain length) of all the fatty acids present. *S. paniculata* oil has a saponification value of 152.22 mg of KOH per gram. It is comparable to *Sapium sebiferum* and *Styrax tonkinesis* oil and lower than peanut oil. Pretreatment processes are needed to improve the oil quality for biodiesel production, and the acid value of the *S. paniculata* oil should especially be reduced before transesterification.

11.3.3.1 ACID PRETREATMENT

The crude oil of *S. paniculata* fruit has a high acid value, which may lead to low efficiency of biodiesel conversion. A pretreatment is required to reduce the acid value of the crude oil to an ideal range before transesterification. Liu et al.[18] conducted an experiment to de-acidify the crude oil produced from

S. paniculata fruit. They mixed the crude oil with methanol at a ratio of 1:2 (w/v) and extracted at 32°C for 10 min, and this process was repeated four times. The acid value could be reduced from 18.59 mg of KOH per gram to 1.5 mg KOH per gram. The quality of oil was improved greatly and it was suitable for biodiesel conversion.

11.3.3.2 FATTY ACID TRANSESTERIFICATION

Liu et al.[18] conducted experiments to convert the *S. paniculata* crude oil into biodiesel. *S. paniculata* crude oil was mixed with methanol at a molar ratio of 1:6 and catalyzed using NaOH at a dosage of 1.2% by mass at 60°C for 2 h. The average transesterification rate was up to 92%. The biodiesel produced had a cetane number of 65.98, density of 0.892 g cm^{-3}, and dynamic viscosity of 3.5 mm^2 s^{-1}. These measurements are similar to the 0# for petroleum diesel and qualify for European standard after biodiesel conversion.[18] Liu et al.[18] compared biodiesel production from *S. Paniculata* fruit oil with European specifications and $0^{\#}$ fossil diesel.[18] The flash point was more than 147°C, which is higher than 0# petroleum diesel (50°C). It is more potent in blending 20% biodiesel made from *S. paniculata* oil into petro-diesel.

11.4 ORNAMENTAL USES

S. paniculata is a great potential ornamental plant with high quality in esthetic appearance.[23] The plant has been introduced to the United States as a hardy ornamental for its beautiful creamy white, fragrant flowers, and sapphire-blue berries.[1] Along with the breeding and selection of *S. paniculata* for biodiesel production, two selections with distinguished sapphire berries have been developed and named as *S. paniculata* "Lan Jingling" and "Zi Qiu".

S. paniculata "Lan Jingling" is selected from a natural population at Daweishan National Park (Liuyang, Hunan) in China. A plant with sapphire-blue fruits and acuminate leaves (Fig. 11.3A) distinguishes itself from other plants around. The population is distributed in the mountain area with an average elevation of 800–1600 m. Provenance test at the experimental plantation at the Hunan Academy of Forestry (Changsha, Hunan) showed that the unique features could be maintained stable and plants can reach 3-m tall in 5 years with a fruit yield of 2.5 kg per tree. It blooms from May to June.

S. paniculata "Zi Qiu" is selected from a pool of clones with high fruit yield at the experimental plantation at the Hunan Academy of Forestry (Changsha, Hunan). The clones are propagated from a plant in the natural population at an elevation of 100–450 m. It is a natural dwarf shrub with a vigorous branching habit. The flowering period is usually from April to May, 1 month earlier than *S. paniculata* "LanJingling". The unique features are grey fruits and acute leaves (Fig. 11.3B).

FIGURE 11.3 Two newly selected *Symplocos paniculata* cultivars (A: "Lan Jingling"; B: "Zi Qiu") for ornamental uses.

However, the exploitation of *S. paniculata* for ornamental uses is still limited. More research is needed to breed and select novel cultivars, test their genetic stability and adaptability, and to develop commercial propagation protocols.

S. paniculata is an ideal plant for bonsai because it grows extremely slow.[35] The special shape and delicate texture of *S. paniculata* roots make it an excellent candidate for root carving.[25] In addition, *S. paniculata* produces a delicate texture of timber, which is ranked one of the finest woods in China. The timber can be processed for upscale furniture, cabinet work, decorative objects, woodcarving, musical instruments, precision molds, religious artifacts, and so on.

11.5 MEDICINAL VALUE

S. paniculata has long been used as a traditional herbal medicine in China. Its roots and leaves have pathogenic heat expelling, detoxifying, detumescence, and anti-inflammatory effects that make it useful in treating mastitis, lymphadenitis, enterodynia, gastric cancer, boils, skin itching, hernia, and urticarial.[31] The flowers can be used to treat fever, stomachache, nausea, vomiting, diarrhea, and burn.[31] The bark has astringent, cooling, and tonic effects, which are useful in the treatment of menorrhagia, bowel complaints, eye diseases, and ulcers.[6] The phytochemicals from stem bark of *S. paniculata* have antimicrobial, analgesic, and anti-inflammatory activities and have been used as traditional medicine for checking abortion in India.[27] Protein Tyrosine Phosphatase-1B (PTP1B) was also extracted from leaves and stems of this species and proposed as a therapy for the treatment of type 2 diabetes and obesity.[24]

11.6 NATURAL DYES

Natural dyes originated from plant materials are a sustainable source of colorants. In contrast to the hazardous effects of synthetic dyes on skin such as allergy, skin cancer, etching, rashes, and so on[13] and severe water and atmospheric pollution resulting from non-degradable byproducts, natural dyes are non-toxic, biodegradable, and ecologically safe. To meet the demand of green chemistry and the use of safer chemicals to minimize the pollution for environmental considerations, the cultivation of dye plants and the development of novel natural dyes have recently drawn considerable attentions. A yellow dye is obtained from its rough yellowish-brown and corky bark.[8] Badoni and Semwal[2] conducted a study to apply the dye extracted from *S. paniculata* leaves and bark on white cotton thread and cloth, particularly to compare the affinity of vegetable mordents (*Euomymus tingens* bark extracts and *Myrica esculenta* leaf extracts) and common synthetic mordents (copper sulfate and stannous chloride) with fiber. They concluded that *S. paniculata* (leaves and bark) dyed with natural mordents (extracts *from E. tingens* bark extracts and *M. esculenta* leaf) produce different color hues with better fastness properties than that of synthetic mordents ($CuSO_4 \cdot 5H_2O$ and $SnCl_2 \cdot 2H_2O$).

Natural dyes have been also used in food industry as safe food additives, in cosmetics, and in pharmaceutical preparations. Chen et al.[4] reported that *S. paniculata* fruits contain abundant red pigment (605 mg/100 g), which can be easily extracted with organic solvent of 50% ethanol and 1.5 mol/L HCl at a volume ratio of 15:85. They further optimize the extraction technique of total flavonoids of *S. paniculata* leaves and found that 4.6% yield of total flavonoids could be extracted using 60% ethanol with application of ultrasonic wave (25 kHz) for 50 min at 80°C.[3] The produced red pigment was identified as water-soluble and alcohol-soluble anthocyanins with high stability in acid solution, in solution containing Na_2SO_3, K^+, Ca^{2+}, Na^+, Mg^{2+}, Al^{3+}, or Zn^{2+}, and under light or heat.[5] High content of glucose maltose citric acid can protect color of the pigment, while H_2O_2, Vitamin C, Cu^{2+}, Mn^{2+}, Fe^{3+}, and high content of sorbic acid can degrade the pigment.

11.7 LIVESTOCK FEED

S. paniculata contains 12.8–13.9% of crude protein and 16.4% of crude fiber in fresh leaves. Seed cake, a byproduct obtained from oil extraction, is also high in protein. In both fresh leaves and seed cake, there are 17 kinds of amino acids such as lysine, methionine, threonine, isoleucine, glutamic acid, alanine, and arginine.[10,36] Among them, seven kinds are essential amino acids for animals, for example: lysine, phenylalanine, methionine, threonine, isoleucine, leucine, and valine. There are only nine kinds of amino acids in rice hull and 10 kinds of amino acids in wheat hull and bean cake. The content of the amino acids in fresh leaves and seed cake of *S. paniculata* is comparable to or higher than that in rice hull, wheat hull, and bean cake.[10,36] In addition, they are also rich in vitamin C and minerals[36] in both the leaves and seed cake of *S. paniculata*. Therefore, the leaves and seed cake would be a valuable livestock protein feed supplement.

11.8 SUMMARY

S. paniculata is a new energy-efficient plant with superior adaptability and ecological benefits. As awareness among people toward eco-friendly natural products increases, it shows great potential for use in biodiesel production. Research activities on the production of biodiesel from the fruit oil are summarized. This chapter also documents the treasures of indigenous knowledge of other multiple uses such as ornamental and medicinal uses, natural dyes production, and livestock feedstock. More detailed research is needed to assess the real potential and availability of this renewable plant species and to optimize production procedures to improve the quality and quantity of biodiesel. Biotechnology and other modern techniques are required to breed better cultivars with high fruit yield and oil content, to develop a rapid propagation protocol to produce this high-demanding species on a commercial scale and to cultivate this species for biodiesel production and other industrial uses.

ACKNOWLEDGMENT

The study was financially supported by China's 12th 5-Year Plan for Key Technologies Research and Development Program, Ministry of Science & Technology of China (2014BAC09B01, Development of Key Technologies for Nonpoint Source Pollution Management and Wetland Ecosystem Restoration around Dongting Lake; and 2015BAD15B02, Selection and Cultivation of Novel Oil Plants for Bioenergy Uses) and the China Scholarship Council (201508430158).

KEYWORDS

- **biodiesel**
- **cooking oil**
- **medicinal value**
- **natural dyes**
- **renewable energy**
- ***Symplocos paniculata***

REFERENCES

1. Arnold Arboretum. *Bulletin of Popular Information*; Arnold Arboretum of Harvard University, 1921, *7*(7), 28.
2. Badoni, R.; Semwal. D. K. Dyeing Performance of *Symplocos paniculata* and *Celtis australis*. *Asian Dyer* **2012,** *8*(6), 51–58.
3. Chen, Z. B. Preliminary Study on Total Flavonoids in Leaves and Red Pigment in Flesh of *Symplocos paniculata* (Thunb.) Miq. Master's Degree Thesis. Fujian Normal University, Fuzhou, China, 2005, 72 p.
4. Chen, Z. B.; Liu, J. Q.; Chen, B. H.; Xu, G. Y.; Xie, R. F. Content Mensuration of Red Pigment in Flesh of *Symplocos paniculata* and Elementary Study on Its Basic Physical and Chemical Properties. *J. Zhangzhou Normal Univ.* **2007,** *1*, 108–112.
5. Chen, Z. B.; Liu, J. Q.; Chen, B. H.; Xu, G. Y.; Xie, R. F. Study on the Stability of Red Pigment in Flesh of *Symplocos paniculata*. *J. Zhangzhou Normal Univ.* **2008,** *1*, 91–96.
6. Chopra, R. N.; Nayar, S. L.; Chopra, I. C.; Asolkar, L. V.; Karkar, K. K.; Chakre, O. J.; Varma, B. S. *Glossary of Indian Medicinal Plants;* Council of Scientific and Industrial Research Publications and Information Directorate: New Delhi, India, 1992; 240 p.
7. Drapcho, C. M.; Nhuan, N. P.; Walker, T. H. *Biofuels Engineering Process Technology.* McGraw-Hill: New York, USA, 2008; 114 p.

8. Gaur, R. D. *Flora of District Garhwal North West Himalaya: With Ethnobotanical Notes*; TransMedia: Srinagar, UP, India, 1999; p 84, 105, 204, 329.
9. Guan, Z. X. *Symplocos paniculata. J. Soil Water Conserv.* **1991,** *8*, 44.
10. Guan, Z. X.; Zhu, T. P.; Chou, T. Q. The Oil and Amino Acid Analysis and Utilization Evaluation of *Symplocos paniculata* Seeds. *Chin. Wild Plant Resour.* **1991,** *2*, 11–14.
11. IEA (International Energy Agency)—OECD (Organization for Economic Cooperation and Development). *World Energy Outlook 2007*, International Energy Agency, Paris, France, 2008, pp 74–77.
12. Ji, C. B.; Zou, K. Seedling Technology of *Symplocos paniculata* in Northern China. *J. Pract. For. Technol.* **2009,** *2*, 28–29.
13. Kamel, M. M.; El-Shishtway, R. R.; Yussef, B. M.; Mashaly, H. Ultrasonic Assisted Dyeing: III. Dyeing of Wook with Lac as a Natural Dye. *Dyes Pigments* **2005,** *65*(2), 103–110.
14. Korbitz, W. Biodiesel Production in Europe and North America, An Encouraging Prospect. *Renew. Energy* **1999,** *16*, 1078–1083.
15. Li, X. F.; Hou, S. L.; Su, M.; Yang, M. F.; Shen, S. H.; Jiang, G. M.; Qi, D. M.; Chen, S. Y.; Liu, G. S. Major Energy Plants and Their Potential for Bioenergy Development in China. *Environ. Manag.* **2010,** *46*, 579–589.
16. Lin, C. S.; Li, Y. Y.; Liu, J. L.; Zhu, W. B.; Chen, X. Diversity of Energy Plant Resources and Its Prospects for the Development and Application. *Henan Agric. Sci.* **2006,** *12*, 17–23.
17. Liu, G. B.; Liu, W. Q.; Huang, C. G.; Du, T. Z.; Hu, D. N.; Huang, Z.; Qiu, Z. B.; Li, B. J.; Shan, T. Physico-chemical Properties and Preparation of Biodiesel with Five Categories of Woody Plants Seeds Oil. *Acta Agric. Univ. Jiangxiensis* **2010,** *32*(2), 339–344.
18. Liu, G. B.; Liu, W. Q.; Huang, C. G.; Du, T. Z.; Huang, Z.; Wen, X. G.; Xia, D. Q.; He, L. Physiochemical Properties and Preparation of Biodiesel by *Symplocos paniculata* Seeds Oil. *J. Chin. Cereals Oils Assoc.* **2011,** *26*(3), 64–67.
19. Liu, J.; Liu, Q.; Jiang, L. J. Diversity and Vertical Distribution Characteristics of *Symplocos* spp. Communities in Daweishan Mountain. *J. Northwest For. Univ.* **2015,** *30*(4), 121–126.
20. Liu, Q.; Li, C. Z.; Jiang, L. J.; Li, H.; Chen, J. Z.; Yi, X. Y. The Oil Accumulation of Oil Plant *Symplocos paniculata. J. Biobased Mater. Bioenergy* **2015,** *9*, 1–5.
21. Liu, Q.; Yang, Y.; Yin, X.; Jiang, L. J. Fruit Morphological Development of Oil Plant *Symplocos paniculata. Chin. Wild Plant Resour.* **2012,** *31*(6), 53–55, 61.
22. Liu, Q.; Yin, X.; Yang, Y.; Chen, J. Z.; Jiang, L. J. Analysis of Genetic Diversity and Genetic Structure in Natural Populations of *Symplocos paniculata. J. Plant Genetic Resour.* **2015,** *16*(4), 751–758.
23. Ma, Q.; Jiang, L. J.; Li, C. Z. The Energy Plants Introduction—12th Species: *Symplocos paniculata. Solar Energy* **2009,** *12*, 27–28.
24. Na, M. K.; Yang, S.; He, L.; Oh, H.; Kim, B. S.; Oh, W. K.; Kim, B. Y.; Ahn, J. S. Inhibition of Protein Tyrosine Phosphatase 1B by Ursane-Type Triterpenes Isolated from *Symplocos paniculata. Planta Medica* **2006,** *72*, 261–263.
25. Peng, C. S.; Zhu, T. B.; Ding, Y. Z. A Survey of Root Carving Art in China. *J. Beijing For. Univ.* **1989,** *11*(1), 93–98.
26. Pyter, R.; Voigt, T.; Heaton, E.; Dohleman, F; Long, S. Giant Miscanthus: Biomass Crop for Illinois. In *Issues in New Crops and New Uses;* ASHS Press: Alexandria, VA, USA, 2007; pp 39–42.

27. Ruchi, B. S.; Deepak, K. S.; Ravindra, S. Chemical Constituents from the Stem Bark of *Symplocos paniculata* Thunb, with Antimicrobial, Analgesic and Anti-inflammatory Activities. *J. Ethnopharmacol.* **2011,** *135*, 78–87.
28. Schmer, M. R.; Vogel, K. P.; Mitchell, R. B; Perrin, R. K. Net Energy of Cellulosic Ethanol from Switchgrass. *Proc. Natl. Acad. Sci. USA* **2008,** *105*(2), 464–469.
29. U.S. Energy Information Administration. *Annual Energy Outlook 2014 Early Release Overview*; Washington, DC, USA, 2014; p ES-4.
30. Wu, Z. Y.; Raven, P. H. *Flora of China*; Missouri Botanical Garden Science Press: St Louis, USA, 1996; 387 p.
31. Xie, Z. W.; Yu, Y. C. *Chinese Herbal Medicine Dictionary*; People's Medical Publishing House: Beijing, China, 1996; p 667.
32. Yang, Y.; Jiang, L. J.; Li, C. Z.; Li, P. W.; Chen, J. Z.; Xu, Q. Investigation and Analysis of Wild *Symplocos paniculata* Resources in Dawei Mountain. *Hunan For. Sci. Technol.* **2011,** *38*(6), 36–38.
33. Yin, X.; Yang, Y.; Liu, Q.; Jiang, L. J. Phenophase Period and Morphological Diversity of Different *Symplocos paniculata* Populations. *Nonwood For. Res.* **2012,** *30*(3), 55–60.
34. Zhang, J. L.; Zeng, X. F.; Liu, X. Q.; Qi, R.; Deng, X. Z.; Xiong, Y. P.; Lu, S. Q.; Chen, Z. X.; Xue, J. C. Study on the Selection of Fireproof Trees in Hubei Province. *J. HuaZhong Agric. Univ.* **2000,** *19*(1), 84–90.
35. Zhu, J. Y.; Lu, J. M.; Xiao, Z. Structures of the Stem Secondary Xylem of *Symplocos paniculata*. *Bull. Bot. Res.* **2006,** *26*(5), 563–564.
36. Zuo, C. Q.; Zhu, T. P.; Du, J. R. Study on Value of Exploitation and Utilization of *Symplocos paniculata*. *Jiangxi Hydraulic Sci. Technol.* **1994,** *20*(1), 68–72.

CHAPTER 12

SALT-TOLERANT BIOENERGY CROPS FOR IMPROVING PRODUCTIVITY OF MARGINAL LANDS

YOUPING SUN[1,*], GENHUA NIU[1], PEIWANG LI[2], and JOE MASABNI[3]

[1]*Assistant Professor, Department of Plants, Soils and Climate Utah State Uni 4820 Old Main Hill Logan UT 84322, USA*

[2]*Hunan Academy of Forestry, 658 South Shaoshan Road, Changsha, 41004 Hunan, China*

[3]*Texas A&M AgriLife Research and Extension at Overton, Texas A&M University, 1710 FM 3053 N, Overton, TX 75684, USA*

[*]*Corresponding author. E-mail: youping.sun@usu.edu*

CONTENTS

ABSTRACT

Development of an efficient, economically viable bioenergy crop production industry in the marginal lands will take time. It is essential that breeding programs pay close attention to screening salt-, drought-, and heat-tolerant genotypes of multipurpose bioenergy crops from available germplasm. In addition, development of cultural practices and irrigation technology for marginal lands with low-quality water is imperative. Finally, development of optimal energy crop agronomic systems will likely require the use of modern molecular biology tools to generate plants with elite traits for biofuel production.

12.1 INTRODUCTION

As the world population and global economy expands, the demand to meet the ever-increasing energy consumption continues to increase. The primary energy consumption of the world increased from 10,557 million tons oil equivalent in 2004 to 12,928 million tons in 2014.[8] China surpassed the USA in 2009 to become the largest energy consumer of the world. Since then, China has increased its primary energy consumption up to 2972 million tons oil equivalent in 2014, a 28.5% increase.[8] This trend of increasing energy consumption will continue in the future in other developing countries. In contrast, discoverable fossil-fuel reserves are being quickly depleted.[32] Fossil fuel has been the major contributor to greenhouse gas emission and global warming. Additionally, use of fossil fuel such as oil and coal has resulted in many environmental and health problems.[11]

Renewable energy (bioenergy) has the potential to reduce greenhouse gas emissions and is expected to make a significant contribution to meet global energy needs. The U.S. Energy Information Administration[77] expected that renewable energy, excluding hydropower, would account for 28% of the overall growth in electricity generation from 2012 to 2040. Biofuel, one of the most important types of renewable energy, is gaining popularity and the demand for biofuel is increasing. The global demand for liquid biofuel has tripled between 2000 and 2007.[31] In 2011, a total of 110 billion liters of biofuel were produced worldwide, among which bioethanol and biodiesel accounted for 78.7% and 21.3%, respectively.[76] Bioethanol is produced from agricultural feedstocks such as corn, *miscanthus* (silver grass), sweet potato, sugar cane, sorghum, and switchgrass,[19,63,67] while biodiesel, alkyl esters of fatty acids, is made from edible animal fats and vegetable oils derived from coconut, linseed, palm seed, rapeseed, soybean, and sunflower.[39] Some biofuel feedstocks are food crops. As a consequence, biodiesel feedstock production may compete in the long run with food supply and/or with food crops for arable land.

There is a growing interest in the use of agricultural land to produce biomass for bioenergy. However, it is widely recognized that availability of productive or arable land to grow bioenergy crops will become a limiting factor in future development of bioenergy. Agricultural production is already facing challenges due to decreasing arable land per capita and depleting freshwater resources. Supply of high-quality water and arable land is falling short to the demand of an increasing world population. Growing bioenergy crops on marginal lands would be an alternative way to conserve water, reduce fossil fuel pollution, and secure food safety.

Marginal lands refer to lands that have low inherent productivity and that have been abandoned or degraded.[73] Marginal lands are primarily located in arid and semi-arid regions where soil salinity is too high for most common economic crops and where high salinity groundwater is the primary water source. FAO[23] estimated that 397 million ha of land throughout the world are affected by salinity. Most of the salt-affected areas are marginal lands.[11]

With proper genotype selection and crop diversity, growing salt- and drought-tolerant bioenergy crops in salt-affected marginal lands may be the only option to enhance productivity without competing with food production and to increase income in rural communities. Due to limited supply and poor water quality, plants grown in marginal lands must be salt- and drought-tolerant. Research is necessary to identify suitable species and crop diversity with specific soil and climate conditions.

This chapter focuses on salt-tolerant bioenergy crops for improving productivity of marginal lands.

12.2 EFFECTS OF SALINITY ON PLANTS

Effective use of marginal lands and saline groundwater to grow bioenergy crops requires knowledge of salt tolerance of potential bioenergy crops. Salt tolerance is defined as the ability of a plant to withstand the effects of high salinity without significant adverse effects to growth, yield reduction, or foliar damage.[56,69] Salinity decreases soil water potential, thereby making water less available to plants. This causes reductions in growth rate, alongside a suite of metabolic changes identical to those caused by water stress.[53] Typical plant responses to salinity include slower plant growth, smaller size of whole plants, and foliar injury such as leaf burn, scorch, necrosis, premature defoliation, or even plant death.[53]

Salinity may also induce a series of metabolic dysfunctions in plants including absorption of excessive minerals, nutrient imbalance, and inhibition of plant photosynthesis and stomatal conductance.[54] In the greenhouse study by authors, excessive minerals including Na, Ca, and Cl accumulated in the leaf tissue of castor plants irrigated with saline solutions compared to their respective control.[71] Salinity also resulted in nutrient imbalance. For example, salinity decreased the concentration of K, Fe, and Al, but increased Mn and Zn concentrations in castor leaf tissue. Additionally, salinity negatively impacted the leaf net photosynthesis (Pn) of castor cultivars "Energia," "Hale," "HCastor," "Memphis," and "Ultra dwarf," but not on "Brigham."

The negative effects of salinity on plant growth and physiological processes often depend on salinity level and length of exposure. The actual response of a plant to salinity is often affected by climatic conditions, type of substrate or soil, and irrigation management.[56,69] For example, salt injury symptoms are more evident under hot, dry, and windy conditions than in a cool, humid environment.[56] The response of a plant to salinity also varies largely with species or even cultivars within a species.[59] In addition, salt tolerance of a plant is dependent on the growth stage. In general, seedlings are more susceptible to high salinity than mature plants.[56] At each salinity level, it is possible to find a selection of plants capable to tolerate and grow normally. Obviously, as salinity levels increase, fewer plants can tolerate the increasing salt stress.

12.3 SALT TOLERANCE OF POTENTIAL BIOENERGY CROPS

12.3.1 AROMATIC LITSEA

Aromatic litsea or May Chang [*Litsea cubeba* (Lour.) Pers.] belongs to the *Lauraceae* family and is native to China, Indonesia, and other parts of Southeast Asia.[80] It is an evergreen tree or shrub that can grow 5–12 m tall. It produces fruits with 3–5% lemony essential oil that can be used for the synthesis of vitamin A and violet-like fragrances.[14] Aromatic *litsea* is mainly cultivated in China with production estimates between 500 and 1500 tons of oil per annum. The kernel oil content is 27.8%.[29] It is also selected as a candidate feedstock for biodiesel production in China because of high fruit yield and oil content. Zhang et al.[83] reported that biodiesel (alkyl esters of fatty acids) could be produced by mixing aromatic *litsea* kernel oil with ethanol at a molar ratio of 1:16 and 10% $SO_4^{2-}/Fe_2O_3–TiO_2$ solid acid catalyst (w/w) at 78°C for 8 h. The yield of biodiesel was above 45%. In another study, a transesterification rate of 97.6% was achieved when aromatic litsea kernel oil was mixed with methanol at a molar ratio of 1:6, 1% NaOH, and 0.5% of hexadecyl-trimethyl-ammonium bromide (w/w) and reacted at 25°C for 15 min.[10]

Salt tolerance of aromatic *litsea* still remains unclear. However, relative plant species in the same family, *Lauraceae*, have been studied for salt tolerance. Avocado (*Persea americana* Mill.) is known to be the most salt-sensitive cultivated fruit tree. Bernstein et al.[6] observed even low levels of salt (15 mM) inhibited tree growth and decreased productivity. They also found that root growth of avocado is more sensitive to salinity than shoot growth. Further studies are needed to quantify the relative salt tolerance of aromatic *litsea* in order to cultivate this special plant species in marginal lands to provide more feedstocks for bioenergy usages.

12.3.2 CASTOR

Castor (*Ricinus communis* L.) is a flowering plant species in the spurge family, *Euphorbiaceae*. It is indigenous to the southeastern Mediterranean basin, eastern Africa, and India, but is widespread throughout tropical regions and widely grown elsewhere as an ornamental plant.[79] Castor is an ideal plant for industrial oil production and bioenergy use because it has high yields and a unique fatty acid composition.[71] FAO[24] reported a worldwide production of 1.54 million ha and 1.76 million tons of seeds in 2010. The seed contains 40–60% oil rich in triglycerides. Castor oil accounts for 0.15% of vegetable oil production in the world.[68] Many varieties have been selected for oil production and bioenergy usage. For instance, "Brigham" is a variety with 10-fold reduction in ricin content adapted for Texas.[3] "Hale" is a dwarf variety with multiple racemes.[9]

Salt tolerance of castor has been documented.[33,40,64] Pinheiro et al.[61] conducted a greenhouse study to evaluate the salt tolerance of most commonly planted castor cultivars in Brazil. No difference in germination was observed between the control (soil salinity of 1.8 dS/m^{-1}) and salt treatment (soil salinity of 8.4 $dS\ m^{-1}$). However, salt treatment decreased predawn leaf water potential by 42%, stomatal conductance to water vapor by 36%, and net carbon assimilation rate by 24% at 38 days after germination. They concluded that castor does not tolerate this level of salt stress at initial growth stages because of significant reduction in dry mass accumulation.

However, considerable variability in salt tolerance exists among genotypes or cultivars of castor.[40,64] Raghavalah et al.[64] screened salt tolerance of 20 genotypes in terms of seed germination, growth parameters, and ion content of leaves. Only nine genotypes exhibited tolerance to increased salinity. Sun et al.[71] also found that "Memphis" is more tolerant to salt stress than "Brigham," "Energia," "Hale," "Ultra dwarf," and "HCastor."

12.3.3 COTTON

Cotton (*Gossypium* spp.) is a perennial shrub native to tropical and subtropical regions around the world. Cotton is the primary natural fiber for textile industry. It is also a major oilseed crop for oil industry as well as a main protein source for animal feed. Cotton is one of the most economically important crops with a total commercial production of more than 11.0 million planted acres and 16.1 million bales of cotton harvested in the USA in 2014.[55] The demand for cotton fiber is steadily increasing worldwide. Along with cotton fiber and seed production, cotton wastes (e.g., residues from fields and gins) have been converted into pellets, ethanol, methane, and pyrolytic products for bioenergy usage in recent years.[70]

Cotton is a moderately salt-tolerant crop with a threshold salinity of 7.7 dS m^{-1}.[48] It is a good candidate crop for salt-affected lands or marginal lands. However, reductions in cotton growth, yield, and fiber quality due to high salinity in soil or irrigation water have been reported.[18,49] Many studies have revealed the existence of genetic variations in salt tolerance among cotton genotypes.[36] Three *Gossypium hirsutum* genotypes (DN1, DP491, and FM 989) and two *Gossypium barbadense* (Cobalt and Pima S-7) were compared for salt tolerance under NaCl or Na_2SO_4 salinity conditions at similar osmotic potentials (100 mM NaCl vs. 70 mM Na_2SO_4 and 150 mM NaCl vs. 111 mM Na_2SO_4).[57] They found that DP491 was more salt tolerant than other genotypes. Ashraf[2] observed that seed emergence and young seedlings are more sensitive to salinity compared to mature cotton plants.

12.3.4 DOGWOOD

Wilson's dogwood or guangpishu [*Cornus wilsoniana* (Wangerin) Soják, syn. *Swida wilsoniana*], a member of *Cornaceae* family, is a deciduous tree that can grow up to 8–10 m tall.[25] It has green leaves, cream white flowers, purple brown fruits, and beautiful exfoliate bark. It is native to the forest land in China with an elevation ranging from 100 to 1000 m, an average temperature of 18–25°C, and an annual precipitation of 1000–1570 mm.[42] Wilson's dogwood is a potential biodiesel feedstock plant because it is a fast-growing plant with high oil content in fruit and high adaptability to marginal land.[42] It can maintain maximum productivity for over 50 years with an average of 50 kg dry fruit per year produced on a mature tree. The fruit oil content can reach up to 33–36%.[82] In 1997–2007, six high-yield and stress-tolerant Wilson's dogwood cultivars ("Xianglin G1," "Xianglin G2," "Xianglin G3," "Xianglin G4," "Xianglin G5," "Xianglin G6") were selected using conventional breeding protocol such as mass selection, elite-tree selection, and clone testing.[42,43] These six cultivars are important breeding materials for the development and improvement of this special plant for biodiesel usage. Biodiesel production from Wilson's dogwood fruit oil at bench scale has been investigated for 10 years by Hunan Academy of Forestry. Recently, this process was scaled up, and about 3000 tons of biodiesel is being produced from dogwood fruit oil every year (Changzhu Li, personal communication).

Dogwoods are generally considered intolerant to salts.[28] For example, red osier dogwood (*C. sericea* L., syn. *C. stolonifera*) has low tolerance to salts, whether on roots or applied aerially to foliage.[5,15] Tartarian dogwood (*C. alba* L.), cornelian-cherry (*C. mas* L.), gray dogwood (*C. racemosa* Lam.),[5,15] and roughleaf dogwood (*C. drummondii* C.A.Mey.)[27] are sensitive to foliar salt solution spray. Flowering dogwood (*C. florida* L.) is sensitive to salts applied to their foliage,[27] but also sensitive to salt at less than 3 dS m^{-1} in soil.[52] However, no previous reports are available on the salt tolerance of Wilson's dogwood or guangpishu.

12.3.5 JATROPHA

Jatropha (*Jatropha curcas* L.), a C3 perennial plant in *Euphorbiaceae* family, is native to tropical America but now thrives in many parts of the tropics and sub-tropics in Africa and Asia. It grows well in areas with low rainfall and harsh climatic conditions and can alleviate soil degradation, desertification, and deforestation.[26,35] Jatropha has received special attention in many countries and is one of the main crops promoted for biodiesel production in marginal lands.[37,41]

Jatropha has been listed as a slightly salt-tolerant plant.[7] Matsumoto et al.[51] reported that jatropha is highly sensitive to Na accumulation, especially in the root zone. However, jatropha is classified as a moderate salt-tolerant plant as its seedlings can grow with irrigation water at EC of up to 4 dS m^{-1} (30 mM NaCl).[16] The salt tolerance of jatropha has been documented. Jatropha plants were evaluated for salt tolerance in a greenhouse by irrigating with saline solution at four salinity levels, namely electrical conductivity (EC) of 1.6 (control, no addition of salts to nutrient solution), 3, 6, and 9 dS m^{-1}.[58] The nutrient solution contained 0.5 g L^{-1} 20 N–8.6 P–16.7 K (Peters 20-20-20, Scotts). Saline solutions were prepared by dissolving sodium chloride (NaCl), magnesium sulfate heptahydrate ($MgSO_4 \cdot 7H_2O$), and calcium chloride ($CaCl_2$) at 87:8:5 (w/w) to the nutrient solution. Plants were watered daily and the experiment terminated after 54 days.

Typical symptoms of salinity stress such as leaf edge yellowing were observed in all elevated salinity treatments on the lower leaves and the degree of foliar salt damage increased with increasing salinity of the solution. Total dry weight of jatropha plants was reduced by 30%, 30%, and 50% when irrigated with saline solutions at EC of 3, 6, and 9 dS m^{-1}, respectively, compared to the control treatment.[58] Leaf Na concentration was much higher than that observed in most glycophytes (any plant that will only grow normally in soils with a low sodium salt content). Leaf Cl concentrations were also high.[58] Although salt tolerance was not tested at the fruiting stage, the reduced vegetative growth indicates that yield may be affected if jatropha plants are irrigated with water of elevated salinity or if jatropha plants are grown in land with high salinity (3 dS m^{-1} or higher). Further studies are needed to quantify the effect of salinity on jatropha oil yield.

12.3.6 SAPPHIRE BERRY

Sapphire berry or Asiatic sweet leaf [*Symplocos paniculata* (Thunb.) Miq.], a member of *Symplocaceae* family, is native to China and Japan.[25] It is a deciduous large shrub or small, low-branched tree. It can reach 3–6 m in height and width and is widespreading at maturity. The leaves are alternate and dark green, and fruits are bright blue and mature in September. Sapphire berry has many uses in the ornamental, food, and medicinal industries. It was introduced and planted as an ornamental specimen in the states of Connecticut, New Jersey, New York, Pennsylvania, Washington DC, and Ohio.[75] The fruit is used in jams, jellies, and sauces.[20] A yellow or red dye is extracted from its leaves and bark.[62] The bark is astringent that is useful in the treatment of menorrhagia, bowel complaints, eye diseases, and ulcers.[13] The fruit oil content can reach up to 36.6%, enriched with unsaturated fatty acid.[81] It has recently been promoted as an ideal feedstock for biodiesel production in China due to high fruit yield and oil content. Liu et al.[45] documented that

92.0% transesterification rate was obtained by mixing seed oil with methanol at a molar ratio of 1:6 and 1.2% catalyst (w/w) at 60°C for 2 h. The fuel properties of the biodiesel made from sapphire berry oil are similar to the 0# petroleum diesel.[44]

Sapphire berry has low tolerance to drought conditions. However, its tolerance to salinity in soils is yet to be known. More research work is needed to determine the relative salt tolerance of sapphire berry. This research-based information would help to produce sapphire berry in marginal lands for bioenergy usages.

12.3.7 SORGHUM

Sorghum [*Sorghum bicolor* (L.) Moench] is a grass species cultivated for its grain used for human and animal food, forage, and ethanol production. Sorghum originated in northern Africa and is now cultivated widely in tropical and subtropical regions. Sorghum is a bioenergy crop that can produce ethanol from the grain (starch), juice (sugar), or plant biomass (lignocellulose). In the USA, sorghum has been produced historically as a feed grain. A recent estimate of approximately 30% of US sorghum grain production is currently used for ethanol production. Since sorghum is not a primary food crop, increasing sorghum production for bioenergy will not compete with food use. Sorghum is also a water-use efficient plant and is well adapted to semi-arid areas.

Sorghum is moderately tolerant to saline conditions.[48] Sorghum is more sensitive to salinity at the seedling emergence stage than any other stage[50] and salt tolerance is cultivar-dependent.[4] We recently determined the salt tolerance of four sorghum hybrids (SS304, NK7829, Sordan79, and KS585) by irrigating with nutrient solution at EC of 1.5 dS m^{-1} or saline solution at EC of 8.0 dS m^{-1}.[60] After 40 days of treatment, KS585 had the most severe leaf edge burn and leaf yellowing, followed by NK7929; SS304 had minor leaf edge burn, while Sordan79 looked healthy without any salt damage. Dry weight of shoots was reduced by 51%, 56%, 56%, and 76% in SS304, NK7829, Sordan79, and KS585, respectively. These results indicated that Sordan79 was the most tolerant genotype, and KS585 was the least tolerant. In another greenhouse study, 10 sorghum varieties ("1790E," "BTx642," "Desert Maize," "Macia," "RTx430," "Schrock," "Shallu," "Tx2783," "Tx7078," and "Wheatland") were evaluated for salt tolerance at two stages.[72] At seedling emergence stage, sorghum were sown in substrates moistened with either nutrient solution [control, 1 g L^{-1} 15N–2.2P–12.5K

(Peters 15-5-15; Scotts)] at an EC of 1.2 dS m^{-1} or salt solution at EC 5, 10, or 17 dS m^{-1}. The salt solution was prepared by adding sodium chloride (NaCl) and calcium chloride (CaC12) at 2:1 (molar ratio) to the nutrient solution. Seedling emergence decreased in all varieties at EC 17 dS m^{-1} compared to the control. Seedling emergence of sorghum "Macia" and "1790E" irrigated with salt solution at EC 17 dS m^{-1} decreased by 50% and 51%, while that of "RTx430" reduced by 97%, and reduction in other varieties ranged from 64% to 90%. Compared to control, both salt solutions at EC 5 and 10 dS m^{-1} reduced the dry weight of sorghum seedlings by 29% and 72%, respectively, on average across varieties. At seedling growth stage, plants were irrigated with nutrient solution at EC 1.2 dS m^{-1} or salt solution at EC 5.0 or 10.0 dS m^{-1} for 30 days. Salt solution at EC 5.0 and 10.0 dS m^{-1} had similar influence on dry weight of all sorghum varieties except "Tx2783." The relative dry weight of "Shallu," "Desert Maize," and "1790E" irrigated with salt solution at EC 10 dS m^{-1} were over 67%, those of "Macia," "Schrock," and "RTx430" ranged from 30% to 33%, and other varieties were 45–59%. Foliar salt damage was observed on all salt-treated sorghum varieties except for "Shallu," which had the lowest shoot dry weight reduction and greatest visual score. These results indicated that salt tolerance of sorghum varied with plant growth stage and varieties. "Shallu," "Desert Maize," and "1790E" were the most salt tolerant varieties, while "Schrock" and "RTx430" showed the least salt tolerance at both stages.

12.3.8 SWITCHGRASS

Switchgrass (*Panicum virgatum* L.) is a native perennial warm season grass that is primarily used for soil conservation, forage production, and ornamental grass. It has been identified as a sustainable source of biomass feedstock for energy production with the potential to produce about 380 L of ethanol per metric ton.[30] Substantial efforts are being made in developing switchgrass for forage production, cellulosic ethanol production, biogas, and direct combustion for thermal energy applications. For instance, "Alamo" is a tetraploid lowland variety with coarser foliage and late maturity date.[66] "Kanlow" is a tetraploid lowland variety that is suited for poorly drained sites or areas subject to periodic flooding.[66] Switchgrass has broad adaptability, it tolerates water and nutrient limitations, and has the ability to produce moderate-to-high biomass yields on marginal lands.[34,65,78]

Salinity causes reduction in seed germination,[12,38,46] seedling emergence,[17] seedling growth,[17,22,38] and yield of switchgrass.[1] It also modifies plants' physiological and biochemical processes.[54] According to Tober et al.,[74] switchgrass is moderately sensitive to salt conditions at EC 5–10 dS m^{-1}. Switchgrass is more sensitive to salinity at the seedling emergence stage than at any other stage.[74] Additionally, the salt tolerance of switchgrass varies with different varieties.[21,47]

In 2013, switchgrass "Alamo," "Cimarron," "Kanlow," "NL 94C2-3," "NSL 2009-1," and "NSL 2009-2" were evaluated at two stages in a greenhouse study. At seedling emergence stage, switchgrass seeds were sown in substrates moistened with nutrient solution [control, 1 g L^{-1} 15N–2.2P–12.5K (Peters 15-5-15; Scotts)] at EC 1.2 dS m^{-1} or salt solution at EC 5, 10, or 20 dS m^{-1}. Salt solution at EC 5 dS m^{-1} did not inhibit the seedling emergence of all switchgrass varieties (Table 12.1). While salt solution at EC 10 dS m^{-1} did not impact the seedling emergence of switchgrass "Kanlow," "NSL 2009-1," and "NSL 2009-2," but reduced "Alamo," "Cimarron," and "NL 94C2-3" by 44%, 33%, and 82%, respectively, compared to control.

TABLE 12.1 Seedling Emergence of Switchgrass Varieties Subirrigated with Salt Solutions at Electrical Conductivity (EC) of 5.0 dS m^{-1} (EC 5) or 10.0 dS m^{-1} (EC 10) in the Greenhouse.

Variety	Seedling emergence (%)		
	Control	EC 5	EC 10
Alamo	41.7[a†]	35.0 (16)[ab]	23.4 (44)[b]
Cimarron	57.5[ab]	63.4 (0)[a]	38.4 (33)[b]
Kanlow	63.4[a]	61.7 (3)[a]	47.5 (25)[a]
NL 94C2-3	51.7[a]	46.7 (10)[a]	9.2 (82)[b]
NSL 2009-1	50.8[a]	34.2 (33)[a]	26.7 (48)[a]
NSL 2009-2	22.5[a]	25.9 (0)[a]	10.8 (52)[a]

Nutrient solution (EC = 1.2 dS m^{-1}) was used as the control. Data for the treatment of EC of 19.8 dS m^{-1} (EC 20) are not presented since no seedlings emerged. Relative reduction (%) in seedling emergence was calculated as percent of control and presented in parentheses.

[†]For each variety, means with same letters are not significantly different among treatments by Tukey's HSD multiple comparison at $P < 0.05$.

At seedling growth stage, switchgrass seedlings were irrigated with aforementioned nutrient solution at EC 1.2 dS m^{-1} (control) or salt solution at EC 5.0 or 10.0 dS m^{-1} for 36 days. Salt solution at EC 5 dS m^{-1} tended to inhibit tiller formation of all switchgrass varieties except "Cimarron," whereas salt solution at EC 10 dS m^{-1} reduced the number of tillers by 32–37% for all switchgrass varieties except "Kanlow" (Table 12.2). Dry weight of switchgrass "Cimarron" and "Kanlow" irrigated with salt solution at EC 5 dS m^{-1}

decreased by 42% and 28%, respectively. All switchgrass varieties irrigated with salt solution at EC 10 dS m^{-1} had a significant reduction of 50–63% in dry weight. These results showed that switchgrass has a moderate tolerance to the salinity at EC 5 dS m^{-1} at both seedling emergence and growth stages.

TABLE 12.2 Number of Tillers and Dry Weight of Switchgrass Varieties Irrigated with Salt Solutions at Electrical Conductivity (EC) of 5.0 dS m^{-1} (EC 5) or 10.0 dS m^{-1} (EC 10) in the Greenhouse.

Variety	Number of tillers			Dry weight (g)		
	Control	EC 5	EC 10	Control	EC 5	EC 10
Alamo	44.6[a]	37.4 (16)[ab]	29.8 (33)[b]	43.1[a]	34.4 (20)[a]	20.8 (52)[b]
Cimarron	50.6[a]	37.8 (25)[b]	34.6 (32)[b]	42.0[a]	24.6 (42)[b]	18.9 (55)[b]
Kanlow'	39.6[a]	32.8 (17)[a]	29.2 (26)[a]	55.2[a]	39.7 (28)[b]	24.5 (56)[b]
NL 94C2-3	44.8[a]	46.4 (0)[a]	30.4 (32)[b]	52.1[a]	41.9 (20)[a]	19.5 (63)[b]
NSL 2009-1	58[a]	44.2 (24)[ab]	36.8 (37)b	51.0[a]	40.0 (22)[ab]	25.4 (50)[b]
NSL 2009-2	55.8[a]	48.6 (13)[ab]	38.0 (32)b	70.5[a]	54.9 (22)[a]	26.7 (62)[b]

Nutrient solution (EC = 1.2 dS m^{-1}) was used as the control. Relative reduction (%) in number of tillers and dry weight were calculated as percent of control and presented in parentheses.
†For each variety, means with same letters are not significantly different among treatments by Tukey's HSD multiple comparison at $P < 0.05$.

12.4 SUMMARY

As population continues to expand, more food and energy are needed to meet the increasing demand. Renewable energy or bioenergy has drawn great attention worldwide because of depleting fossil-fuel reserves and environmental consequences of exhaust gases from fossil fuel. Bioenergy is produced from agricultural feedstocks (e.g., corn, miscanthus, sorghum, sugar cane, sweet potato, and switchgrass) and edible animal fats and vegetable oils derived from coconut, linseed, rapeseed, palm seed, soybean, and sunflower. Many of them are food crops. Therefore, the production and use of bioenergy may compete in the long term with food supply and with food crops for arable land. It would seem necessary to use marginal lands to grow bioenergy crops and save quality arable land for food production. Marginal lands have low inherent productivity, poor soil, or other undesirable characteristics. The primary challenge for growing bioenergy crops on marginal lands is the elevated salinity in the soil and irrigation water. This paper reviews the salt tolerance of several potential bioenergy crops such as aromatic litsea, castor, cotton, dogwood, jatropha, sapphire berry, sorghum, and switchgrass.

ACKNOWLEDGMENT

This work was partially supported by the National Science and Technology Support Program, Ministry of Science and Technology, China (2015BAD15B02). The content is solely the responsibility of the authors and does not necessarily represent the official views of the funding agency.

KEYWORDS

- **aromatic *Litsea***
- **bioenergy**
- **bioethanol**
- **dogwood**
- ***Jatropha***
- **renewable energy**
- **sapphire berry**

REFERENCES

1. Anderson, E.; Voigt, T.; Lee, D. *Salt Tolerance in Panicum virgatum and Spartina pectinata*, American Society of Agronomy, Crop Science Society of America, and Soil Science Society of America International Annual Meetings, Cincinnati, OH, October 21–24, 2012 (abstract).
2. Ashraf, M. Salt Tolerance of Cotton: Some New Advances. *Crit. Rev. Plant Sci.* **2012,** *21*, 1–30.
3. Auld, D. L.; Rolfe, R. D.; McKeon, T. A. Development of Castor with Reduced Toxicity. *J. New Seeds* **2001,** *3*, 61–69.
4. Azhar, F. M.; McNeilly, T. Variability for Slat Tolerance in *Sorghum bicolor* (L.) Moench. Under Hydroponic Conditions. *J. Agron. Crop Sci.* **1987,** *159*, 269–277.
5. Beckerman, J.; Lerner, B. R. *Salt Damage in Landscape Plants*. Purdue Extension, Purdue University, Bulletin ID-412-W, 2009, p 10.
6. Bernstein, N.; Meiri, A.; Zilberstaine, M. Root Growth of Avocado is More Sensitive to Salinity than Shoot Growth. *J. Am. Soc. Hort. Sci.* **2004,** *129*(2), 188–192.

7. Black, R. J. *Salt Tolerant Plants for Florida.* IFAS Extension, University of Florida, ENH26, 2003, p 7.
8. *BP Statistical Review of World Energy;* BP: London, 2015, www.bp.com (accessed December 15, 2016).
9. Brigham, R. D. Registration of Castor Variety Hale. *Crop Sci.* **1970,** *10*, 457.
10. Cai, H.; Zhong, S.; Ai, H. Preparation of Biodiesel from *Litsea cubeba* Kernel Oil. *J. Central South Univ. (Sci. Technol.)* **2009,** *40*(6), 1517–1521.
11. Cai, X.; Zhang, X.; Wang, D. L. Availability for Biofuel Production. *Environ. Sci. Technol.* **2011,** *45*(1), 334–339.
12. Carson, M. A.; Morris, A. N. Germination of *Panicum virgatum* Cultivars in a NaCl Gradient. *BIOS* **2012,** *83*(3), 90–96.
13. Chopra, R. N.; Nayar, S. L.; Chopra I. C. *Glossary of Indian Medicinal Plants (Including the Supplement);* Council of Scientific and Industrial Research: New Delhi, 1986; p 233.
14. Coppen, J. J. W. *Non-Wood Forest Products: Flavors and Fragrances of Plant Origin*; FAO: Rome, 1995; p 3.
15. Davidson, H. *Tree and Shrub Tolerance to Deicing Salt Spray*. Michigan State University Extension, Michigan State University, Horticulture Bulletin HM-95, 1996, p 11.
16. Díaz-López, L.; Gimeno, V.; Lidón, V.; Simón, I.; Martínez, V.; García-Sánchez, F. The Tolerance of *Jatropha curcas* Seedlings to NaCl: An Ecophysiological Analysis. *Plant Physiol. Biochem.* **2012,** *54*, 34–42.
17. Dkhili, M.; Anderson B. In *Salt Effects on Seedling Growth of Switchgrass and Big Bluestem,* Proceedings of the Twelfth North American Prairie Conference, 1990, pp 13–16.
18. Dong, H. Technology and Field Management for Controlling Soil Salinity Effects on Cotton. *Aust. J. Crop Sci.* **2012,** *6*, 333–341.
19. Drapcho, C. M.; Nhuan, N. P.; Walker, T. H. *Biofuels Engineering Process Technology;* McGraw-Hill: New York, 2008; p 388.
20. Facciola. S. *Cornucopia: A Source Book of Edible Plants*; Kampong Publications: Vista, CA, 1990; p 205.
21. Fan, X.; Hou, X.; Wu, J.; Zhu, Y. In *Relative Salt Tolerance of Switchgrass (Panicum Virgatum) Varieties During Germination Development*, Proceedings of the 19th European Biomass Conference and Exhibition, 2011, pp 149–151.
22. Fan, X.; Hou, X.; Zhu, Y.; Wu, J. Impacts of Salt Stress on the Growth and Physiological Characteristics of *Panicum virgatum* Seedlings. *Chin. J. Appl. Ecol.* **2012,** *23*(6), 1476–1480.
23. FAO. *Global Network on Integrated Soil Management for Sustainable Use of Salt-Affected Soils*; FAO Land and Plant Nutrient Management Service: Rome, 2015. http://www.fao.org/ag/agl/agll/spush/intro.htm (accessed December 15, 2016).
24. FAO. *World Crop Production Statistics;* United Nations Food and Agriculture Organization: Rome, 2010. http://faostat.fao.org/site/567 (accessed December 15, 2016).
25. Flora of China Editorial Committee. *Flora of China;* Science Press: Beijing, 1994; p 400.
26. Francis, G.; Edinger, R.; Becker, K. A Concept for Simultaneous Wasteland Reclamation, Fuel Production, and Socio-Economic Development in Degraded Areas in India: Need, Potential and Perspectives of *Jatropha* Plantations. *Natl. Resour. Forum* **2005,** *29*(1), 12–24.
27. Gilman, E. F.; Watson, D. G. *Cornus drummondii (Roughleaf Dogwood)*. Florida Cooperative Extension Service, University of Florida. Fact Sheet ST-184, 1993, p 12.
28. Gilman, E. F.; Watson, D. G. *Cornus florida (Flowering Dogwood)*. Florida Cooperative Extension Service, University of Florida. Fact Sheet ST-185, 1993, p 6.

29. He, Y. A New Extraction Protocol for the Kernel Oil of *Litsea cubeba*. *Econ. For. Res.* **1986,** *4*(2), 69–71.
30. Hull, T. *Switchgrass: Native American Powerhouse? Renewable Energy Resources*, 2007. http://www.prognog.com/driving/ethanol/switchgrass-native-american-powerhouse.html (accessed December 15, 2016).
31. IEA (International Energy Agency); OECD (Organization for Economic Cooperation and Development). *World Energy Outlook 2007;* International Energy Agency: Paris, 2008; p 199.
32. IEA (International Energy Agency), OECD (Organization for Economic Cooperation and Development). *World Energy Outlook 2014*; International Energy Agency: Paris, 2014; p 180.
33. Janmohammadi, M.; Abbasi, A.; Sabaghnia, N. Influence of NaCl Treatments on Growth and Biochemical Parameters of Castor Bean (*Ricinus communis* L.). *Acta Agric. Slovenica* **2011,** *99*(1), 31–40.
34. Jimmy, C. Plant Materials Center. *Plant Fact Sheet for Switchgrass (Panicum virgatum L.)*; USDA-Natural Resources Conservation Service: Americus, GA, 2011; p 2.
35. Jingura, R. M. Technical Options for Optimization of Production of *Jatropha* as a Biofuel Feedstock in Arid and Semi-arid Areas of Zimbabwe. *Biomass Bioenergy* **2011,** *35*(5), 2127–2132.
36. Khan, T. M.; Saeed, M.; Mukhtar, M. S.; Khan. A. M. Salt Tolerance of Some Cotton Hybrids at Seedling Stage. *Int. J. Agric. Biol.* **2001,** *3*, 188–191.
37. Kheira, A. A. A.; Atta, N. M. M. Response of *Jatropha curcas* L. to Water Deficits: Yield, Water Use Efficiency and Oilseed Characteristics. *Biomass Bioenergy* **2009,** *33*(10), 1343–1350.
38. Kim, S.; Rayburn, A. L.; Voigt, T.; Parrish, A.; Lee, D. K. Salinity Effects on Germination and Plant Growth of Prairie Cordgrass and Switchgrass. *Bioenergy Res.* **2012,** *5*(1), 225–235.
39. Korbitz, W. Biodiesel Production in Europe and North America: An Encouraging Prospect. *Renew. Energy* **1999,** *16*, 1078–1083.
40. Kumar, D.; Daulay, H. S.; Sharma, P. C. Tolerance of Castor to Soil Salinity. *Ann. Arid Zone* **1989,** *28*(3–4), 249–255.
41. Kumar, D.; Singh, S.; Sharma, R.; Kumar, V.; Chandra, H.; Malhotra, K. Above-ground Morphological Predictors of Rooting Success in Rooted Cuttings of *Jatropha curcas* L. *Biomass Bioenergy* **2011,** *35*(9), 3891–3895.
42. Li, C. Genetic Diversity of Plus Tree and Its Fruit Fatty Acid in *Swida wilsoniana*. Beijing Forestry University, Beijing, PhD Dissertation, 2010, p 188.
43. Li, C.; Liu, Y.; Luo, J.; Li, R.; Yuan, R.; Liu, C. Study on *Cornus wilsoniana* Oil Refining. *Cereals Oils Process.* 2007, *11*(3), 76–78.
44. Liu, G.; Liu, Y.; Huang, C.; Du, T.; Hu, D.; Huang, Z.; Qiu, Z.; Li, B.; Shan, T. Physiochemical Properties and Preparation of Biodiesel with Seed Oil of Five Woody Plants. *Acta Agric. Univ. Jiangxiensis* **2010,** *32*(2), 339–344.
45. Liu, G.; Liu, Y.; Huang, C.; Du, T.; Huang, Z.; Qiu, Z.; Wen, X.; Xia, D.; He, L. Physiochemical Properties and Preparation of Biodiesel by *Symplocos paniculata* Seeds Oil. *J. Chin. Cereals Oils Assoc.* **2011,** *26*(3), 64–67.
46. Liu, Y.; Wang, Q.; Zhang, Y.; Cui, J.; Chen, G.; Xie, B.; Wu, C.; Liu, H. Synergistic and Antagonistic Effects of Salinity and pH on Germination in Switchgrass (*Panicum virgatum* L.). *PLoS ONE* **2014,** *9*(1), 1–10.

47. Liu, Y.; Zhang, X.; Zhao, B.; Childs, K.; Buell, C.; Kim, J. In *Relative Salt Tolerance of 33 Switchgrass Cultivars*; American Society of Agronomy, Crop Science Society of America, and Soil Science Society of America International Annual Meetings, Tampa, FL, November 3–6, 2013 (abstract).
48. Maas, E. V. Salt Tolerance of Plants. *Appl. Agric. Res.* **1986,** *1*, 12–26.
49. Maas, E. V.; Hoffman, G. J. Crop Salt Tolerance—Current Assessment. *J. Irrig. Drain. Div.* **1977,** *103*, 115–134.
50. Macharia, J. M.; Kamau, J.; Gituanja, J. N.; Matu, E. W. Effects of Sodium Salinity on Seed Germination and Seedling Root and Shoot Extension of Four Sorghum [*Sorghum bicolor* (L.) Moench] Cultivars. *Int. Sorghum Millets Newslett.* **1994,** *35*, 124–125.
51. Matsumoto, H.; Yeasmin, R.; Kalemelawa, F.; Watanabe, T.; Aranami, M.; Nishihara, E. Evaluation of NaCl Tolerance in the Physical Reduction of *Jatropha curcas* L. Seedlings. *Agric. Sci.* **2014,** *2*(3), 23–35.
52. Miyamoto, S.; Martinez, I.; Padilla, M.; Portillo, A.; Ornelas, D. *Landscape Plant Lists for Salt Tolerance Assessment;* Texas A&M AgriLife Research Center: El Paso, TX, 2004; p 3.
53. Munns, R. Comparative Physiology of Salt and Water Stress. *Plant Cell Environ.* **2002,** *25*, 239–250.
54. Munns, R.; Tester M. Mechanisms of Salinity Tolerance. *Annu. Rev. Plant Biol.* **2008,** *59*, 651–681.
55. National Cotton Council of America. *Production and Acreage Information*; National Cotton Council of America: Washington, DC; 2015. http://www.cotton.org/econ/cropinfo/production/production.cfm (accessed December 15, 2016).
56. Niu, G.; Cabrera, R. I. Growth and Physiological Responses to Landscape Plants to Saline Water Irrigation—A Review. *HortScience* **2010,** *45*(11), 1605–1609.
57. Niu, G.; Rodriguez, D.; Dever, J.; Zhang, J. Growth and Physiological Responses of Five Cotton Genotypes to Sodium Chloride and Sodium Sulfate Saline Water Irrigation. *J. Cotton Sci.* **2013,** *17*, 233–244.
58. Niu, G.; Rodriguez, D.; Mendoza, M.; Jifon, J.; Ganjegunte, G. Responses of *Jatropha curcas* to Salt and Drought Stresses. *Int. J. Agron.* **2012,** 1–7 (Article ID 632026).
59. Niu, G.; Starman, T.; Byrne, D. Responses of Growth and Mineral Nutrition of Garden Roses to Saline Water Irrigation. *HortScience* **2013,** *48*, 756–761.
60. Niu, G.; Xu, W.; Rodriguez, D.; Sun, Y. Growth and Physiological Responses of Maize and Sorghum Genotypes to Salt Stress. *ISRN Agron.* **2012,** 1–12, (Article ID 145072).
61. Pinheiro, H. A.; Silva, J. V.; Endres, L.; Ferreira, V. M.; Câmara, C. A.; Cabral, F. F.; Oliveira, J. F.; Carvalho, L. W. T.; Santos, J. M.; Filho, B. G. S. Leaf Gas Exchange, Chloroplastic Pigments and Dry Matter Accumulation in Castor Bean (*Ricinus communis* L) Seedlings Subjected to Salt Stress Conditions. *Ind. Crops Products* **2008,** *27*, 385–392.
62. Polunin, O.; Stainton, A. *Flowers of the Himalayas*; Oxford University Press: Oxford, UK, 1997; p 300.
63. Pyter, R.; Voigt, T.; Heaton, E.; Dohleman, F.; Long, S. Giant Miscanthus: Biomass Crop for Illinois. In *Issues in New Crops and New Uses*; Janick, J.; Whipkey, A., Eds.; ASHS Press: Alexandria, VA, 2007; pp 39–42.
64. Raghavaiah, C. V.; Lavanya, C.; Kumaran, S.; Jeevanroyal, T. J. Screening Castor (*Ricinus communis*) Genotypes for Salinity Tolerance in Terms of Germination, Growth and Plant Ion Composition. *Indian J. Agric. Sci.* **2006,** *76*(3), 196–199.
65. Sanderson, M. A.; Reed, R. L.; Mclaughlin, S. B.; Wullschleger, S. D.; Conger, B. V.; Parrish, D.; Wolf, D. D.; Taliaferro, C. M.; Hopkins, A. A.; Ocumpaugh, W. R.; Hussey,

M. A.; Read, J. C.; Tishler, C. R. Switchgrass as a Sustainable Bioenergy Crop. *Bioresour. Technol.* **1996,** *56*, 83–93.

66. Sanderson, M. A.; Reed, R. L.; Ocumpaugh, W. R.; Hussey, M. A.; Van Esbroeck, G.; Read, J. C.; Tischler, C. R.; Hons, F. M. Switchgrass Cultivars and Germplasm for Biomass Feedstock Production in Texas. *Bioresour. Technol.* **1999,** *67*, 209–219.
67. Schmer, M. R.; Vogel, K. P.; Mitchell, R. B.; Perrin, R. K. Net Energy of Cellulosic Ethanol from Switchgrass. *PNAS* **2008,** *105*(2), 464–469.
68. Scholz, V.; Silva, J. N. Prospects and Risks of the Use of Castor Oil as a Fuel. *Biomass Bioenergy* **2008,** *32*, 95–100.
69. Shannon, M. C.; Grieve, C. M.; Francois, L. E. Whole-plant Response to Salinity. In *Plant Environment Interaction*; Wilkinson, R. E., Ed.; Marcel Dekker: New York, 1994; pp 199–244.
70. Sharm-Shivappa, R. R.; Chen, Y. Conversion of Cotton Wastes to Bioenergy and Value-Added Products. *Trans. ASABE* **2008,** *51*, 2239–2246.
71. Sun, Y.; Niu, G.; Osuna, P.; Ganjegunte, G.; Auld, D.; Zhao, L.; Peralta-Videa, J. R.; Gardea-Torresdey, J. L. Seedling Emergence, Growth, and Leaf Mineral Nutrition of *Ricinus communis* L. Cultivars Irrigated with Saline Solution. *Ind. Crops Products* **2013,** *49*, 75–80.
72. Sun, Y.; Niu, G.; Osuna, P.; Zhao, L.; Ganjegunte, G.; Peterson, G.; Peralta, J.; Gardea-Torresdey, J. Variability in Salt Tolerance of *Sorghum bicolor* L. *Agric. Sci.* **2014,** *2*(1), 9–21.
73. Tang, Y.; Xie, J.; Geng, S. Marginal Land-based Biomass Energy Production in China. *J. Integr. Plant Biol.* **2010,** *52*(1), 112–121.
74. Tober, D.; Duckwitz, W.; Sieler, S. *Plant Materials for Salt-affected Sites in the Northern Great Plains;* United States Department of Agriculture-Natural Resources Conservation Service. Plant Materials Center: Bismarck, ND, 2007; p 4.
75. U.S. Department of Agriculture. *PLANTS Database: Symplocos paniculata (Thunb.) Miq. Sapphire-Berry*; Natural Resources Conservation Service: Washington, DC, 2015. http://plants.usda.gov/core/profile?symbol=sypa12 (accessed December 15, 2016).
76. U.S. Energy Information Administration. *International Energy Statistics;* U.S. Department of Energy: Washington, DC, 2011. http://www.eia.gov/ (accessed December 15, 2016).
77. U.S. Energy Information Administration. Annual Energy Outlook 2014 Early Release Overview, Washington, DC, 2014.
78. Vogel, K. P. Energy Production from Forage. *J. Soil Water Conserv.* **1996,** *51*, 137–139.
79. Weiss, E. A. *Oilseed Crops*, 2nd edition; Blackwell Science: Oxford, UK, 2000; p 324.
80. Wikipedia, the Free Encyclopedia. *Litsea cubeba*. Wikimedia Foundation, Inc.: San Francisco, CA, 2015. https://en.wikipedia.org/wiki/Litsea_cubeba (accessed December 15, 2016).
81. Yang, Y. Genetic Diversity and Fruit Fatty Acid of *Symplocos paniculata* in Hunan Areas. Central South University of Forestry and Technology, Hunan, China, PHD Dissertation, 2013, p 150.
82. Zeng, H.; Fang, F.; Su, J.; Li C.; Jiang, L. Extraction of Seed Oil of *Swida wilsoniana* by Supercritical CO_2, Microwave, and Ultrasound. *J. Chin. Cereals Oils* **2005,** *20*(2), 67–70.
83. Zhang, Q.; Zhou, K.; Ma, P. Preparation of Biodiesel from *Litsea cubeba* Kernel Oil. *Chem. Eng. Oil Gas* **2015,** *2*, 14–17.

GLOSSARY OF TECHNICAL TERMS

Application efficiency is the amount of water stored in the root zone that is available to meet crop transpiration needs in relation to the amount of irrigation water applied to the field.

Automation is the process of performing operations without the need for constant human involvement except for periodic inspections and routine maintenance. A classic example of automation in surface systems is automatic opening and closing of border/bay inlets by use of actuators.

Basin irrigation refers to irrigation of land by surrounding it with embankment and flooding it water.

Border irrigation refers to irrigation controlled or directed by short dikes around areas treated.

Coherency analysis measures the significance of the spatial correlation between two sets of observations for various frequencies.

Crop coefficient is a coefficient of specific crop growth under standard conditions (well wetted free from disease crop land). The coefficient integrates the differences in the evapotranspiration rate between the crop and the grass reference surface.

Crop evapotranspiration (mm): $ET_c = K_c\ ET_o$

Crop stress factor: The value of K_s is in the range of FC and θ_t. Below θ_t the value of Ks decreases linearly between 1 at θ_t and zero at WP.

Crop water requirement is the water demand of a given crop for its metabolic activity and evapotranspiration needs for specific location and planting period.

Dielectric constant refers to a quantity measuring the ability of a substance to store electrical energy in an electric field.

Evapotranspiration is the combination of evaporative losses from the soil surface and transpiration from plant surfaces.

Feedback control refers to decisions (for example time to cut off the flow into a furrow) made based on some form of measurement or feedback from the irrigation process. For instance, water sensors may be placed anywhere along the furrow to provide feedback on the time of arrival of the water front.

A common form of feedback control practiced by irrigators is cutting off the flow of water when it has reached the end of the field.

Field capacity (%) is the maximum water content that soil can hold by means of capillary forces. The conventional value of soil water tension at field capacity is 1/3 bar.

Furrow irrigation is a method of irrigation where water is applied to sloping furrows at one end of the field and the water advances along the field with infiltration during the process through the sides and base of the furrow.

Gates refer to metallic or concrete structures that are used for the control of water into borders and basins, as well as controlling the level of water in head itches.

Gravimetric water content of soil refers to moisture content in soil on a mass basis and is the ratio of the mass of the liquid phase in the given soil sample to the mass of the solid material.

Hydraulics refers to the scientific study of water and other liquids, in particular their behavior under the influence of mechanical forces and their related uses in engineering.

Infiltration is the process by which water on the ground surface enters the soil.

Inflow rate refers to the rate of flow from direct connections to the collection system such as basins or furrows.

Irrigation is the controlled application of water to arable lands in order to supply crops with the water requirements not satisfied by natural precipitation.

Irrigation modeling in surface irrigation systems is the process of mathematically describing the hydraulic characteristics of water as it flows from one end of the field to the other. Modelling make it possible to evaluate the components of water balance that is difficult or practically impossible to measure, such as the distribution of the applied depths. Models frequently come in the form computer software.

Irrigation scheduling is the irrigation application program that indicates when and how much water is to be applied to the crop area.

Management allowed deficit: Published values of MAD vary between 0.2 and 0.7. MAD represents the fraction of the total available water that can be extracted from the soil before crop water stress occurs.

Optimization is referred to in irrigation is the process of manipulating the various design and management variables affecting the irrigation process with the aim of achieving the best or optimal outcome possible. Optimization can be achieved through trial and error, irrigator experience, or simulation models.

Real-time control in surface irrigation implies that measurements taken during an irrigation event are processed and used for the modification and optimization of the same irrigation event. Real-time control in surface systems is feasible when the control process is automated so that the feedback can be implemented rapidly. An example is when an irrigation event is monitored and the feedback is implemented while the irrigation is still underway.

Reference evapotranspiration is evapotranspiration from a reference surface, which is hypothetical grass reference crop with specific characteristics not short of water and uniform growth.

Roughness coefficient refers to the hydraulic resistance.

Semivariogram is a graph of how semivariance changes as the distance between observations changes. Semivariograms are used for measuring the degree of dissimilarity between observations as a function of distance.

Smart irrigation refers to an irrigation system which involves the use intelligent (computerized) devices and systems that monitor the site conditions in real time and provide information to be used in the optimization process. Smart irrigation may be seen as a more advanced form of an automated irrigation system.

Soil moisture stress refers to a term when water supply does not meet the crop water requirements. This will be done strategically to safe water with minimum or no significant yield reduction.

Spatial autocorrelation measures the correlation of a variable with itself through space.

Spatial dependence refers to the tendency for observations close together in space to be more highly correlated than those that are further apart. Also is called spatial autocorrelation. Spatial dependence imputes that up to some distance apart from each other, two observations at different locations are not statistically independent.

Spatial statistics is the field of study concerning statistical methods that use space and spatial relationships (such as distance, area, volume, length, height, orientation, centrality and/or other spatial characteristics of data) directly in their mathematical computations.

Spatial variability of soil property refers to the quantity that is measured at different spatial locations exhibits values that differ across the locations.

Spectral analysis is a tool for the exploratory analysis of spatial patterns.

Surface irrigation is the irrigation water application method in which water is conveyed over the field surface by gravitation force. In this case, the soil acts both as a means of conveying water and the surface through which infiltration occurs. The most common configurations of surface irrigation are basin, bay and furrow.

Surface irrigation refers to the group of application techniques where water is applied and distributed over the soil surface by gravity.

Telemetry refers to accessing and/or transmitting data and control a system remotely. Telemetry systems are vital components of automatic surface irrigation methods for they allow measurement of various parameters (for example inflow, advance time and soil moisture) from a remote location and the results conveyed to a central location mainly via some form of radio communication.

Temporal stability of soil property refers to the time invariant association between spatial location and statistical parameters of soil properties.

Threshold moisture content (%) is defined as the [FC – RAW].

Time to cut off refers to the time at which the irrigation application is cut-off.

Total available water is the water content above field capacity cannot be held against the forces of gravity and as the water content below wilting point cannot be extracted by plant roots. Total available water (%): TAW = FC – WP

Volumetric water content in the soil represents the fraction of the total volume of soil that is occupied by the water contained in the soil.

Water productivity is the amount of product or yield gained from unit volume of water application. It measures how a system converts water into goods and services.

Water use efficiency refers to the effectiveness of a water application system. The two most common efficiency measures of an irrigation system are application efficiency (AE) and requirement efficiency (RE). AE is the ratio of the volume of water added to the root zone to the total volume of water applied. RE is the ratio of the volume of water added to the root zone to the water deficit prior to irrigation. Water use efficiency may also be characterized in terms of the uniformity or evenness of the applied water across the field.

Wetted perimeter refers to the surface of the channel bottom and sides in direct contact with the aqueous body in open channels. It is a ratio of wetted area to the hydraulic radius.

Wilting point (%) is the moisture content at which plants can no longer remove water from the soil. The conventional value of soil water tension at wilting point is 15 bars.

Yield reduction is the relationship between the ratio of yield reduce from due to soil moisture stress to the ratio of the reduce amount of water applied.

Yield response factor addresses the relationship between crop yield and water use, where relative yield reduction is related to the corresponding relative reduction in evapotranspiration (ET).

INDEX

R

S

T

V

W

Printed and bound by CPI Group (UK) Ltd, Croydon, CR0 4YY

23/10/2024

01777701-0005